Stochastic Equations and Differential Geometry

Mathematics and Its Applications (*Soviet Series*)

Volume 30

Ya. I. BELOPOLSKAYA
Ukrainian Academy of Sciences, Kiev, U.S.S.R.
and
Yu. L. DALECKY
Polytechnic Institute, Kiev, U.S.S.R.

Stochastic Equations and Differential Geometry

Springer-Science+Business Media, B.V.

Library of Congress Cataloging-in-Publication Data

CIP

Belopolskaya, Ya. I., 1943-
Stochastic equations and differential geometry.

(Mathematics and its applications. Soviet series)
Translated from the Russian original.
Bibliography: p.
Includes index.
1. Stochastic differential equations. 2. Geometry, Differential. I. Daletskiĭ, I͡U. L. (I͡Uriĭ L'vovich) II. Title. III. Series: Mathematics and its applications (Kluwer Academic Publishers). Soviet series.
QA274.23.B45 1989 519.2 88-13476
ISBN 978-94-010-7493-3 ISBN 978-94-009-2215-0 (eBook)
DOI 10.1007/978-94-009-2215-0

printed on acid free paper

Table of Contents

SERIES EDITOR'S PREFACE

'Et moi, ..., si j'avait su comment en revenir, je n'y serais point allé.'

Jules Verne

The series is divergent; therefore we may be able to do something with it.

O. Heaviside

One service mathematics has rendered the human race. It has put common sense back where it belongs, on the topmost shelf next to the dusty canister labelled 'discarded nonsense'.

Eric T. Bell

Mathematics is a tool for thought. A highly necessary tool in a world where both feedback and nonlinearities abound. Similarly, all kinds of parts of mathematics serve as tools for other parts and for other sciences.

Applying a simple rewriting rule to the quote on the right above one finds such statements as: 'One service topology has rendered mathematical physics ...'; 'One service logic has rendered computer science ...'; 'One service category theory has rendered mathematics ...'. All arguably true. And all statements obtainable this way form part of the raison d'être of this series.

This series, *Mathematics and Its Applications*, started in 1977. Now that over one hundred volumes have appeared it seems opportune to reexamine its scope. At the time I wrote

> "Growing specialization and diversification have brought a host of monographs and textbooks on increasingly specialized topics. However, the 'tree' of knowledge of mathematics and related fields does not grow only by putting forth new branches. It also happens, quite often in fact, that branches which were thought to be completely disparate are suddenly seen to be related. Further, the kind and level of sophistication of mathematics applied in various sciences has changed drastically in recent years: measure theory is used (non-trivially) in regional and theoretical economics; algebraic geometry interacts with physics; the Minkowsky lemma, coding theory and the structure of water meet one another in packing and covering theory; quantum fields, crystal defects and mathematical programming profit from homotopy theory; Lie algebras are relevant to filtering; and prediction and electrical engineering can use Stein spaces. And in addition to this there are such new emerging subdisciplines as 'experimental mathematics', 'CFD', 'completely integrable systems', 'chaos, synergetics and large-scale order', which are almost impossible to fit into the existing classification schemes. They draw upon widely different sections of mathematics."

By and large, all this still applies today. It is still truc that at first sight mathematics seems rather fragmented and that to find, see, and exploit the deeper underlying interrelations more effort is needed and so are books that can help mathematicians and scientists do so. Accordingly MIA will continue to try to make such books available.

If anything, the description I gave in 1977 is now an understatement. To the examples of interaction areas one should add string theory where Riemann surfaces, algebraic geometry, modular functions, knots, quantum field theory, Kac-Moody algebras, monstrous moonshine (and more) all come together. And to the examples of things which can be usefully applied let me add the topic 'finite geometry'; a combination of words which sounds like it might not even exist, let alone be applicable. And yet it is being applied: to statistics via designs, to radar/sonar detection arrays (via finite projective planes), and to bus connections of VLSI chips (via difference sets). There seems to be no part of (so-called pure) mathematics that is not in immediate danger of being applied. And, accordingly, the applied mathematician needs to be aware of much more. Besides analysis and numerics, the traditional workhorses, he may need all kinds of combinatorics, algebra, probability, and so on.

In addition, the applied scientist needs to cope increasingly with the nonlinear world and the extra mathematical sophistication that this requires. For that is where the rewards are. Linear models are honest and a bit sad and depressing: proportional efforts and results. It is in the non-linear world that infinitesimal inputs may result in macroscopic outputs (or vice versa). To appreciate what I am hinting at: if electronics were linear we would have no fun with transistors and computers; we would have no TV; in fact you would not be reading these lines.

There is also no safety in ignoring such outlandish things as nonstandard analysis, superspace and anticommuting integration, p-adic and ultrametric space. All three have applications in both electrical engineering and physics. Once, complex numbers were equally outlandish, but they frequently proved the shortest path between 'real' results. Similarly, the first two topics named have already provided a number of 'wormhole' paths. There is no telling where all this is leading - fortunately.

Thus the original scope of the series, which for various (sound) reasons now comprises five subseries: white (Japan), yellow (China), red (USSR), blue (Eastern Europe), and green (everything else), still applies. It has been enlarged a bit to include books treating of the tools from one subdiscipline which are used in others. Thus the series still aims at books dealing with:

- a central concept which plays an important role in several different mathematical and/or scientific specialization areas;
- new applications of the results and ideas from one area of scientific endeavour into another;
- influences which the results, problems and concepts of one field of enquiry have, and have had, on the development of another.

In the beginning, there were differential equations in the form of 'when differentiation (a differential operator) is applied to a certain function f, there should result a certain function g', then came the coordinate free (or invariant) approach: differential equations on manifolds. For stochastic (Ito) equations, the local coordinate version is well established, of vast importance, and immensely rich in applications. With the continuing spectacular rise of the importance of all things stochastic within the general framework of mathematics and its applications, it has become quite important to have an appropriate global version available. That now exists and this volume, based among many other sources, on the two authors' own researches of the last 20 years, is a comprehensive account encompasssing both the finite- and infinite-dimensional cases.

Let me backtrack a little bit. The study of global differential equations, i.e. of expressions like $\dot{x} = f(x)$, where now x lives on a global differentiable manifold like a sphere, involved the realization that the right-hand side of this equation is not a function at all but something quite different, namely a vector-field. This was, and is, a far from trivial and important conceptual advance, which even now is sometimes felt to be a bit painful. All the same, nowadays, it is hardly necessary anymore to argue the advantages of the coordinate-free point of view. True, if actual numbers are desired, it may at some point become necessary to reintroduce coordinates - though this is by no means as often the case as is commonly thought -, but according to current wisdom, this should be postponed as long as possible (and possibly even a bit longer).

The stochastic case is harder than the deterministic one, and the appropriate notion is not something like a stochastic vector-field; moreover, as expected, there are, in the infinite-dimensional, case a number of measure-theoretical matters to be resolved. This has all been handled and it is with pleasure that I can advise: study the results contained in the volume you have currently in your hands.

The shortest path between two truths in the real domain passes through the complex domain.

J. Hadamard

La physique ne nous donne pas seulement l'occasion de résoudre des problèmes ... elle nous fait pressentir la solution.

H. Poincaré

Never lend books, for no one ever returns them; the only books I have in my library are books that other folk have lent me.

Anatole France

The function of an expert is not to be more right than other people, but to be wrong for more sophisticated reasons.

David Butler

Bussum, September 1989

Michiel Hazewinkel

Introduction

Up to recent times, stochastic differential equation theory has been treated only as a special section of probability theory. Nevertheless, due to its important applications in analysis and mathematical physics, it nowadays appears to be located at the crosspoint of different mathematical and physical theories. That is why one may see terms such as 'stochastic analysis' and 'stochastic differential geometry' in the pages of modern works.

Nowadays, there are a number of monographs devoted to different aspects of stochastic analysis and various connections between stochastic equations and partial differential equations both of parabolic and elliptic types. Among them are the manuscripts by I.I. Gichman and A.V. Skorohod [45], [46], N. Ikeda and S. Watanabe [48], Yu. L. Dalecky and S.V. Fomin [28], T. Bismut [16], K. Elworthy [37] and others.

The main goal of our book is an invariant description of Ito's stochastic equations on a smooth manifold (both of finite and infinite dimension). This description is supposed to be compatible with some additional structures of a differential geometric or algebraic nature which the manifold may be equipped with. The solutions of stochastic equations which are diffusion processes on the manifold are used afterward to construct Cauchy problem solutions for parabolic equations with respect to both functions and measures on the manifold, as well as sections of vector bundles over it.

The first work devoted to the theory of diffusion processes on a manifold was the article [49] by K. Ito. Some peculiarities of the construction of a diffusion process on a nonlocal manifold were revealed in this work. First, was the fact that one cannot construct a global process by patching together solutions of local equations without additional assumptions. The obstacle here is the fact that a diffusion process can leave whatever small neighborhood of an initial point during whatever small time with a positive probability.

Next, an Ito differential equation cannot be treated as a vector field (even a random one) in contrast with an ordinary differential equation. This happens since, under coordinate transformations, stochastic equation coefficients are transformed according to Ito's formula, rather than according to the usual transformation rules of differential calculus.

It had been pointed out in works by H. McKean [62] and R. Gangolli [41] that Ito's formula has a differential geometric nature and may be connected with an exponential map on the manifold generated by a given connection.

Finally, there is another specific feature of the situation considered. Namely, one cannot define on an arbitrary manifold a fundamental notion of a linear theory such as a stochastic integral.

Keeping in mind all these peculiarities it is, nevertheless, possible to develop an invariant theory of stochastic differential equations. We describe such a theory in a way which permits us to include both the finite and infinite dimensional case. Moreover, we subsequently try to deal with a differential geometric approach based on a connection theory and using an exponential map. From this point of view, it is natural to use probability constructions only at a local level. To overcome the difficulties which arise while changing local constructions into global ones, we formally introduce the stochastic differential as the germ of local diffusion processes and consider a special bundle of stochastic differentials – an Ito bundle – which had been presented as a ghost in the above cited work by K. Ito [49]. The sections of this bundle are naturally called stochastic differential equations. Remember that sections of a tangent bundle are called ordinary differential equations.

The intention to extend considerations to an infinite dimensional case is motivated by the deeper and deeper penetration of infinite dimensional manifolds into both modern analysis and mathematical physics (see, for example, works by V.I. Arnold [2] D. Ebin and T. Marsden [33] and others).

The crucial role played by measure theory in the infinite dimensional case, had been pointed out in a number of works on analysis infinite dimensional linear spaces. Unlike the finite dimensional case, the absence of an invariant measure analogous to a Lebesgue measure does not permit us to treat distributions as generalized functions if we want to develop a harmonic analysis. Notice that a weak version of a harmonic analysis may be developed by using a Gaussian measure instead of a Lebesgue one. Nevertheless, if we are going to deal with measures on a manifold, we lose this opportunity as well. This leads to the necessity to construct a large enough class of measures compatible with the smooth structure of the manifold as well as with those algebraic structures which the manifold may be equipped with.

A collection of measures which are differentiable along vector fields constitutes such a class. Obviously, the investigation of those measurable properties is an important task.

A natural new object arises in this way on an infinite dimensional manifold. This object is a differential equation with respect to measures which must be studied along with differential equations with respect to functions.

An example of a parabolic equation with respect to a measure given, in particular, by a forward Kolmogorov equation. Unlike the finite dimensional case, one may not reduce this equation to an equation with respect to a function which gives a density of the transition probability with respect to Lebesgue measure.

The choice of subjects to be discussed in the book is motivated by the goal declared above. In the course of the exposition, we had to deal with objects which have very little in common from a traditional point of view, namely with random processes connections on infinite dimensional manifolds and bundles, and finally parabolic equations with respect to both functions and measures. To make the number of potential readers of this book as large as possible, we give some necessary information concerning those objects. We hope that the book will be of use to specialists in stochastic process theory who are interested in possible applications of their theory. Next, we have tried to do our best to demonstrate that stochastic equations may be used as a powerful tool for dealing with various problems of global analysis. To this purpose we have been trying to make both the language and way of exposition as little 'probablistic' as possible.

The first chapter of the book is devoted to an exposition of some known facts and notions concerning analysis both on functions and measures in Banach spaces. The reader may find more details in the book by Yu. L. Dalecky and S. V. Fomin [28]. Nevertheless, the investigation of properties of measures differentiable along vector fields leads to some new results.

The theory of connections for vector bundles is the main topic of the second chapter. The foundation of this theory may be found in the book by S. Lang [58]. Afterwards, we investigate properties of smooth measures on manifolds and construct covariant differential operations for functions and measures.

It should be mentioned that the reader who wants to go through the main results of this book cannot omit those two chapters. Even those who are specialists in differential geometry must look them through, treating their content as a collection of necessary notations and definitions.

The third chapter gives a more or less detailed review of necessary results from the theory of stochastic equations, both in Hilbert spaces and in Banach spaces with smooth norms. This point of view on the content of the chapter is natural for a specialist in random process theory: though an extension of the theory to the infinite dimensional case is connected with some difficulties; in addition we give an exposition of the theory which differs from the traditional one.

The reader who is not a specialist in probability theory will find all the informations needed in this chapter. Next, he will not come in contact practically with specific probabilistic considerations. Nevertheless to enter more deeply into the subject, more detailed information is needed about probabilistic constructions.. This information may be found in a number of books. See for example [45 – 48], [55], or [64], where the infinite dimensional case is treated.

The last three chapters contain an exposition of the main results of this book.

In Chapter 4, we construct a stochastic differential equation on a smooth Banach manifold and solve the Cauchy problem for it. In this book we deal only with manifolds without boundaries. Moreover, we assume that both the manifold structure and coefficient properties grant the existence of a global solution of the Cauchy problem. We pay special attention to manifolds equipped with vector bundle total space structures and describe processes compatible with those structures. In particular, we investigate a tensor stochastic parallel displacement generating a process of this type. A differential lift of a stochastic equation based on changing the equation on the manifold to an equation on its tangent bundle whose coefficients are derivatives of the initial coefficients, gives an example of the above-mentioned processes as well. The investigation of differential lifts of the stochastic equation gives the possibility to state conditions which guarantees the smoothness of the solution of a stochastic differential equation, and to explain the very sense of this notion. In constructing a global solution of a stochastic equation on a manifold, a crucial part is played by the 'principle of localization'. This principle states that under some conditions the solution may be constructed as the limit of a certain sequence of processes such that the probability that the sequence is stabilized tends to 1.

A solution of a stochastic equation with nonrandom coefficients gives a Markov stochastic process on the considered manifold. Like any Markov process, it generates a pair of linear evolution operator families acting in a pair of dual spaces, namely the space of functions and the space of measures defined on the manifold. Those family generators are second-order elliptic operators on functions and measures. This approach permits us to investigate the Cauchy problem for some parabolic equations and, in particular, for both forward and backward Kolmogorov equations connected with the under study Markov process. The correspondent results may be seen in Chapter 5.

While dealing with a manifold equipped with a vector bundle total space structure and a stochastic equation compatible with this structure, we construct a Markov process on the basis of the bundle and a special linear map acting in its fibres, which is called a multiplicative functional of the Markov process. This linear map generates an evolution family acting on vector bundle sections and, thus, we may investigate the Cauchy problem for an equation with respect to vector bundle sections. In a local trivialization, this equation has the form of a parabolic equations system. This approach is interesting, even in the finite dimensional case, since it permits us to solve equations with degenerated main symbols.

If one considers stochastic equation with coefficients depending on the equation distribution solution, it is realized that in this way we may investigate quasilinear and even nonlinear differential parabolic equations on smooth manifolds and vector bundles. A natural way to solve such equations is to construct successive approximations of the desired solution by means of solving linearized parabolic equations at each step. In order to prove the convergence of those approximations, we need some special estimates which are derived thanks to the investigation of corresponding stochastic equations. Such an approach, initially proposed by H. McKean [63] and M.I. Freidlin [39 – 40], is developed in the second section of the fifth chapter.

Different ways to prove smoothness properties of a diffusion process transition probability have been discussed in recent years in connection with works by P. Malliavin [59 – 61] (see J. Bismut [18], D. Stroock [73], and others). All those authors treated a stochastic equation solution transition probability as the image of a Gaussian measure (in the trajectory space of a Wiener process) under a map which maps a Wiener process into a stochastic equation solution. Notice that in the finite dimensional case, the statement that a measure is smooth means that it has a smooth density with respect to the Lebesgue measure. The theory developed in this book permits us to consider the infinite dimensional case as well if one calls smooth a measure which possesses a smooth logarithmic derivative.

In the sixth and last chapter, we deal with manifolds with Lie groups acting on them, and particularly with a Lie group itself or with a principal bundle.

First, we describe the principal bundle connection theory. Next, we consider stochastic equations with invariant coefficients and state different properties of their solutions.

Finally, we shed light on connections between measure smooth properties and a 'quasi-invariance' property, that is the existence of a large enough collection of shifts which transform a measure into a measure which is absolutely continuous with respect to the initial one.

Here we stop before meeting a large and important problem connected with quite different fields, namely infinite dimensional Lie group theory and representation theory for Lie groups, since difficulties which arise here are of an essentially different nature.

Another problem which we do not touch upon here is the investigation of the behaviour of a stochastic equation solution on small times and for small diffusion coefficients, as well as the behaviour of the solution distribution. In the finite dimensional case, this problem has been studies by A. D. Wentzel and M. I. Freidlin [75], S. A. Molchanov [66], K. Elworthy [38], M. Pinsky [70], and others.

The authors began their investigations in the field of stochastic differential geometry at the end of the sixties, when the first results had been announced (see Yu. L. Dalecky, Ya. I. Schnaiderman [30]). Our exposition is based mainly on a couple of reviews published in 'Uspechi Mathematicheskych nauk' (see Ya. I. Belopolskaya and Yu. L. Dalecky [12], and Yu. L. Dalecky [24]) and a paper published in 'Trudach Moskovskogo Mathematicheskogo obschestwa' (proceedings of Moscow Mathematical Society) (see Ya. I. Belopolskaya and Yu. L. Dalecky [9]) and contains detailed proofs of the results of those articles as well as their extension and development.

Some problems discussed in the book have previously been reported in Vilnius International conferences and seminars on stochastic differential equations as well as in the seminar on stochastic processes in function spaces of the Mathematical Institute of the Science Academy of UkrSSR.

We thank all colleagues who took part in discussions of those problems for their critical remarks which improved the exposition. We are especially indebted to A.V. Skorohod whose influence on our understanding of stochastic process theory problems cannot be overestimated.

CHAPTER 1

Functions and Measures in Linear Spaces

In this chapter we introduce the main notions of differential calculus for functions of infinite-dimensional arguments defined in a region of a Banach space. Next, we consider measures in Banach spaces and develop a theory of measure differentiation along vector fields.

1. Spaces, Mappings, Differential Operations

1.1. SPACES AND OPERATORS

By a Banach space we mean a real separable complete linear normed space. The appearance of complex spaces will be specially mentioned. Denote $\| x \|_B$ as the norm of the element $x \in B$, omitting the index B if it will not lead to any misunderstanding.

Given a pair of Banach spaces, denote $L(B_1, B_2)$ as the Banach space of linear continuous mappings from B_1 into B_2 (operators) with usual uniform norm

$$\| A \| = \sup_{\| x \|_{B_1} = 1} \| Ax \|_{B_2}$$

for $A \in L(B_1, B_2)$. For the sake of briefness, we shall write $L(B)$ instead of $L(B, B)$.

A category is defined if we have listed its objects $(\ldots, X, Y, \ldots)$ and described a set of morphisms $\mathrm{Mor}(X, Y)$ for each pair X, Y as well as an associative composition $\circ : \mathrm{Mor}(X, Y) \circ \mathrm{Mor}(Y, Z) \to \mathrm{Mor}(X, Z)$ with identity element $i_X \in \mathrm{Mor}(X, X)$

$$i_y \circ f = f \circ i_X, \quad f \in \mathrm{Mor}(X, Y).$$

Let $\mathcal{Z}$ denote a category, whose objects are Banach spaces and whose morphisms are $\mathrm{Mor}(B_1, B_2) = L(B_1, B_2)$. The part of unit element is played in this case by the identity map of the corresponding space $i_X \equiv I_X$.

Let $B, B_1, \ldots, B_n$ be Banach spaces. Denote $L_n(B_1 \times \ldots \times B_n, B)$, $n = 2, 3, \ldots$ the Banach space of multilinear maps

$$\varphi : B_1 \times \ldots \times B_n \to B$$

with norm

$$\| \varphi \| = \sup_{\substack{\|x_k\|_{B_k} \le 1 \\ 1 \le K \le n}} \| \varphi(x_1, \ldots, x_n) \|_B .$$

In particular, we denote $L_n(B_1, B) = L_n(B_1 \times \ldots \times B_n, B)$ and $L_n(B) = L_n(B, B)$. There exists an isomorphism

$$L_{n-k}(B_1 \times \ldots \times B_{n-k}, L_k(B_{n-k-1} \times \ldots \times B_n), B) = L_n(B_1 \times \ldots \times B_n, B).$$

A functor is a mapping of categories which maps morphism composition into morphism composition and either preserves the arrow directions (covariant functor) or changes them to the inverse (contravariant functor).

The simplest functors in the Z category are those generated by the L symbol.

Given a Banach space B consider a map

$$X \to L(B, X)$$

and put in correspondence to $\varphi \in \mathrm{Mor}(X, Y)$ a morphism

$$L(B, \varphi) : L(B, X) \to L(B, Y)$$

acting as

$$L(B, \varphi)\, \psi = \varphi \circ \psi, \quad \psi \in L(B, X).$$

Evidently in this way we obtain a covariant functor $L(B, \cdot)$. In the same way one may obtain a contravariant functor $L(\cdot, B)$.

$$X \to L(B, X),$$

$$\mathrm{Mor}(X, Y) \ni \varphi \to L(\varphi, B) \in \mathrm{Mor}(L(Y, B), L(X, B)),$$

$$L(\varphi, B)\, \psi = \psi \circ \varphi.$$

Notice that the contravariant functor $* = L(\cdot, R^1)$ is a particular case of the above construction. It puts in correspondence to each Banach space X its dual space X^* and to each $\varphi \in L(X, Y)$ its adjoint map $\varphi^* \in L(Y^*, X^*)$. By definition, the space X^* consists of all continuous linear functionals on X. Denote by $\langle x, x^* \rangle_X = x^*(x)$ a canonical bilinear form $X \times X^* \to R^1$ (to be short we sometimes omit the subscript x).

One may easily check that the symbol L_n gives rise to the (n + 1-multiple) functor. We shall use the following notations

$$L_n(\varphi_1 \times \ldots \times \varphi_n, \varphi)\, \psi : x_1 \times \ldots \times x_n \to \varphi \circ \psi(\varphi_1(x_1), \ldots, \varphi_n(x_n)),$$

where

$$\psi \in L_n(X_1 \times \ldots \times X_n, Y), \quad \varphi \in L(Y, \tilde{Y}), \quad \varphi_j \in L(\tilde{X}_j, X_j),$$

and

$$L_n(\varphi_1 \times \ldots \times \varphi_n, \varphi) : L_n(X_1 \times \ldots \times X_n, Y) \to L_n(\tilde{X}_1 \times \ldots \times \tilde{X}_n, \tilde{Y}). \tag{1.1}$$

Notice that sometimes it will be convenient to take as a Banach space (in a wide sense) a complete linear topological space with topology, defined by a certain Banach norm. That is why we shall identify Banach spaces with equivalent norms.

A Hilbert space H is a Banach space with an inner product $(x, y)_H$ which gives rise to a norm $\| x \|_H^2 = (x, x)_H$. The dual space H^* may be identified with H by putting

$$\langle x, y \rangle = (x, y).$$

In the above-mentioned wide sense, there exists only one (separable) Hilbert space. Hilbert space structure permits us to define some special map classes. Let $H_1, \dots, H_n, H$ be Hilbert spaces, $\{e_j^k\}_{j=1}^{\infty}$, $k = 1, \dots, n$ be an orthobasis in H_k. Denote $L_{n2}(H_1 \times \dots \times H_n, H)$ as a subspace of $L_n(H_1 \times \dots \times H_n, H)$ consisting of maps φ such that

$$\sigma_2^2(\varphi) = \sum_{j_1 \dots j_n = 1}^{\infty} \| \varphi(e_{j_1}^{(1)}, \dots, e_{j_n}^{(n)}) \|_H^2 < \infty .$$

Those maps are called Hilbert–Schmidt (H S) operators. The set of H S operators is a Hilbert space with norm σ and inner product

$$\sigma_2(\varphi, \psi) = \sum_{j_1 \dots j_n = 1}^{\infty} \left(\varphi(e_{j_1}^{(1)}, \dots, e_{j_n}^{(n)})\ \psi(e_{j_1}^{(1)}, \dots, e_{j_n}^{(n)}) \right) .$$

It is easy to check that $\sigma_2(\varphi, \psi)$ does not depend on the choice of a basis.

Using notations (1.1), we obtain the relation

$$L_n(\varphi_1 \times \dots \times \varphi_n, \varphi) : L_{n2}(X_1 \times \dots \times X_n, Y) \to L_{n2}(\tilde{X}_1 \times \dots \times \tilde{X}_n, \tilde{Y}). \tag{1.2}$$

An extension of the set of Hilbert–Schmidt operators leads to the notion of 2-absolutely summing operators. For a Banach space B, a mapping $\varphi \in L_n(H_1 \times \dots \times H_n, B)$ is called a 2-absolutely summing operator if the following estimate is valid:

$$\sum_j \| \psi(x_{j_1}^{(1)}, \dots, x_{j_n}^{(n)}) \|_B \le C^2 \sup_{\|y_k\|_{H_k} \le 1} \prod_{k=1}^{n} \sum_{j_k} | (x_{j_k}, y_{j_k})_{H_k} |^2 \tag{1.3}$$

for an arbitrary set $\{x_{j_k}^{(k)}\} \in H_k$, $k = 1, \dots, n$.

Denote $\sigma_2(\psi)$ as the minimal constant C for which (1.3) takes place. The space $L_{n2}(H_1 \times \dots \times H_n, B)$ of all 2-absolutely summing operators is a Banach space with norm σ_2. An estimate

$$\sum_j \| \psi(e_{j_1}^{(1)}, \dots, e_{j_n}^{(n)}) \|_B^2 \le \sigma_2^2(\psi) \tag{1.4}$$

evidently follows (1.3). Moreover, (1.2) is valid as well as the estimate

$$\sigma_2(L_n(\varphi_1 \times \dots \times \varphi_n, \varphi)\, \varphi) \le \sigma_2(\psi) \| \varphi_1 \| \dots \| \varphi_n \| \, \| \varphi \| . \tag{1.5}$$

Let H be a Hilbert space. An operator A is called a nuclear operator if the series $\sum_{j=1}^{\infty} (\varphi(e_j), e_j)$ converges for any orthonormal basis in H. It is known that the sum of the series $\mathrm{Tr}\,\varphi$, which is called the trace of the operator φ, does not depend on the basis choice. As is easy to see, among nuclear operators there are operators of the form $\varphi = \psi_1 \circ \psi_2$ with $\psi_2 \in L_{12}(H, B)$, $\psi_1 \in L(B, H)$, and $\psi_1^* \in L_{12}(H, B^*)$.

An extension of the nuclear operators class leads to the notion of 1-absolutely summing operators. For a Banach space B, we call $\varphi \in L_n(H_1, \dots, H_n, B)$ a 1-absolutely summing operator if the estimate

$$\Sigma \, \| (\psi(x^{(1)}_{j_1}, \dots, x^{(n)}_{j_n}) \|_B \le C \sup_{\|y_*\|_{H_k} \le 1} \prod_{k=1}^{n} \sum_{j_k} |(x_{j_k}, y_{j_k})| \tag{1.6}$$

holds for any set $\{x^{(k)}_{j_k}\} \in H_k$, $j_k = 1, \dots, m_k$. Denote $\sigma_1(\psi)$ as the minimal constant C for which (1.6) holds.

The spaces $L_{n1}(H_1 \times \dots \times H_n, B)$ of all 1-absolutely summing operators is, once again, a Banach space with norm σ_1. It follows from (1.6) that

$$\Sigma \, \| (\psi(e^{(1)}_{j_1}, \dots, e^{(n)}_{j_n}) \|_B \le \sigma_1(\psi).$$

One may also prove the estimate

$$\sigma_1(L_n(\varphi_1 \times \dots \times \varphi_n, \varphi)\, \psi) \le \sigma_1(\psi) \, \|\varphi_1\| \dots \|\varphi_n\| \, \|\varphi\|. \tag{1.7}$$

Consider next a class $N(H, B) \subset L_2(H, B)$ of bilinear 1-absolutely summing operators, acting from a Hilbert space H into a Banach space B, possessing the following property: for any orthonormal basis $\{e_k\}$ in H the series $\Sigma_k G(e_k, e_k) = \mathrm{Tr}\, G \in B$ converges absolutely and its sum (the trace of the mapping G) does not depend on the basis choice.

PROPOSITION 1. *Let* $\psi \in L_2(B_1 \times B_2, B)$ *and* $\varphi_k \in L_{12}(H, B_k)$, $k = 1, 2$. *Then* $L_2(\varphi_1 \times \varphi_2, B) \subset N(H, B)$.

Proof. The absolute convergence of the series

$$\mathrm{Tr}\, L_2(\varphi_1 \times \varphi_2, B)\, \psi = \sum_j \psi(\varphi_1(e_j), \varphi_2(e_j))$$

in an easy consequence of the estimate (1.4) for $\varphi_1 \in L_{12}(H, B_1)$, $\varphi_2 \in L_{12}(H, B_2)$. Let now $y \in B^*$. The value $\langle \psi(x_1, x_2), y \rangle$ defines a mapping $g(y) \in L(B_1, B_2^*)$ in agreement with the formula

$$\langle x_2, g(y)\, x_1 \rangle = \langle \psi(x_2, x_1), y \rangle .$$

Notice that for an arbitrarily chosen orthonormal basis $\{e_k\}_{k=1}^{\infty}$, there exists

$$\begin{aligned} \mathrm{Tr}\, \varphi_2^* \, g(y)\, \varphi_1 &= \sum_{k=1}^{\infty} \langle \varphi_2(e_k), \; g(y)\, \varphi_1(e_k) \rangle \\ &= \sum_{k=1}^{\infty} \langle \psi(\varphi_2(e_k), \; \varphi_1(e_k)) \rangle \\ &= \langle \mathrm{Tr}\, L_2(\varphi_2 \times \varphi_1, B)\, \psi, y \rangle . \end{aligned} \tag{1.8}$$

Thus, both the left- and right-hand sides of (1.8) do not depend on the basis choice.

Let H be a Hilbert space, B a Banach space, and there exists a continuous mapping

$J \in L(H, B)$ which imbeds H into B:

$$\text{Ker } J = \{x \in H : Jx = 0\} = 0.$$

Let $\text{Im } J = \{y \in B : y = Jx\}$ be dense in B. Then the adjoint map $J^* \in L(B^*, H^*)$ is an embedding as well and its image $\text{Im } J^*$ is dense in H. Identifying H and H^*, we obtain a triple of spaces

$$B^* \overset{J^*}{\to} H \overset{J}{\to} B \tag{1.9}$$

with dense embedding and a simple correspondence between the linear product in H and canonical form defined on $B \times B^*$

$$(h, J_y^*)_H = \langle J h, y \rangle_B, \quad y \in B^*, \ h \in H. \tag{1.10}$$

We call the triple (1.9) the rigged Hilbert space associated with J in a canonical way, if this mapping is 2-absolutely summing. We shall say, that the embedding J generates a Hilbert–Schmidt structure in B and denote this structure (B, J). By definition, a morphism $\varphi : (B_1, J_1) \to (B_2, J_2)$ of H S structures is a mapping $\varphi \in L(B_1, B_2)$ such that $\varphi \circ J_1 = J_2 \circ \varphi$ and $\varphi \circ J_1 \in L(H_1, B_2)$.

In this case, for any $A \in L(B, H)$ both the operator $A J$ and its adjoint $(A J)^* = J^* A^*$ are Hilbert–Schmidt operators. Hence, a product $J^* A_1^* A_2 J$ is a nuclear operator in H

$$\sum_{k=1}^{\infty} (J^* A_1^* A_2 J e_k, e_k) = \sum_{k=1}^{\infty} (A_1 J e_k, A_1 J e_k),$$

and, in general, an operator $J^* \Phi J$ is nuclear if $\Phi \in L(B, B^*)$ because

$$\text{Tr } J^* \Phi J \leq \| \Phi \| \, \sigma_2^2 (J).$$

In what follows, we shall often identify the spaces B^* and H with their images with respect to embedding (1.9), omitting for simplicity the embedding operators. According to this convention, a rigged Hilbert space is a triple

$$B^* \subset H \subset B. \tag{1.11}$$

and $A J = A|_H$ is a restriction of the operator A to a dense linear subset of H. Thus, a restriction to H of $A \in L(B, H)$ is a H S operator and a restriction of $\Phi \in L(B, B^*)$ is a nuclear operator in H.

We call the triple (1.9) a nuclear rigging of a Hilbert space, if the mapping J is a 1-absolutely summing operator. In this case, we say that J generates a nuclear structure in B and denote it (B, J) as well.

The relation (1.10) in this situation has the form

$$(h, y)_H = \langle h, y \rangle, \quad h \in B^*, \ y \in H$$

and shows that the corresponding restriction of the canonical form to $B \times B^*$ coincides with the inner product in H.

It often happens that both B and B^* are Hilbert spaces. In these cases, we shall use the notations $B = H_-$, $B^* = H_+$. In the rigged Hilbert space

$$H_+ \subset H \subset H_- , \tag{1.12}$$

a pairing of H_+ and H_- is given by inner product in H.

We describe now a special construction of the rigged Hilbert space (1.12). Let $S \in L_{12}(H)$ be an operator in H with unbounded densely defined inverse $S^{-1} = K$. By introducing the norm $\| x \|_+ = \| Kx \|_H$ in the domain $\mathcal{D}_K$, we equip it with a Hilbert space structure. Next, we define in H another norm $\| x \|_- = \| S^* x \|_H$ and denote H_- the completion of H in this norm. The operator S^* may be continuously extended to the isometric mapping $\tilde{S}^* : H_- \to H$ and one may define a form

$$\langle \varphi, \xi \rangle_{H_-} = (K\varphi, \tilde{S}^* \xi)_H , \quad \varphi \in H_+ , \ \xi \in H_-$$

which, evidently, satisfies (1.10). In this case, we shall say that (1.12) is generated by the operator S.

1.2. DIFFERENTIABLE FUNCTIONS

Consider a function $f(x)$, valued in a Banach space B and defined on a certain neighborhood U_x of a point $x \in B$. Let B_1 be a Banach space densely embedded in B (possibly coinciding with B). The function f is (strongly) differentiable along B_1 at the point x, if there exists $A \in L\ (B_1, B_2)$ such that

$$f(x+h) - f(x) = Ah + o(\| h \|_{B_1}), \ x + h \in U_x, \ h \in B_1 .$$

The linear map $A : B_1 \to B_2$ is usually denoted $A = f'(x)$ and called the strong derivative of the function $f(x)$ along B_1 at the point x. We shall omit the indication of the subspace B_1 if it is evident from the context (in particular, if $B_1 = B$).

A function f is differentiable in a region $G \subset B$ if it is differentiable at any point of this region.

Together with strong derivative, one may define a weak derivative along a direction $h \in B_1$, which is a linear mapping $f'_w(x) : B_1 \to B_2$ given by the formula

$$f'_w(x)\, h = \frac{d}{d\mathcal{E}} f(x + \mathcal{E}h) \Big|_{\mathcal{E}=0} . \tag{1.13}$$

If a weak derivative $f'_w(x)$ along h does exist in U_x and $f'_w(x) : U_x \to L(B_1, B_2)$ is a continuous map, then both the existence of the strong derivative and an identity $f'(x) = f'_w(x)$ follow from the relation

$$f(x+h) - f(x) = \int_0^1 f'\ (x + \tau h)\, d\tau .$$

In what follows, we always suppose these conditions, granted the above identification, are fulfilled and treat the above relation as a convenient way of calculating derivatives.

Let the B_2-valued function φ be defined in a neighbourhood $U_{f(x)}$ and differentiable along B. Then a composition $\varphi \circ f'(x) = \varphi'(f(x))\, f'(x)$.

One may define higher derivatives in a recurrent way $f^k = (f^{(k-1)})'$. Taylor series gives another way to define them. In the above notations, let there exist multilinear symmetric maps $A_k \in L_k(B_1, B_2)$ such that

$$f(x+h) - f(x) = \sum_{k=1}^{\infty} A_k(h, \dots, h) + \sigma(x, h)$$

with $\sigma(x, h) = o\left(\| h \|_{B_1}^n\right)$. We call $A_k = f^{(k)}(x)$ the kth order derivative of the function f at the point x. One may prove the equivalence of the above two definitions by differentiating the last relation, taking into account the identification

$$L_k(B_1, B) = L\left(B_1, L_{k-1}(B_1, B)\right).$$

For functions, defined in a region $G \subset B$, we introduce some simpler linear differential operators.

Denote $C_k(G, B_1, B_2)$ the class of functions $u : G \to B_2$ possessing k continuous and bounded in G, derivatives along directions out of a given linear set $B_1 \subset B$. If $k = 0$, then $C_0(G, B_2)$ is the space of continuous functions. Denote $C_{k2}(G, B_1, B_2)$ the class of functions $u : G \to B_2$ possessing k derivatives along B_1, which belong $L_{k2}(B_1, B_2)$.

First, suppose, that $u \in C_1(G, B_1, R^1)$. Then for any point

$$x \in G, \ u'(x) \in L(B_1, R^1) = B_1^*.$$

Consider a function $g \in C_0(G, B_1)$ that is a B_1-valued vector field on G. It makes sense to consider the expression $\langle g(x), u'(x) \rangle_{B_1}$ and we call it the derivative of the function u along the vector field g. Using the vector field integral flow $S_t^g \circ x$, which will be defined in the next section, one may prove that

$$\langle g, u'(x) \rangle = \frac{\mathrm{d}}{\mathrm{d}\, t}\, g\, \left(S_t^g \circ x\right) \Big|_{t=0}.$$

Denote $D = D^{(B_1)}$ the differentiation along the subspace $B_1 \subset B$ and notice that in the algebraic sense, D behaves as an element of the adjoint space B_1^*. Now for $u \in C_1(G, B_1, R^1)$ put by definition

$$u'(x) = D\, u(x), \quad D g = \langle g, D \rangle_{B_1}.$$

To extend this notation to vector functions, consider a linear operator

$$g \otimes b : c \mapsto \langle e, g \rangle\, b$$

acting from B_1 to B_2, $g \in B_1^*$, $b \in B_2$. It is easy to check that the adjoint operator $(g \otimes b)^* \in L(B_2^*, B_1^*)$ acts according to the rule

$$(g \otimes b)^* \varphi = (\varphi \otimes g)\, b = \langle b, \varphi \rangle_{B_2}\, g.$$

For $\upsilon \in C_1(G, B_1, B_2)$, $G \subset B$, it is natural to denote

$$\upsilon'(x) = D \otimes v(x)$$

and next

$$D_{g(x)} \upsilon(x) = \upsilon'(x)\, g(x) = (D \otimes \upsilon(x))\, g(x)$$
$$= \langle g(x), D \rangle\, \upsilon(x).$$

In these cases, when all products, written down below, are correctly defined, the operation D acts like a differentiation:

$$(D \otimes \alpha\, \upsilon) = D\, \alpha \otimes \upsilon + \alpha\, (D \otimes \upsilon),$$

$$\langle g,\ D \rangle \langle \upsilon, w \rangle$$
$$= \langle \langle g, D \rangle\, \upsilon, w \rangle + \langle \upsilon, \langle g, D \rangle\, w \rangle$$
$$= \langle (D \otimes \upsilon)\, g, w \rangle + \langle \upsilon, (D \otimes w)\, g \rangle,$$

that is

$$D \langle \upsilon, w \rangle = (D \otimes \upsilon)^*\, w + (D \otimes w)^*\, \upsilon.$$

In what follows B, as a rule, possesses the H S structure $B \supset H \supset B^*$ and the role of B_1 is played either by B or by $H.$

Let, in addition, be $g \in C_1(G, B, B^*)$. Then $g'(x) \in L(B, B^*), x \in G$ and, as it has been shown above, this operator has a nuclear restriction to H. Its trace is called the divergence of the vector field g

$$\operatorname{div} g(x) = \operatorname{Tr} g'(x)$$

and the above arguments make it natural to denote

$$\operatorname{div} g(x) = (D, g)(x)$$

because, for an orthonormal basis $\{e_k\}_{k=1}^{\infty}$, in H with $e_k \in B^*$, we have

$$\Sigma\, ((D \otimes g)\, e_k, e_k)$$
$$= \Sigma\, ((e_k, D)\, g, e_k)$$
$$= \Sigma\, (e_k, D)\, (g, e_k).$$

One may construct more general differential expressions of the first order by inserting the symbol D into a bilinear form $\alpha \in L_2(H \times H, B_2)$

$$\alpha(D, g) = \Sigma\, \alpha\, (e_k, (e_k, D)\, g)$$
$$= \Sigma\, \alpha\, (e_k, (D \otimes g)\, e_k) \overset{\text{def}}{=} \operatorname{Tr}\ \alpha\, (\cdot\,, (D \otimes g)\,\cdot\,).$$

In particular, for a bilinear functional $\alpha(\varphi, \psi) = (C\varphi, \psi)$ and $C \in L(H)$, we have

$$(C D, \upsilon) = \operatorname{Tr}\, (C\,\cdot\,, \upsilon'\,\cdot\,) = \operatorname{Tr}\ C^* \upsilon'.$$

Consider next differential operations of the second order.

Let $u \in C_2(G, B, R^1)$. Then

$$\upsilon = u' \in C_1(G, B, B^*),\ \upsilon' = u'' \in C_0(G, L(B, B^*)).$$

For each $x \in G$, the restriction of υ' to H is nuclear and, thus, for $C \in L(H)$ the expression

$$D_C^{(2)} u = \operatorname{Tr} C^* \upsilon' = (C D, \upsilon) = (C D, Du) = (C D, D)\, u$$

makes sense. The operator $\Delta u = (D, D)\, u = D_u^{(2)} = \operatorname{Tr} u''(x)$ is called the Laplace operator.

In particular, let $A \in L_{12}(\mathcal{H}, H)$, where $\mathcal{H}$ is a Hilbert space as well. Then

$$(A A^* D, D)\, u = \operatorname{Tr}_{\mathcal{H}} A A^* u'' = \operatorname{Tr}_{\mathcal{H}} A^* u'' A.$$

Notice that this is correctly defined as well if $u \in C_2(G, H, R^1)$ and

$$A \in L_{11}(B, H) \subset L_{12}(B, H).$$

We may say the same about $D_C^{(2)}$ if $C \in L(B, B^*)$. Let, in particular, $C = a \otimes b$ with $a, b \in H$. Then

$$D_C^{(2)} u = ((a \otimes b)\, D, D)_H\, u = (a, D)_H\, (b, D)_H\, u = D_a D_b\, u = u''(a, b).$$

This equality does not hold if a and b are H-valued vector fields depending on x, because in that case

$$(a, D)_H\, (b, D)_H = (b\, (a, D), D\, u) + ((a, D)\, b, D)\, u$$

and thus

$$D_{a \otimes b}^{(2)} u = \left(D_a D_b - D_{D_a b}\right) u = u''(a, b).$$

In the sequel, we shall need one more variant of the above expressions. For operators C_1, C_2 having H S restrictions to H define an inner product

$$(C_1, C_2) = \Sigma\, (C_1 e_k, C_2 e_k) = \operatorname{Tr} C_2^* C_1 = \operatorname{Tr} C_1^* C_2.$$

In particular, for $a_1, b_1, a_2, b_2 \in H$, one has

$$(C_1, a_2 \otimes b_2) = \Sigma\, (C_1 e_k, (e_k, a_2)\, b_2) = (C_1, a_2, b_2)$$

and

$$(a_1 \otimes b_1, a_2 \otimes b_2) = (a_1, a_2)\, (b_1, b_2).$$

Using those relations, we may write

$$(C D, D) = (C, D \otimes D)$$

and , in general, for an arbitrary bilinear functional α

$$\alpha\, (D, D) = (\alpha, D \otimes D)\, .$$

In the same way, we may define differential operations of higher order

$$C\, (D, \ldots, D) = (C, D \otimes \ldots \otimes D)$$

with C being an n-linear functional in H. For a pair of those functionals, we have

$$(C_1, C_2) = \sum_{j_1, \ldots, j_n} C_1 \left(e_{j_1}, \ldots, e_{j_n}\right) C_2 \left(e_{j_1}, \ldots, e_{j_n}\right).$$

As an example, let us the calculate the differential operator

$$\sigma(z_1, z_2) = \operatorname{div} D_{z_1} z_2 - D_{z_1} \operatorname{div} z_2$$

for a pair of vector fields $z_j \in C_1(B, B, B^*)$, $j = 1, 2$. Consider the expansions

$$z_1 = \sum_k z_1^k e_k, \quad z_2 = \sum_k z_2^k e_k,$$

where $\{e_k\}$ is an orthonormal basis in H and assume first that those expansions have only a finite number of nonzero terms.

In this situation

$$D_{z_1} z_2 = \sum e_k D_{z_1} z_2^k, \quad \operatorname{div} z_2 = \sum_k D_{e_k} z_2^k$$

and thus

$$\operatorname{div} D_{z_1} z_2 = \sum_k D_{e_k} D_{z_1} z_2^k, \; D_{z_1} \operatorname{div} z_2 = \sum_k D_{z_1} D_{e_k} z_2^k.$$

From this we have

$$\begin{aligned}\sigma(z_1, z_2) &= \sum_k (D_{e_k} D_{z_1} - D_{z_1} D_{e_k}) z_2^k \\ &= \sum_{k,j} (D_{e_k} z_1^j D_{e_j} - z_1^j D_{e_j} D_{e_k}) z_2^k \\ &= \sum_{k,j} (D_{e_k} z_1^j) D_{e_j} z_2^j \\ &= \sum_{k,j} (z'_1 e_k, e_j)(z'_2 e_j, e_k) = \operatorname{Tr}_H z'_1 z'_2.\end{aligned}$$

Next, we take into account that for the above vector fields z'_1 and z'_2 are nuclear operators and we may omit the assumption about the number of nonzero terms in z_1 and z_2 expansions. Moreover, it is evident that the above calculation of $\sigma(z_1, z_2)$ is correct as well in the case when z'_1 and z'_2 are Hilbert–Schmidt operators in H.

1.3. DIFFERENTIAL EQUATIONS

Let $f: G \to B$ be a given function in a region $G \subset B$ which satisfies a Lipschitz condition

$$\| f(x) - f(y) \| \le C \| x - y \|, \quad x, y \in G. \tag{1.14}$$

Consider an ordinary differential equation

$$\frac{dx}{dt} = f(x). \tag{1.15}$$

For any point $x_0 \in G$ there exists $\sigma > 0$ such that for $|t - t_0| < \sigma$ one may define the unique integral path $x(t)$ starting from the point $x_0 = x(t_0)$ and satisfying (1.15).

If (1.14) is valid, then the integral path may be defined up to the first exit moment. In particular, for $G = B$ it is defined for each x_0 and all $t \in R^1$. In this way, one may define a one-parameter group of mappings

$$S_t : B \to B, \quad S_0 = \mathrm{id}_B, \quad S_t \circ S_\tau = S_{t+\tau}, \quad \tau, t \in R^1,$$

satisfying the following relation

$$\frac{\mathrm{d}\, S_t \circ x}{\mathrm{d}\, t} = f\,(S_t \circ x).$$

If $f \in C_k\,(G, B, B)$, then $S_\tau \in C_{k+1}\,(B)$. The set of mappings S_t, $t \in R^1$ is called the integral flow of the vector field f.

All these statements are valid, if one considers the equation

$$\frac{\mathrm{d}\, x}{\mathrm{d}\, t} = f\,(t, x) \tag{1.16}$$

with the right hand side term f continuously depending on the parameter t and satisfying (1.14) uniformly in t. In this case the one-parameter group of mappings S_t must be changed to the two-parameter family of mappings $S_{t,\tau} : x\,(\tau) \mapsto x\,(t)$ once again possessing the evolution property

$$S_{t,\tau} = S_{t,\theta} \circ S_{\theta,\tau}, \quad S_{t,t} = \mathrm{id}, \tag{1.17}$$

$$\tau < \theta < t.$$

Consider, in particular, the linear equation

$$\frac{\mathrm{d}\, x}{\mathrm{d}\, t} = A\,(t)\, x$$

with $A\,(t) \in L\,(B)$, $0 \le t \le T$. If $A\,(t) \equiv A$ does not depend on t, then there exists a representation

$$S_t = e^{At},$$

where the exponent function may be defined by the absolutely converging series

$$e^{At} = \sum_{k=0}^{\infty} \frac{1}{k!}\, t^k A^k$$

for all $t \in R^1$.

In the nonautonomous case, it is not difficult to obtain a multiplicative representation

$$S_{t,\tau} = \lim_{\max|\Delta_k t| \to 0} \overset{\curvearrowleft}{\prod_{k=1}^{n}} e^{A(t_{k-1})\Delta_k t}, \tag{1.18}$$

where

$$\tau = t_0 < t_1 < \ldots < t_n = t, \quad \Delta_k t = t_k - t_{k-1}$$

and the arrow indicates the order of multiplication. The limit in (1.18) is understood in the sense of uniform operator norm convergence. The formula (1.18) justifies the notation

$$S_{t,\tau} = \overset{\frown}{\exp} \int_{\tau}^{t} A\,(\theta)\,\mathrm{d}\,\theta \tag{1.19}$$

and the name 'multiplicative integral' for it.

Notice that the composition in (1.18) is nothing than a linear operator multiplication.

Let $\varphi\,(t)$ be a continuous B-valued function. The solution of the nonuniform equation

$$\frac{\mathrm{d}\,x}{\mathrm{d}\,t} = A\,(t)\,x + \varphi\,(t)$$

may be written in the form

$$x\,(t) = S_{t,\tau} \circ x + \int_{\tau}^{t} S_{t,\theta} \circ \varphi\,(\theta)\,\mathrm{d}\,\theta. \tag{1.20}$$

Let us now return to treating the nonlinear equation (1.16) which is equivalent to the integral equation

$$x_{\tau}\,(t) = x + \int_{\tau}^{t} f\,(\theta, x_{\tau}\,(\theta))\,\mathrm{d}\,\theta.$$

If f is smooth, then by differentiating the last relation with respect to t, we obtain the integral equation

$$\frac{\mathrm{d}}{\mathrm{d}\,\tau}\, x_{\tau}\,(t) = -f\,(\tau, x) + \int_{\tau}^{t} f'\,(\theta, x_{\tau}\,(\theta))\,\frac{\mathrm{d}}{\mathrm{d}\,\tau}\, x_{\tau}\,(\theta)\;\mathrm{d}\,\theta$$

and, thus, a linear differential equation for

$$\upsilon\,(t, \tau; x) = \frac{\mathrm{d}}{\mathrm{d}\,\tau}\, S_{t,\tau} \circ x,$$

$$\frac{\mathrm{d}\,\upsilon\,(t\;\tau;\, x)}{\mathrm{d}\,t} = f'\,(t, S_{t,\tau} \circ x)\, v\,(t, \tau; x),$$

$$\upsilon\,(\tau, \tau; x) = -f\,(\tau, x).$$

At last (1.19) yields the representation

$$\upsilon\,(t, \tau; x) = -\overset{\frown}{\exp}\left\{ \int_{\tau}^{t} f'\;(\theta,\, S_{\theta,\tau} \circ x)\;\mathrm{d}\;\theta \right\}.$$

A function $u\,(t, x)$ which is smooth with respect to a pair of variables $(t, x) \in R^1 \times B$, is called an integral of Equation (1.16) if

$$u\,(t, x\,(t)) = \mathrm{const} \tag{1.21}$$

holds for solutions of (1.16).

Putting $x\,(t) = S_{t,\tau} \circ x$ and differentiating (1.21) with respect to t for $t = \tau$, we obtain for $u\,(t, x)$ a first-order partial differential equation

$$\frac{\partial u(t,x)}{\partial t} + \langle f(t,x,u'(t,x))\rangle = 0. \tag{1.22}$$

It is easy to see that, to the contrary, each solution of (1.22) is an integral of (1.16). Solutions of (1.16) are called characteristics of (1.22). It is possible to construct a solution to the Cauchy problem for Equation (1.22) under the condition

$$u(\tau,x) = \varphi(x). \tag{1.23}$$

In fact, it follows from (1.21) that given (t,x) and a solution of the problem (1.22), (1.23), the relation

$$u(t,x) = u(\tau, S_{t,\tau}\circ x) = \varphi(S_{t,\tau}\circ x) \tag{1.24}$$

is valid. In this way, we have proved the uniqueness of the solution of this problem. Let us show that the function (1.24) is in fact a solution of (1.22), (1.23) under the condition that $\varphi(x)$ is smooth. Due to the uniqueness of the solution, we may take any t as a starting moment, hence it is enough to put $t=\tau$. Now a simple calculation gives

$$\left.\frac{\partial u(t,x)}{\partial t}\right|_{t=\tau} = \varphi'(S_{t,\tau}\circ x)\frac{d}{dt}S_{t,\tau}\circ x|_{t=\tau}$$

$$= -\langle f(x),\varphi'(x)\rangle = -\langle f(x), u'(\tau,x)\rangle.$$

Consider now a more general linear first order equation for a map $u: R^1\times B\to B_1$, $u:(t,x)\mapsto B_1$, $u:(t,x)$, where B_1 is a Banach space as well:

$$\frac{\partial u(t,x)}{\partial t} + u'(t,x)\,a(t,x) + b(t,x)\,u(t,x) = 0, \tag{1.25}$$

with

$$a(t,x)\in B,\quad b(t,x)\in L(B_1),\quad x\in B,\ t\in R^1.$$

We shall now show a trick which later will be used in much more general situations.

Let $y\in B_1^*$ and $\upsilon(t,x,y) = \langle u(t,x), y\rangle$ be a scalar function on $\widetilde{B} = B\times B_1^*$ for each t.

Let us calculate the partial derivatives of this function

$$\upsilon'_y(t;x,y) = u(t,x)\in B_1,$$

$$\upsilon'_x(t;x,y) = [u'_x(t,x)]^*,\quad y\in B_1^*.$$

As easy consequence of these relations and (1.22) is the equation

$$\frac{\partial \upsilon(t;x,y)}{\partial t} + \langle c(t;x,y), \upsilon'(t;x,y)\rangle = 0, \tag{1.26}$$

with

$$c(t;x,y) = \{a(t,x),\quad b^*(t,x)\,y\}\in\widetilde{B}$$

and with the space $\widetilde{B}^* = B^*\times B_1$ being treated as a space dual to $\widetilde{B} = B\times B_1^*$ under the pairing

$$\langle (a, b^*y), (\beta,u)\rangle = \langle a,\beta\rangle + \langle b^*y, u\rangle,\quad (\beta,u)\in\widetilde{B}^*.$$

As a result, we now obtain the scalar equation (1.26). The equation for its characteristics may be written in the form of the system

$$\frac{\mathrm{d}\, x}{\mathrm{d}\, t} = a\,(t, x),$$

$$\frac{\mathrm{d}\, y}{\mathrm{d}\, t} = b\,(t, x)\, y.$$

If $S_{t,\tau} \circ x$ is a solution of the first equation, then one may give a solution of the second equation in the form

$$\overleftarrow{\exp} \int_{\tau}^{t} b^{*}\,(\theta, S_{\theta,t} \circ x)\,\mathrm{d}\,\theta.$$

Next, due to (1.24), the solution of (1.26) corresponding to the initial data $u\,(\tau, x) = \varphi\,(x)$ may be given in the form

$$\begin{aligned} \upsilon\,(t; x, y) &\\ &= \langle u\,(t, x), y\rangle \\ &= \left\langle \varphi\,(S_{t,\tau} \circ x) \left[\exp \int_{\tau}^{t} b^{*}\,(\theta, S_{\theta,t} \circ x)\,\mathrm{d}\,\theta\right] y \right\rangle . \end{aligned}$$

Thus, we obtain the following representation for the solution of the Cauchy problem (1.25), (1.23)

$$\begin{aligned} u\,(t, x) &\\ &= \left[\overrightarrow{\exp} \int_{\tau}^{t} b^{*}\,(\theta, S_{\theta,t} \circ x)\,\mathrm{d}\,\theta\right] \varphi\,(S_{t,\tau} \circ x) \\ &= \left[\overleftarrow{\exp}\left(-\int_{\tau}^{t} b^{*}\,(\theta, S_{\theta,t} \circ x)\,\mathrm{d}\,\theta\right)\right] \varphi\,(S_{t,\tau} \circ x). \end{aligned} \tag{1.27}$$

2. Measures and Integrals

2.1. MEASURE SPACES, INTEGRALS

A pair $(X, \mathcal{Z})$ with X a set and $\mathcal{Z}$ a σ-algebra of subsets of X is called a measurable space. The role of morphisms in the measurable space category is played by measurable mappings

$$f : (X, \mathcal{Z}) \rightarrow (X_1, \mathcal{Z}_1)\,.$$

By definition, f is measurable if the inverse image of a measurable set is measurable, that is, belongs to the corresponding σ-algebra $\mathcal{Z}$:

$$A_1 \in \mathcal{Z}_1 \Rightarrow f^{-1}(A_1) = \{x \in X : f(x) \in A_1\} \in \mathcal{Z}.$$

A real function $\mu : \mathcal{Z} \to R^1$ is called a submeasure if it is finitely additive

$$\mu\left(\bigcup_{n}^{k=1} A_k\right) = \sum_{k=1}^{n} \mu(A_k), \quad A_k \in \mathcal{Z},\ A_k \cap A_j = \varnothing,\ k \neq j,\ k, j = 1, \dots, n.$$

We will deal only with bounded submeasures. In this case, the additional requirement of submeasure continuity:

$$A_n \underset{n\to\infty}{\to} \varnothing \Rightarrow \mu(A_n) \underset{n\to\infty}{\to} 0$$

is sufficient for μ to be σ-additive

$$\mu\left(\bigcup_{n}^{k=1} A_k\right) = \sum_{k=1}^{n} \mu(A_k), \quad A_k \in \mathcal{Z},\ A_k \cap A_j = \varnothing,\ k = j,$$

where the series is absolutely convergent by definition.

A (finite) measure is a σ-additive submeasure. It is called positive, if $\mu(A) \geq 0\ (\forall A \in \mathcal{Z})$ and a probability measure if, moreover, $\mu(X) = 1$.

A triple $(X, \mathcal{Z}, \mu)$ with μ a measure on $(X, \mathcal{Z})$ is called a measure space.

For any measure μ one may define two positive measures on the same measurable space

$$\mu^+(E) = \sup_{A \subset E} \mu(A),$$

$$\mu^-(E) = \inf_{A \subset E} \mu(A)$$

which satisfy the relation

$$\mu(E) = \mu^+(E) + \mu^-(E).$$

Moreover, there exists a measurable set X_0, such that

$$\mu^+(E) = \mu(E \setminus X_0), \quad \mu^-(E) = -\mu\big(E \cap (X \setminus X_0)\big).$$

The measure $|\mu| = \mu^+ + \mu^-$ is called a full variation of the measure μ.

By introducing the norm

$$\|\mu\| = |\mu|(X),$$

one may give a structure of the Banach space $M(X, \mathcal{Z})$ to the linear space of all measures defined on $(X, \mathcal{Z})$.

If a sequence of measures $\{\mu_n\}_{n=1}^{\infty}$ converges in the topology of this space, we say that it converges in variations. Evidently a convergence of μ_n in the variation implies a convergence of the real-valued sequence $\{\mu_n(A)\}_{n=1}^{\infty}$, for each $A \in \mathcal{Z}$. Vice-versa, it is known that the convergence of the last sequence implies the existence of a measure μ on $\mathcal{Z}$, which satisfies

$$\mu(A) = \lim_{n} \mu_n(A).$$

Let $(X, \mathcal{Z}, \mu)$ be a measure space with a positive measure μ. Consider the linear space $\mathcal{L} = \mathcal{L}_0(X, \mathcal{Z})$ of simple real functions $f(x) = \Sigma f_k j_{A_k}(x)$, where $j_A(x)$ is the characteristic function of the set A. (We suppose that R^1 is equipped with a measurable space structure $(R^1, \mathcal{Z}_{R^1})$ with $\mathcal{Z}_{R^1}$ a Borel-σ-algebra.) On the space $\mathcal{L}$ one may define the integral

$$\int_X f(x)\, \mu(d\,x) = \sum_k f_k\, \mu(A_k).$$

This is a positive linear functional

$$f(x) \geq 0, \quad (x \in X) \Rightarrow \int_X f(x)\, \mu(d\,x) \geq 0$$

and continuously monotonous

$$f_n(x) \underset{n \to \infty}{\to} 0 \Rightarrow \int_X f_n(x)\, \mu(d\,x) \to 0.$$

One may introduce the norm

$$\| f \| = \int_X | f(x) |\, \mu(d\,x)$$

in the space $\mathcal{L}$. Completing $\mathcal{L}$ with respect to this norm, we obtain the Banach space $\mathcal{L}_1(X, \mu)$ of absolutely integrable (with respect to the measure μ) functions (or more precisely, classes of functions, which are mod μ identified). We may extend our integral to this space preserving all properties described above.

In the same way, one may define a space $\mathcal{L}_p(X, \mu)$ of the pth power absolutely integrable, with respect to μ, functions with norm

$$\| f \|_p = \int_X | f(x) |^p\, \mu(d\,x).$$

Consider a measure

$$\mu_f(A) = \int_A f(x)\, \mu(d\,x) = \int_X f(x)\, j_A(x)\, \mu(d\,x)$$

for $f \in \mathcal{L}_1(X, \mu)$. The measure μ_f on $(X, \mathcal{Z})$ is called an indefinite integral of a function (with respect to μ). We denote $\mu_f = f \cdot \mu$. The indefinite integral is absolutely continuous with respect to μ, $\mu_f < \mu$. Recall that $\nu < \mu$ means that

$$\mu(A) = 0 \Rightarrow \nu(A) = 0, \quad A \in \mathcal{Z}.$$

On the contrary, if $\nu < \mu$, then there exists a function $f \in \mathcal{L}_1(X, \mu)$ such that $\nu = \mu_f$. This function is called the Radon–Nikodim derivative (density) of ν with respect to μ and denoted

$$f(x) = \frac{\nu(d\,x)}{\mu(d\,x)}.$$

Two measures μ_1, μ_2 are called equivalent $\mu_1 \sim \mu_2$ if $\mu_1 < \mu_2$ and $\mu_2 < \mu_1$.

Consider a sequence of measurable functions $f_n(x)$ on $(X, \mathcal{Z}, \mu)$. A natural notion of convergence is the almost everywhere (a, e) convergence

$$\mu(x : f_n(x) \mapsto f(x)) = 1.$$

If, in addition, the simple condition

$$|f_n(x)| \le g(x)$$

holds for an integrable function g,

$$\int_X g(x)\,\mu(dx) < \infty ,$$

then the convergence in the mean sense

$$\|f - f_n\|_{\mathcal{L}_1} = \int_X |f(x) - f_n(x)|\,\mu(dx) \to 0$$

is granted for the almost everywhere convergent sequence f_n. In particular, this guarantees the possibility of taking the limit under the integral sign

$$\int_X f(x)\,\mu(dx) = \lim_{n\to\infty} \int_X f_n(x)\,\mu(dx).$$

In the same way, one may define integrals with respect to nonpositive measures. It is easy to see that each measure is absolutely continuous with respect to its full variation $\mu < |\mu|$. Moreover, its density

$$\rho(x) = \frac{\mu(dx)}{|\mu|(dx)}$$

possesses the property $\|\rho(x)\| = 1$. That is why properties of an integral with respect to μ may be studied by investigating the integral with respect to $|\mu|$

$$\int_X f(x)\,\mu(dx) = \int_X f(x)\,\rho(x)\,|\mu|\,(dx).$$

Let $(X, \mathcal{Z}, \mu)$ be a measure space and

$$f : (X, \mathcal{Z}) \to (X_1, \mathcal{Z}_1).$$

The mapping f generates a measure on $(X_1, \mathcal{Z}_1)$

$$f(\mu) = \mu^f = \mu \circ f^{-1},\ \mu^f(A) = \mu(f^{-1}(A)).$$

Notice that the following formula holds when changing the integration variable.

$$\int_{X_1} \varphi(x_1)\,\mu^f(dx_1) = \int_X \varphi(f(x))\,\mu(dx).$$

For simple functions on $(X_1, \mathcal{Z}_1)$, this formula is merely a definition of μ^f. For arbitrary functions on $(X_1, \mathcal{Z}_1)$, one may prove it by a limit procedure.

Let X be a topological space and the σ-algebra $\mathcal{Z}$ generated by open sets. A sequence of measures defined on $(X, \mathcal{Z})$ is called weakly convergent if for $n \to \infty$

$$\int_X \varphi(x)\,\mu_n(dx) \to \int_X \varphi(x)\,\mu(dx)$$

for any function φ, continuous on X. Evidently, convergence in variation implies weak measure convergence. Let us point out that the weak convergence described above does not coincide with weak convergence in Banach space $\mathcal{M}(X, \mathcal{Z})$ topology. The latter is equivalent to the convergence of a bounded measure sequence μ_n on each $A \in Z$, $\mu_n(A) \to \mu(A)$.

2.2. MEASURES IN LINEAR SPACES

Let X be a linear space, Y be a set of linear functionals on X, total in the sense that

$$\langle x, y \rangle = 0, \quad \forall\, y \in Y \Rightarrow x = 0.$$

Any finite set $\tau = \{y_1, \ldots, y_n\} \in Y^n$ gives rise to a linear mapping

$$x \overset{\tilde{\tau}}{\to} (\langle x, y_1 \rangle, \ldots, \langle x, y_n \rangle)$$

from the space X into the n-dimensional space R_τ. Moreover, inverse images of Borel sets form a σ-algebra Z_τ in X. A union algebra $\overset{0}{Z}_Y = \cup Z_\tau$ is an algebra of sets which are called cylindrical. Denote Z_Y the minimal σ-algebra which includes all cylindrical sets.

Any measure μ defined on (X, Z_Y) generates a set of measures $\mu_\tau = \mu \circ \tau^{-1}$ defined on finite-dimensional spaces R_τ. This set is called a system of finite-dimensional distributions, corresponding to μ. Finite-dimensional distributions satisfy a consistency condition: if $\tau_1 \subset \tau$ and $\tau_1 = P \circ \tilde{\tau}$, where $P : R_\tau \to R_{\tau_1}$ is a projection, then $\mu_{\tau_1} = \mu_\tau \circ P^{-1}$. Vice-versa, a set of measures $\{\mu_\tau\}$ defined on R_τ, satisfying the above consistency condition, generates a submeasure μ, correctly defined on Z_Y by

$$\mu\big|_{Z_\tau} \circ \tau^{-1} = \mu_\tau .$$

This submeasure is usually called a cylindrical measure. Given a measure μ, define a characteristic functional χ_μ, which is a function on Y, by the formula

$$\chi_\mu(y) = \int_X e^{i\langle x, y\rangle} \mu(\mathrm{d}\,x) = \int_{-\infty}^{\infty} e^{iz} \mu_y(\mathrm{d}\,z),$$

where $\mu_y(\cdot)$ is the one-dimensional distribution corresponding to $\tau = \{y\}$. Evidently, to calculate a characteristic functional, one needs only a system of finite-dimensional distributions. On the other hand, this system may be reconstructed in a unique way if one knows the characteristic functional, since

$$\chi_\mu\Big(\sum_k \alpha_k y_k\Big) = \int_X e^{i \sum_k \alpha_k \langle x, y_k\rangle} \mu(\mathrm{d}\,x)$$

$$= \int_{R_\tau} e^{i \sum_k \alpha_k z_k} \mu_\tau(\mathrm{d}\,z),$$

and the measure μ_τ on the finite-dimensional space is defined by its Fourier transformation in a unique way.

Thus, the characteristic functional generates the cylindrical measure and vice versa. Nevertheless, generally speaking, without some additional assumptions (which may be quite complicated) a cylindrical measure on $\overset{0}{Z}_Y$ could correspond to no measure defined on Z_Y.

We consider now an important case of positive measure μ. In this case, the characteristic functional is positively defined, that is

$$\sum \chi_\mu (y_j - y_k)\, c_j \bar{c}_k \geq 0$$

for any set $\{y_j\}_{j=1}^n$ and complex numbers $c_1, \ldots, c_n$.

In what follows we shall be interested mainly in measures defined on a Banach space equipped with a Hilbert–Schmidt structure

$$X = B \supset H \supset B^* = Y. \tag{2.1}$$

In this situation we always choose for $\mathcal{Z}$ a Borel σ-algebra in X.

It is well known that any definite positive function χ from B^* into R^1 continuous at the origin in Hilbert space H topology is the characteristic function of some positive measure on B.

The function

$$\chi(y) = e^{-\frac{1}{2}\|y\|_H^2}, \quad y \in H,$$

possessing all properties described above, gives an example which will be crucial in the following. The measure on B, corresponding to it, is called the Gaussian measure canonically associated with rigging (2.1) (or, what is the same, with a Hilbert–Schmidt structure).

Let now μ be an arbitrary measure on B. A moment (of the nth order) of the measure μ is an integral of the form

$$\int_X \langle x, y_1\rangle \ldots \langle x, y_n\rangle\, \mu(\mathrm{d}x) = S_\mu(y_1, \ldots, y_n).$$

The condition which guarantees the existence of the nth order moment, which is a continuous multilinear functional on B^*, is the estimate

$$\int_X \|x\|^n \mu(\mathrm{d}x) < \infty.$$

It is an easy task to prove that under this condition the characteristic functional χ_μ is differentiable up to the nth order and

$$S_\mu(y_1, \ldots, y_n) = (-1)^n \chi_\mu^{(n)}(0)(y_1, \ldots, y_n).$$

In particular, the linear functional

$$S_\mu(y) = \int_X \langle x, y\rangle\, \mu(\mathrm{d}x) = -i\chi'_\mu(0)$$

is called the mean of the measure μ, while the bilinear functional

$$S_\mu(y_1, y_2) = \int_X \langle x, y_1\rangle \langle x, y_2\rangle\, \mu(\mathrm{d}x) = -\chi''_\mu(0)(y_1, y_n)$$

is called its correlation functional. If

$$S_\mu(y_1, y_2) = \langle \Phi_{y_1}, y_2\rangle,$$

where $\Phi \in L(B^*, B)$, then Φ is called the measure correlation operator.

It is easy to see that the canonical Gaussian measure has zero mean, while its correlation operator is the identity operator. Thus, the following formula

$$\int_X \langle x, y\rangle^2 \mu(\mathrm{d}x) = \| y \|_h^2, \quad (y \in B^*)$$

holds. It shows that the mapping

$$y \mapsto \langle x, y\rangle$$

which puts in correspondence a linear functional ℓ_y on B to any $y \in B^*$ may be extended to an isometric isomorphism between H and a subspace of the space $\mathcal{L}_2(B^*, \mu)$. The elements of this subspace are called measurable linear functionals on B. Sometimes we will preserve the notation $\ell_y = \langle x, y\rangle$, even in the case $y \overline{\in} B^*$.

Notice that

$$\int_X e^{i\ell_y(x)} \mu(\mathrm{d}x) = e^{-\frac{1}{2}\|y\|^2}.$$

Along with the measurable linear functional, we shall need later measurable linear operators.

A measurable function $A(x)$ on B which is the limit of an almost everywhere convergent sequence $A_n(x)$, $A_n \in L(B, B_1)$

$$A(x) = \lim A_n(x) \quad (\mathrm{mod}\ \mu) \tag{2.2}$$

is called a measurable linear operator from B to B_1.

Evidently a linear operator continuous on B gives rise to a measurable operator A. If B is equipped with a H S structure (B, J), then $A_0 = AJ \in L_{12}(H, B_1)$, as $J \in L_{12}(H, B)$.

On the other hand, as the convergence in (2.2) takes place on a positive measure set which contains H, one may uniquely define the operator $A \in \mathcal{L}(H, B_1)$

$$Ax = \lim A_n x, \quad x \in H.$$

So, a measurable operator on B has a continuous restriction to H.

3. Measure Differentiation

3.1. LOGARITHMIC DERIVATIVES

Consider a smooth vector field z which maps a Banach space B into itself, $z: B \to B$, and possesses a bounded derivative $z'(x)$, $\sup \| z'(x) \| < \infty$. The integral flow S_t $(-\infty < t < \infty)$ corresponding to it is a family of operators that maps B into itself and preserves the Borel σ-algebra $\mathcal{Z}$. That is why for $\mu \in \mathcal{M}(B, \mathcal{Z})$ the relation

$$\mu_t \in S_t(\mu) = \mu \circ S_t^{-1} \in \mathcal{M}(B, \mathcal{Z})$$

makes sense.

DEFINITION 3.1. We say a measure μ is differentiable along the vector field z if there exists a measure

$$D_z \mu = -\frac{d}{dt} \mu_t \Big|_{t=0}$$

the derivative of μ along z in the following sense: for any $\varphi \in C_1(B, B, R^1)$

$$\int_B \varphi(x) (D_z \mu)(dx) = -\lim_{t \to 0} \int_B \varphi(x) \frac{\mu_t - \mu}{t} (dx). \tag{3.1}$$

Notice that after a simple transformation, this relation is changed to the following

$$\int_B \varphi(x) (D_z \mu)(dx) = -\int_B (D_z \varphi)(x) \mu(dx). \tag{3.2}$$

The notation $D_z \varphi = (z, D) \varphi$ which has been used above, makes the corresponding notation $D_z \mu = (D, z) \mu$ natural.

DEFINITION 3.2. Let μ be differentiable along z and let its derivative $D_z \mu$ be absolutely continuous with respect to μ, $D_z \mu \prec \mu$. Then the density

$$\rho_\mu(z; x) = \frac{D_z \mu(dx)}{\mu(dx)} = \frac{((D, z)\mu)(dx)}{\mu(dx)}$$

is called the logarithmic derivative of the measure μ along z.

If the measure μ has the logarithmic derivative along z, one may rewrite formula (3.2) (integration by parts) in the form

$$\int_B (D_z \varphi)(x) \mu(dx) = -\int_B \varphi(x) \rho_\mu(z; x) \mu(dx). \tag{3.3}$$

There is a useful way for calculating logarithmic derivatives.

PROPOSITION 3.1. *Let* $\mu_t \prec \mu$ *and the density*

$$\beta(t, x) = \frac{\mu_t(dx)}{\mu(dx)}$$

be differentiable with respect to t at $t = 0$ *as an element of the space* $L_1(B, \mu)$. *Then there exists the logarithmic derivative*

$$\rho_\mu(z; x) = -\frac{\partial \beta(t, x)}{\partial t} \Big|_{t=0}.$$

Proof. Under the above condition, it is easy to check that

$$\int_B \varphi(x) \frac{\mu_t(dx) - \mu(dx)}{t}$$

$$= \int_B \varphi(x) \frac{\beta(t, x) - 1}{t} \mu(dx) \to$$

$$\to \int_B \varphi(x) \frac{\partial \beta(t, x)}{\partial t} \Big|_{t=0} \mu(dx).$$

EXAMPLE. Let μ be a canonical Gaussian measure for the triple $B \supset H \supset B^*$, $z(x) = h \in H$ be a constant vector field and $S_{-t} x = x - t h$, the corresponding shift transformation. It is known that in this case $\mu_t \prec \mu$ and

$$\beta(t, x) = \exp\left\{t(x, h) - \frac{t^2}{2}\|h\|^2\right\}.$$

Thus, there exists $D_h \mu$ and $\rho_\mu(h; x) = -(x, h)$.

Let us investigate both structure and properties of logarithmic derivatives.

Integration by parts formula lead to the relations

$$D_z(\alpha, \mu) = \alpha D_z \mu + (D_z \alpha)\mu = D_{\alpha z} \mu, \tag{3.4}$$

that is

$$(D, z)(\alpha \mu) = \alpha(D_z \alpha) + (z, D)\alpha \mu$$

and

$$\rho_\mu(\alpha z; x) = \alpha(x)\rho_\mu(z; x) + D_z \alpha. \tag{3.5}$$

If $\alpha(x) \neq 0$. (μ – a.e.) then

$$\rho_{\alpha\mu}(z; x) = \frac{1}{\alpha(x)} \rho_\mu(\alpha z; x) = \rho_\mu(z; x) + D_z \quad n\,\alpha(x). \tag{3.6}$$

Those formulas show that a map $z \mapsto \rho_\mu(z; \cdot)$ acts on a set of vector fields, along which μ is differentiable as a linear first-order differential operator. In particular, for a constant vector field $z(x) \equiv h$, $\rho(h; x)$ is a linear functional for each $x \in B$.

Consider now a space with a Hilbert–Schmidt structure $B \supset H \supset B^*$.

Denote $\mathcal{M}_0(B)$ a class of measures, possessing logarithmic derivatives along any constant vector field $z(x) \equiv h \in B^*$

$$\rho_\mu(h; x) = \langle \lambda(x), h \rangle, \tag{3.7}$$

where $\lambda = \lambda_\mu : B \to B$ (μ – a.e.) is a weakly measurable mapping. We call it the (vector) logarithmic derivative of the measure μ.

EXAMPLE. The canonical Gaussian measure $\mu \in \mathcal{M}_0(B)$ and $\lambda(x) = -x$.

Let $\sigma_1(B^*)$ be the space of vector fields $z : B \to B^*$ belonging to $C_1(B, B, B^*)$ and satisfying the condition

$$\|z\|_1 = \sup_{x \in B} \left\{\|z'(x)\|_{L(B, B^*)}, \|z(x)\|_B\right\} < \infty.$$

THEOREM 3.1. *Let $\mu \in \mathcal{M}_0(B)$ and $z \in \sigma_1(B^*)$. Then μ has logarithmic derivative along z and the equality*

$$\rho_\mu(z; x) = \langle \lambda(x), z(x) \rangle + \operatorname{div} z(x) \tag{3.8}$$

holds, that is

$$(D, z)\mu = \langle D + \lambda(x), z(x) \rangle \mu. \tag{3.9}$$

Proof. First, put $z(x) = \alpha(x) h$, where $h \in B^*$ and $\alpha(x)$ is a scalar function of class C_1. Then

$$z'(x)\,h_1 = \langle h, \alpha'(x)\rangle\, h_1,$$

$$\operatorname{div} z'(x) = \operatorname{Tr} z'(x) = \sum_k \langle e_k, \alpha'(x)\rangle\,(h, e_k)$$

$$= \langle \alpha'(x), h\rangle = D_h\,\alpha$$

and (3.8) follows now from (3.5) and (3.7).

One may prove the assertion of the theorem in the general case by a limiting procedure in the formula

$$\int_B D_z\,\varphi(x)\,\mu(d\,x)$$

$$= -\int_B \varphi(x)\,[\langle\lambda(x), z(x)\rangle + \operatorname{Tr} z'(x)]\,\mu(d\,x), \tag{3.10}$$

which is justified by the possibility of approximating $z \in \sigma_1(B^*)$ (along with its derivative) by linear combinations of expressions like $\alpha(x)\,h$.

3.2. ESTIMATES OF LOGARITHMIC DERIVATIVE POWERS

Let us now estimate integrals of the form

$$\int_B |\,\rho(z; x)\,|^{2m}\,\mu(d\,x)$$

with respect to a positive measure μ. The estimate, as we shall see below, will be closely connected with estimates for higher derivatives of the measure.

Consider the class $\mathcal{M}_1(B, H) \subset \mathcal{M}_0(B)$ of those measures which have (a.e.) vector logarithmic derivatives along H such that $\lambda'(x) \in L(H)$ and $\sup \|\,\lambda'(x)\,\|_{L(H)} < \infty$.

First, choose a vector field z for which the function $\rho_\mu(z; x)$ is bounded together with its first-order derivative. Next, we shall weaken the latter requirement by means of vector-field approximation and a limiting procedure. Now let us estimate the integral

$$\int_B |\,\rho_\mu(z; x)\,|^2\,\mu(d\,x)$$

$$= \int_B \rho_\mu(z; x)\,(D_z\,\mu)(d\,x)$$

$$= -\int_B D_z\,\rho_\mu(z; x)\,\mu(d\,x). \tag{3.11}$$

The expression under the integral sign may be calculated, using (1.13):

$$D_z\,\rho_\mu = (z, D)\,[\langle\lambda(x), z(x)\rangle + \operatorname{div} z(x)]$$

$$= \langle D_z\,\lambda(x), z(x)\rangle + \langle\lambda(x), D_z\,z(x)\rangle +$$

$$+ D_z\,\operatorname{div} z(x) = \rho\,(D_z\,z; x) +$$

$$+ \langle D_z\,\lambda(x), z(x)\rangle - \operatorname{Tr}\,[z'(x)]^2. \tag{3.12}$$

Substituting the expression in (3.11), we obtain the formula

$$\int_B |\rho_\mu(z; x)|^2 \mu(\mathrm{d}x)$$

$$= \int_B \{\mathrm{Tr}\,[z'(x)]^2 - \langle \lambda'(x)\, z(x), z(x)\rangle\}\, \mu(\mathrm{d}x), \tag{3.13}$$

since

$$\int_B \rho(D_z z; x)\, \mu(\mathrm{d}x) = (D_{D_z z}\mu)(B) = 0.$$

Notice, moreover, that the right-hand side of (3.12) makes sense under much more general assumptions about the vector field.

Consider the Hilbert space $\mathcal{H}_1^2(B, H, \mu)$ of vector functions $z: B \to H$ which is the closure of $C_1(B, H, H)$ with respect to the norm

$$\sigma_{12}(z) = \left\{\int_B \left[\| z(x)\|^2 + \mathrm{tr}\, z'(x)\,(z'(x))^*\right] \mu(\mathrm{d}x)\right\}^{1/2}.$$

Evidently, the right-hand side of (3.13) is defined for $z \in \mathcal{H}_1^2(B, H, \mu)$ and for vector fields in the class $\sigma_1(B^*)$ described above, which is dense in $\mathcal{H}_1^2(B, H, \mu)$. Hence, the estimate

$$\int_B |\rho_\mu(z; x)|^2 \mu(\mathrm{d}x) \le \mathrm{const}\, \sigma_{12}^2(z) \tag{3.14}$$

holds.

This proves that the linear differential operator

$$\rho_\mu : z \mapsto \langle \lambda(x), z(x)\rangle + \mathrm{div}\, z(x)$$

is a continuous map

$$\rho_\mu : \sigma_1(B^*) \to \mathcal{L}_1(B, \mu)$$

and thus, by continuity, may be extended to the whole $\mathcal{H}_1^2(B, H, \mu)$. We keep the previous notation for this extension.

It remains to be noted that in formula (3.13) one may take the limit for $z_n \to z$ with $z_n \in \sigma_1(B^*)$, $z \in \mathcal{H}_1^2$, which leads to the following result.

THEOREM 3.2. *Let* $\mu \in \mathcal{M}_1(B, H)$ *and* $z \in \mathcal{H}_1^2(B, H, \mu)$. *Then* μ *is differentiable along* z *and its logarithmic derivative does exist and satisfies* (3.13) *and* (3.14).

Remark 1. Let $z(x) \equiv h \in H$. Then for μ almost all x the logarithmic derivative $\rho_\mu(h; x)$ is the extension of the linear functional $\langle \lambda(x), h\rangle_B$ defined on B and satisfying the estimate

$$\int_B |\rho_\mu(h; x)|^2 \mu(\mathrm{d}x) \le \mathrm{const}\, \| h \|_H^2 .$$

Remark 2. Consider in particular a canonical Gaussian measure. Then we obtain $\lambda'(x) = -1$ and formula (3.13) takes the form

$$\int_B |\rho_\mu(z; x)|^2 \mu(\mathrm{d}x)$$

$$= \int_B \{(\| z(x) \|_H^2 + \mathrm{Tr}\,[z'(x)]^2\}\, \mu(\mathrm{d}\,x), \tag{3.15}$$

where for $z \in C_1(B, B, B^*)$,

$$\rho_\mu(z, x) = \mathrm{tr}\, z'(x) - \langle x, z(x)\rangle_B\,. \tag{3.16}$$

For $z(x) \equiv h \in B^*$ this leads to the equality $\rho_\mu(z; x) = -\langle x, h\rangle_B$ where the right-hand side term is the continuous linear functional on B considered above.

As it follows from the equality

$$\int_B |\langle x, h\rangle_B|^2\, \mu(\mathrm{d}\,x) = \| h \|_H^2\,,$$

this functional may be extended to the whole H by means of a limiting procedure in a square mean sense. The function constructed in this way is measurable on B, and called a square integrable (with respect to the Gaussian measure) measurable linear functional on B. When it does not cause confusion, we shall use the same notation $\langle x, h\rangle_B$ in the case $h \in H$ (but $h \notin B^*$).

Let us use the above results to estimate higher-order logarithmic derivative integrals.

$$\int_B [\rho(z; x)]^{2n}\, \mu(\mathrm{d}\,x)$$

$$= \int_B [\rho(z; x)]^{2n-1}\, (D_z\, \mu)(\mathrm{d}\,x)$$

$$= -(2n-1) \int_B [\rho(z; x)]^{2n-2}\, (D_z\, \rho)(z; x)\, \mu(\mathrm{d}\,x).$$

Now the Hölder inequality leads to

$$\frac{1}{2n-1} \int_B \rho^{2n}(z; x)\, \mu(\mathrm{d}\,x)$$

$$\leq \left\{\int_B \rho^{2n}(z\,; x)\, \mu(\mathrm{d}\,x)\right\}^{1-1/n} \times$$

$$\times \left\{\int_B [D_z\, \rho(z\,; x)]^n\right\}^{1/n},$$

or

$$\frac{1}{2n-1} \left\{\int_B \rho^{2n}(z; x)\, \mu(\mathrm{d}\,x)\right\}^{1/n}$$

$$\leq \left\{\int_B [D_z\, \rho(z; x)]^n\, \mu(\mathrm{d}\,x)\right\}^{1/n}.$$

Finally, due to (3.12) and the Minkowsky inequality, we obtain the estimate

$$\frac{1}{2n-1} \left\{\int_B \rho^{2n}(z; x)\, \mu(\mathrm{d}\,x)\right\}^{1/n}$$

$$\leq \left\{\int_B \left[\mathrm{tr}\,[z'(x)]^2 - \langle \lambda'(x)\, z(x), z(x)\rangle\right]^n \mu(\mathrm{d}\,x)\right\}^{1/n} +$$

$$+ \left\{ \int_B [\rho\,(D_z z; x)]^n \mu\,(d\,x) \right\}^{1/n} . \tag{3.17}$$

It follows from $z \in C_m(B,B,H)$ that $D_z z \in C_{m-1}(B,B,H)$ so that one may iterate the above estimate.

For example, if $n = 2$, then

$$\left\{ \int_B \rho^4 (z; x) \mu\,(d\,x) \right\}^{1/2}$$

$$\leq 3 \left\{ \int_B \left(\mathrm{tr}\, [z'(x)]^2 - \langle \lambda'(x)\, z(x), z(x) \rangle \right)^2 \mu\,(d\,x) \right\} +$$

$$+ 3 \left\{ \int_B \left(\mathrm{tr} \left[(z'(x)\, z(x))' \right]^2 - \right. \right.$$

$$\left. \left. - \langle \lambda'(x)\, z'(x)\, z(x),\ z'(x)\, z(x) \rangle \right) \mu\,(d\,x) \right\}^{1/2} ,$$

and after some transformations based on the fact that $\lambda'(x)$ is bounded, we get the inequality

$$\left\{ \int_B \rho^4 (z; x) \mu\,(d\,x) \right\}^{1/4}$$

$$\leq \mathrm{const} \left\{ \left(\int_B \| z(x) \|^4 \mu\,(d\,x) \right)^{1/4} + \right.$$

$$+ \left(\int_B \sigma_2^4 (z'(x))\, \mu\,(d\,x) \right)^{1/4} +$$

$$\left. + \left(\int_B \sigma_{22}^4 (z''(x))\, \mu\,(d\,x) \right)^{1/4} \right\} . \tag{3.18}$$

More complicated computations, which would be tedious to report here, permit us to obtain

$$\left\{ \int_B \rho^{2^k} (z; x) \mu\,(d\,x) \right\}^{1/2^k}$$

$$\leq 2^{2k} \sum_{j=1}^{k} \left\{ \int_B \left| \mathrm{Tr}\, [z'_j]^2 (x) - \langle \lambda'(x)\, z_j(x)\, z_j(x) \rangle \right|^{2^{k-j}} \mu\,(d\,x) \right\}^{1/2^k} ,$$

where $z_1 = z_2$, $z_2 = D_{z_1} z_1$, $z_j = D_{z_1} z_{j-1}$ and, at last,

$$\left\{ \int \rho^{2^k} (z; x) \mu\,(d\,x) \right\}^{1/2^k} \leq \mathrm{const}\, \sigma^{\mu}_{k, 2^k} (z), \tag{3.19}$$

where

$$\sigma^{\mu}_{m,n} (z) = \sum_{j=0}^{m} \left\{ \int_B \sigma_{j2}^n \left(z^{(j)} \right) \mu\,(d\,x) \right\}^{1/n} . \tag{3.20}$$

Denote $W_m^n (B, H, \mu)$ the Banach space which is the closure of $C_m (B,B,H)$ with respect to the norm $\sigma^{\mu}_{m,n}$. Now we may state the following result.

THEOREM 3.3. *For* $\mu \in \mathcal{M}_1 (B, H)$ *and* $z \in W_k^{2^k}$ *the logarithmic derivative* $\rho_\mu (z; x)$ *satisfies the estimate* (3.19) *and thus the map*

$$z \mapsto \rho_\mu(z; x)$$

continuously maps $W_k^{2^k}$ *into* $\mathcal{L}_{2^k}(B, \mu)$.

3.3. HIGHER-ORDER DIFFERENTIAL OPERATIONS

Let z_1, z_2 be vector fields belonging to $C_1(B, H, H)$. Given $\varphi \in C_2(B, H, R^1)$, consider a second-order symmetric differential operation

$$\begin{aligned}\overset{(2)}{\mathcal{D}}_{z_1 \otimes z_2} \varphi(x) &= (z_1(x) \otimes z_2(x), D \otimes D)\, \varphi(x) \\ &= \varphi''(x)\,(z_1(x), z_2(x))\end{aligned}$$

and try to calculate the adjoint operation $\left(\overset{(2)}{\mathcal{D}}_{z_1 \otimes z_2}\right)^*$ on measures, defined by the relation

$$\int_B \varphi(x) \left(\overset{(2)}{\mathcal{D}}_{z_1 \otimes z_2}\right)^* \mu(\mathrm{d}x) = \int_B \overset{(2)}{\mathcal{D}}_{z_1 \otimes z_2} \varphi(x)\, \mu(\mathrm{d}x).$$

As far as

$$\overset{(2)}{\mathcal{D}}_{z_1 \otimes z_2} = (z_1, D)\,(z_2, D) - (D_{z_1} z_2, D)$$

one may write down the adjoint operation in the form

$$\begin{aligned}\left(\overset{(2)}{\mathcal{D}}_{z_1 \otimes z_2}\right)^* &= (D, z_2)\,(D, z_2) + (D, D_{z_1} z_2) \\ &= (D \otimes D, z_2 \otimes z_1).\end{aligned}$$

Let us derive an expression for $\left(\overset{(2)}{\mathcal{D}}_{z_1 \otimes z_2}\right)^*$ assuming that $\mu \in \mathcal{M}_1(B, H)$.

$$\begin{aligned}&\left(\overset{(2)}{\mathcal{D}}_{z_1 \otimes z_2}\right)^* \mu \\ &= (D, z_2)\,[\rho(z_1; x)\,\mu] + \rho(D_{z_1} z_2; x)\,\mu = \rho(z_1; x)\,(D, z_2)\,\mu + \\ &\qquad + (z_2, D)\,\rho(z_1; x)\,\mu + \rho\left(D_{z_1} z_2; x\right)\mu \\ &= \left[\rho(z_1; x)\,\rho(z_2; x) + D_{z_1}\,\rho(z_1; x) + \rho(D_{z_1} z_2; x)\right]\mu.\end{aligned}$$

Next,

$$\begin{aligned}&D_{z_2}\,\rho(z_1; x) \\ &= (z_2, D)\,(\lambda(x) + D, z_1) \\ &= \langle \lambda(x) + D, D_{z_2} z_1\rangle + \langle D_{z_2} \lambda(x), z_1\rangle - \mathrm{Tr}\, z'_1\, z'_2\,, \qquad (3.21)\end{aligned}$$

so that

$$\begin{aligned}&\left(\overset{(2)}{\mathcal{D}}_{z_1 \otimes z_2}\right)^* \mu \\ &= \left[\rho(z_1; x)\,\rho(z_2; x) + \rho\,\left(D_{z_1} z_2 + D_{z_2} z_1; x\right) + \right.\end{aligned}$$

$$+ \langle \lambda'(x) z_2, z_1 \rangle] \mu - \operatorname{Tr} z'_1 z'_2 \mu .$$

It follows from Theorem 3.3 that the function

$$\begin{aligned} &\rho_2(z_1, z_2; x) \\ &\quad = \rho(z_1; x)\,\rho(z_2; x) + \rho\left(D_{z_1} z_2 + D_{z_2} z_1; x\right) + \\ &\qquad + \langle \lambda'(x) z_2, z_1 \rangle - \operatorname{Tr} z'_1 z'_2 \end{aligned}$$

belongs to $\mathcal{L}_2(B, \mu)$ if $\mu \in \mathcal{M}_1(B, H)$ and $z_1, z_2 \in W_4^1(B, H, \mu)$.

Under these conditions, $\left(\mathcal{D}^{(2)}_{z_1 \otimes z_2}\right)^* \mu \prec \mu$ and

$$\rho_2(z_1, z_2; x) = \frac{\left(\mathcal{D}^{(2)}_{z_1 \otimes z_2}\right)^* \mu(\mathrm{d}x)}{\mu(\mathrm{d}x)} .$$

Using the expression for $\rho(z, x)$, we can write this function in a more symmetric way:

$$\begin{aligned} &\rho_2(z_1, z_2; x) \\ &\quad = (z_2, D)(\lambda + D, z_1) + \rho(z_2, x)(\lambda + D, z_1) + \\ &\qquad + \left(\lambda + D, D_{z_1} z_2\right) = (\lambda + D, z_2)(\lambda + D, z_1) + \\ &\qquad + \left(\lambda + D, D_{z_1} z_2\right) = ((\lambda + D) \otimes (\lambda + D), z_2 \otimes z_1). \end{aligned}$$

Consider now more general differential operations like

$$\mathcal{D}^{(2)}_C \varphi(x) = (C(x) D, D)\, \varphi(x) = \operatorname{Tr} C^*(x)\, \varphi''(x).$$

We state the following result.

THEOREM 3.4. *Let* $C \in C_2(B, L, (H))$ *and* $\mu \in \mathcal{M}_1(B, H)$. *Then*

$$\rho_C(x) = ((\lambda(x) + D) \otimes (\lambda(x) + D), C(x)) \in \mathcal{L}_2(B, \mu)$$

and

$$\frac{\left(\mathcal{D}^{(2)}_C\right)^* \mu(\mathrm{d}x)}{\mu(\mathrm{d}x)} = \rho_C(x).$$

Thus, for $\varphi \in C_2(B, H, R^1)$, *we arrive at the relation*

$$\begin{aligned} &\int_B (C(x) D, D)\, \varphi(x)\, \mu(\mathrm{d}x) \\ &\quad = \int_B \varphi(x) \left[(D \otimes D, C(x))\, \mu\right](\mathrm{d}x) \\ &\quad = \int_B \varphi(x) \left((D + \lambda(x)) \otimes (D + \lambda(x)), C(x)\right) \mu(\mathrm{d}x) . \end{aligned} \tag{3.22}$$

Proof. All the statements of the theorem are easily justified for operators having the form $z_1(x) \otimes z_2(x)$.

Due to the expansion

$$C(x) = \sum_k e_k \otimes C(x)\, e_k$$

it is enough to prove that we may go to the limit in (3.22). To this purpose, we calculate the expression

$$\int \left| \sum_{k=m}^{m+r} ((\lambda + D) \otimes (\lambda + D),\, e_k \otimes C(x)\, e_k) \right|^2 \mu(dx)$$

$$= \int \left| \left(\lambda + D, \sum_{k=m}^{m+r} e_k (\lambda + D, C(x)\, e_k)\right) \right|^2 \mu(dx)$$

$$\leq \text{const} \int \left\{ \sigma_2^2 \left[D \otimes \sum_{k=m}^{m+r} e_k (\lambda + D, C(x)\, e_k) \right] + \right.$$

$$\left. + \sum_{k=m}^{m+r} |(\lambda + D, C(x)\, e_k)|^2 \right\} \mu(dx). \tag{3.23}$$

For the second summand, the estimate

$$\int \sum_{k=m}^{m+r} |(\lambda + D, C(x)\, e_k)|^2 \mu(dx)$$

$$\leq \sum_{k=m}^{m+r} \int \{(\sigma_2^2 (C'(x)\, e_k) + \| C(x)\, e_k \|^2\} \mu(dx) \tag{3.24}$$

holds.

Moreover, using (3.21) we obtain

$$(e_j, D) \sum_{k=m}^{m+r} e_k (\lambda + D, C(x)\, e_k$$

$$= \sum_{k=m}^{m+r} e_k \left\{ (\lambda + D\ D_{e_j} C(x)\, e_k) + (\lambda'(x)\, e_j,\ C(x)\, e_k) \right\}$$

and, as a consequence,

$$\int \sigma_2^2 \left(D \otimes \sum_{k=m}^{m+r} e_k\ (\lambda + D, C(x)\, e_k) \right) \mu(dx)$$

$$= \sum_j \sum_{k=m}^{m+r} \int |(\lambda + D, D_{e_j} C(x)\, e_k) + (\lambda'(x)\, e_j, C(x)\, e_k)|^2 \mu(dx)$$

$$\leq \text{const} \sum_j \sum_{k=m}^{m+r} \int \{\sigma_2^2 (D \otimes D_{e_j} C(x)\, e_k) + \tag{3.25}$$

$$+ \| D_{e_j} C(x)\, e_k \|^2 + |(\lambda'(x)\, e_j, C(x)\, e_k)|^2\} \mu(dx)$$

$$\leq \text{const} \sum_{k=m}^{m+r} \{ \sigma_{22}^2 (C''(x)\, e_k) + \sigma_2^2 (C'(x)\, e_k) + \| C(x)\, e_k \|^2 \} \,\mu(\mathrm{d}\,x).$$

Estimates (3.24) and (3.25) show that for $m \to \infty$ and an arbitrary $r > 0$, the right-hand side of (3.23) tends to zero.

This leads to a square mean convergence of the above series

$$\sum_{k=m}^{m+r} ((\lambda + D) \otimes (\lambda + D), e_k \otimes C(x)\, e_k) = ((\lambda + D) \otimes (\lambda + D), C(x))$$

for $C \in C_2(B, L(H))$ and to the estimate

$$\int |\rho_c(x)|^2 \mu(\mathrm{d}\,x)$$
$$\leq \text{const} \left\{ \int [\, \sigma_{23}^2 (C''(x)) + \sigma_{22}^2 (C'(x)) + \sigma_2^2 (C(x))] \right\} \mu(\mathrm{d}\,x).$$

Let us now look at higher-order differential operations.

Consider the class $C_{m,2}(B, L_{n_2}(H, R^1))$ of n-linear functionals $K(x) : H \times \ldots \times H \to R^1$ depending on $x \in B$ which are $\mathcal{L}_{n_2}(H)$ mappings together with their derivatives up to the mth order.

For $K \in C_{m,2}(B, L_{n_2}(H, R^1))$ define a differential operator

$$D_K^{(n)} \varphi(x) = K(x)(D, D, \ldots, D)\, \varphi(x) = \mathrm{Tr}\, K(x)\, \varphi^{(n)}(x)\,.$$

Let $\mathcal{M}_n(B, H)$ be the class of those measures in $\mathcal{M}_0(B)$ whose logarithmic derivatives are boundedly differentiable up to the nth order.

THEOREM 3.5. *For $K \in C_{m,2}(B, L_{n_2}(H, R^1))$, $\mu \in \mathcal{M}_1(B, H)$, we have the following equality*

$$\int K(x)(D, \ldots, D)\, \varphi(x)\, \mu(\mathrm{d}\,x)$$
$$= (-1)^n \int \varphi(x)\, (D \otimes \ldots \otimes D, K(x))\, \mu(\mathrm{d}\,x)$$
$$= (-1)^n \int \varphi(x)\, ((D + \lambda) \otimes \ldots \otimes (D + \lambda), K(x))\, \mu(\mathrm{d}\,x) \tag{3.26}$$

and

$$\rho_n(x) = \frac{D_K^{*(n)} \mu(\mathrm{d}\,x)}{\mu(\mathrm{d}\,x)}$$
$$= ((D + \lambda) \otimes \ldots \otimes (D + \lambda), K(x)) \in \mathcal{L}_2(B, \mu). \tag{3.27}$$

Proof. We shall give, for the sake of brevity, only a sketch of the proof which, in principle, is the same as the proof of the previous theorem.

The proof is based on the expansion

$$\mathcal{D}_{\mathcal{K}}^{*(n)} = \sum_{j_1, \ldots, j_n = 1}^{\infty} D_{e_{j_1}} \ldots D_{e_{j_n}}\, \mathcal{K}(x)\, (e_{j_n}, \ldots, e_{j_1})$$

for which (3.26) is easily verified by induction for each summand. To check (3.27), one may use the induction method as well.

COROLLARY. *Given a differential operation*

$$\mathcal{D}^{(n)}_{z_1 \dots z_n} = (z_1, D) \dots (z_n, D) ,$$

where $z_j\,(j, \dots, n)$ *is a vector field belonging to* $C_n(B, H)$, *the adjoint operation is*

$$\mathcal{D}^{*(n)}_{z_1 \dots z_n} = (-1)^n (D, z_n) \dots (D, z_1)$$

and

$$\rho_n = \frac{\mathcal{D}^{*(n)}_{z_1 \dots z_n} \mu\,(\mathrm{d}\,x)}{\mu\,(\mathrm{d}\,x)} = (D + \lambda, z_n) \dots (D + \lambda, z_1) \in \mathcal{L}_2(B, \mu) .$$

3.4. SMOOTH MEASURE MAPPINGS

Let B, B_1 be two Banach spaces and the mapping $f: B \to B_1$ be measurable. For a Borel measure μ defined on B, one may consider the image $\mu^f = \mu \circ f^{-1}$ and try to determine whether the measure μ^f will be smooth when both the measure μ and the mapping f are smooth.

Let f be differentiable along the vector field z. A vector field $z^f(y)$, $(y \in B_1)$, is called f-connected with z if

$$z^f(f(x)) = f'(x)\, z(x).$$

First of all we prove the following simple assertion which means, in particular, that a logarithmic derivative of a smooth measure is invariant under smooth invertible mappings.

PROPOSITION 3.2. *Let* $f \in C_1(B, B, B_1)$ *and the measure* μ *be differentiable along* z. *Then the measure* μ^f *is differentiable along* z^f *and*

$$(D, z^f)\, \mu^f = ((D, z)\, \mu)^f . \tag{3.28}$$

If, moreover, f *is invertible, then*

$$\rho_{\mu^f}(z^f; y) = \rho_\mu(z; f^{-1}(y)) . \tag{3.29}$$

Proof. First of all we point out that for $\varphi \in C_1(B_1, B_1 R^1)$,

$$\begin{aligned} \langle \varphi'(y), z^f(y) \rangle \big|_{y=f(x)} &= \langle \varphi'(f(x)), f'(x)\, z(x) \rangle \\ &= \langle [f'(x)]^* \varphi'(f(x)), z(x) \rangle \\ &= \langle (\varphi \circ f)'(x), z(x) \rangle, \end{aligned}$$

that is for $\varphi \in C_1(B_1, B_1, R^1)$,

$$\langle z^f(y), D_y\rangle \varphi(y)\,|_{y=f(x)} = \langle z(x), D_x\rangle \varphi^f(x). \tag{3.30}$$

Due to this,

$$\begin{aligned}\int_{B_1} &\langle \varphi'(y), z^f(y)\rangle \mu^f(d\,y)\\ &= \int_B \langle(\varphi\circ f)'(x), z(x)\rangle \mu(d\,x)\\ &= -\int_B (\varphi\circ f)'(x)\,(D_z\mu)(d\,x)\\ &= -\int_{B_1} \varphi(y)\,(D_z\mu)^f(d\,y).\end{aligned}$$

which implies (3.28).

Next, if f is invertible, we have the relation

$$\begin{aligned}\int_{B_1} &\langle \varphi(y) D_{z^f}\rangle \mu^f(d\,y)\\ &= -\int_B \varphi(f(x))\,\rho_\mu(z; x)\,\mu(d\,x)\\ &= -\int_B \varphi(y)\,\rho_\mu(z; f^{-1}(y))\,\mu_f\,(d\,y)\end{aligned}$$

which is equivalent to (3.29).

COROLLARY. *Suppose that the spaces considered are equipped with* H S *structures* (B, J) *and* (B_1, J_1) *and the mapping* f *is compatible with these structures*:

$$f \in C_2(B, H, B_1),\quad f: H \to H_1,\quad f'(x): H \to H_1,\quad x \in B.$$

Then for $z(x) \equiv h$ *we obtain the following formula*

$$\begin{aligned}\langle &\lambda^f(f(x)), f'(x)\,h\rangle\\ &= \langle \lambda(x), h\rangle - \operatorname{div} f'(f^{-1}(y))\, z(f^{-1}(y))\end{aligned} \tag{3.31}$$

(*giving the vector logarithmic derivative* λ^f *of* μ^f) *as a consequence of* (3.29) *and* (3.10).

The next result will give the possibility to prove of existence of the logarithm derivative without assumptions on the invertibility of the mapping f.

THEOREM 3.6. *Let* $\mu \in \mathcal{M}_1(B, H)$, $z \in \mathcal{H}_1^2$, (B, H, μ) and $f \in C_1(B, H, B_1)$. *Then the measure* $\mu^f = \mu\circ f^{-1}$ *has logarithmic derivative* ρ_{μ^f} *along the vector field* $\xi = z^f$.

Proof. Given a smooth function φ defined on B_1, we have

$$D_\xi \varphi(y)\,|_{y=f(x)}$$

$$= \langle \xi(y), \varphi'(y)\rangle \big|_{y=f(x)}$$

$$= \langle f'(x)\, z(x), \varphi'(f(x))\rangle$$

$$= \langle z(x), (f'(x))^* \varphi'(f(x))\rangle$$

$$= D_z(\varphi \circ f)(x).$$

Thanks to this,

$$\Big| \int_{B_1} D_\xi \varphi(y)\, \mu^f(dy) \Big|$$

$$= \Big| \int_B D_z(\varphi \circ f)(x)\, \mu(dx) \Big|$$

$$= \Big| \int_B (\varphi \circ f)(x)\, \rho_\mu(z; x)\, \mu(dx) \Big|$$

$$\le \Big[\int_{B_1} \varphi^2(y)\, \mu^f(dy)\Big]^{1/2} \times$$

$$\times \Big[\int_B \rho_\mu^2(z; x)\, \mu(dx)\Big]^{1/2}$$

$$\le C\, \sigma^\mu_{2,1}(z)\, \| \varphi \|_{\mathcal{L}_2(B_1, \mu^f)} .$$

Hence, $\int_B D_\xi \varphi(y)\, \mu^f(dy)$ is a linear continuous functional on $\mathcal{L}_2(B_1, \mu^f)$ and there exists $\beta \in \mathcal{L}_2(B_1, \mu^f)$ such that

$$\int_{B_1} D_\xi \varphi(y)\, \mu^f(dy) = -\int_{B_1} \varphi(y)\, \beta(y)\, \mu^f(dy).$$

Evidently, $\beta(y) = \rho_{\mu^f}(\xi; y)$.

One may extend the method we have to prove the last theorem to calculate higher-order differential operations on measures.

First, consider a second-order operator $D_{z_2} D_{z_1}$, for $z_1, z_2 \in \mathcal{H}_2^2(B, H\,\mu)$ and let ξ_1, ξ_2 be the vector fields respectively, of the class $\mathcal{H}_2(B_1, H_1, \mu^f)$ f-connected with, respectively, z_1 and z_2.

Then, due to (3.30)

$$D_{\xi_2} D_{\xi_1} \varphi(y) \big|_{y=f(x)} = D_{z_2} D_{z_1} (\varphi \circ f)(x) .$$

So, if $\varphi \in C_2(B_1, R^1) \cap \mathcal{L}_2(B_1, \mu^f)$, then

$$\Big| \int_{B_1} D_{\xi_2} D_{\xi_1} \varphi(y)\, \mu^f(dy) \Big|$$

$$= \Big| \int_B D_{z_2} D_{z_1} (\varphi \circ f)(x)\, \mu(dx) \Big|^2$$

$$= \Big| \int_B (\varphi \circ f)(x)\, \rho_2(z_1, z_2; x)\, \mu(dx) \Big|^2$$

$$\le \int_{B_1} |\varphi(y)|^2 \mu^f(dy) \int_B |\rho_2(z_1, z_2; x)|^2 \mu(dx).$$

Using the same arguments as in the proof of the above theorem, we get the following result.

THEOREM 3.7. *If* $z_1, z_2 \in \mathcal{H}_2^2(B, H, \mu)$, $\mu \in \mathcal{M}_1(B, H)$ *then there exists the second logarithmic derivative*

$$\rho_2^f(\xi_1, \xi_2; y) = \frac{(D_{\xi_1} D_{\xi_2} \mu^f)(dy)}{\mu^f(dy)} \in \mathcal{L}_2(B_1, \mu^f)$$

and

$$\int_{B_1} |\rho_2^f(\xi_1, \xi_2; y))| \mu^f(dy) \le \int_B |\rho_2^f(z_1, z_2; x)|^2 \mu(dx) < \infty. \tag{3.32}$$

Remark. In the same way as above, using Theorem 3.6, one may derive a similar estimate for the function

$$\rho_n^f(\xi_1, \dots, \xi_n; y) = \frac{(D_{\xi_1} \dots D_{\xi_n} \mu^f)(dy)}{\mu^f(dy)},$$

where $\xi_j \in \mathcal{H}_2^n(B_1, H, \mu^f)$ are vector fields *f*-connected with $z_j \in \mathcal{H}_2^n(B, H, \mu)$, $\mu \in \mathcal{M}_1(B, H)$ and

$$\int_{B_1} |\rho_n^f(\xi_1, \dots, \xi_n; y)|^2 \mu^f(dy) \le \int_B |\rho_n(z_1, \dots, z_n; x)|^2 \mu^f(dx) < \infty \tag{3.33}$$

if $f \in C_n(B, H, B_1)$.

CHAPTER 2

Functions and Measures on Smooth Manifolds

In this chapter we give some necessary preliminaries concerning differential geometric objects; namely, smooth manifolds, vector bundles and their connections. These notions are used afterwards to construct invariant differential operators in sections of vector bundles. The final topic of this chapter is the theory of smooth measures on manifolds.

1. Smooth Manifolds and Vector Bundles

1.1. BANACH MANIFOLDS

We begin the construction of the manifold category by treating local manifolds.

We shall call a zero neighborhood of the origin U in a Banach space B a local manifold modeled on B.

Let U_1, U_2 be a pair of local manifolds with models B_1 and B_2 and $f: V_1 \to U_2$ be a C_k-mapping defined on a certain zero neighborhood $V_1 \subset U_2$ such that $f(0) = 0$. We call these f morphisms of the class $C_k(U, B_1, B_2)$, $k \geq 1$.

It follows from Banach space smooth mapping properties that in this way we obtain the $\mathcal{M}_{\text{loc}}$ category. Banach spaces are the objects of this category, while smooth mappings preserving the origin are its morphisms. In particular, a smooth invertible mapping stands for an isomorphism.

Let, now, X be a Hausdorff topological space homeomorphic to a certain local manifold $U \subset B$. A homeomorphism

$$\varphi : X \to U$$

(a coordinate mapping) introduces a local manifold structure in X, since it permits us to identify X with U. Moreover, for any point x_0 one may construct, by an additive shift, a coordinate mapping $\varphi_x(x_0) = \varphi(x) - \varphi(x_0)$ which brings x_0 into zero and, hence, equip a neighborhood of the point x_0 with a local manifold (l.m.) structure.

A pair of coordinate mappings

$$\varphi_1 : X \to U_1, \quad \varphi_2 : X \to U_2$$

is called compatible if the mapping $F_{12} = \varphi_2 \circ \varphi_1^{-1}$

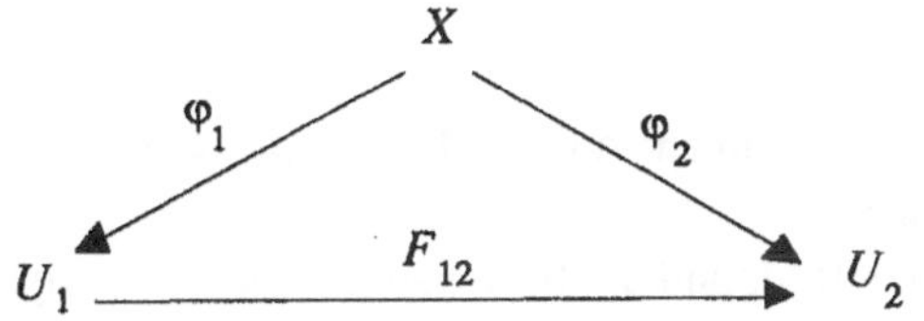

and, as a consequence,
$F_{21} = \varphi_1 \circ \varphi_2^{-1}$ is a morphism in $\mathcal{M}_{\mathrm{loc}}$ category.

Consider a pair of topological spaces X and Y equipped with local manifold structures. The set of C_k-smooth mappings from X into Y is the set of those $g : X \to Y$ for which the mapping $f = \psi \circ g \circ \varphi^{-1}$ corresponding to the given pair φ, ψ of local coordinates

$$\begin{array}{ccc} X & \xrightarrow{\quad g \quad} & Y \\ \varphi \downarrow & & \downarrow \psi \\ U & \xrightarrow{\quad f \quad} & V \end{array}$$

turns out to be a local manifold morphism.

Indeed, changing the chosen pair of coordinate mappings to another one compatible with it, one will obtain a local manifold morphism once again. Thus, a local manifold structure on X defines in a unique way the set of smooth mappings from X (into a Banach space) and conversely it may be uniquely defined by describing this set.

DEFINITION 1.1. A Hausdorff topological space X is called a C_k-smooth manifold if each point $x \in X$ has a neighborhood equipped with a local manifold structure and, for each pair of neighborhoods, those structures may be identified on the neighborhood intersection.

The pair (U, φ) is called a chart of the manifold X if φ is a coordinate mapping which maps the neighborhood U into a model space B.

For a connected X, all the charts have the same model space B (up to isomorphism equivalence). This space is called the model of the manifold X.

A manifold is called a Banach (Hilbert) manifold if its model is a Banach (Hilbert) space.

For a pair of C_k-smooth manifolds X, Y by a morphism $f \in \mathrm{Hom}(X, Y)$ we mean a mapping $f : X \to Y$ belonging to $C_k(X, B, Y)$ which possesses a smooth, in the above sense, restriction to each chart.

Denote by Man the category of Banach manifolds, using the notation Man (B) for the subcategory, whose objects are manifolds with a fixed model B.

1.2. VECTOR BUNDLES

A vector bundle is a manifold with an additional structure. Once again, to describe it we start by treating local objects.

Let X be a smooth manifold, and E be a Banach space. A triple $(\mathcal{E}, X, \pi)$ is called a local vector bundle (l.v.b.) with a base X and fibre E if $\mathcal{E} = X \times E$ is the direct product of the base and the fibre, called the total space of the vector bundle, and $\pi : \mathcal{E} \to X$ is the canonical projection onto the first factor

$$\pi : (x, a) \mapsto x, \quad x \in X, a \in E .$$

Evidently, $\pi \in \text{Hom}\,(\mathcal{E}, X)$. For each point $x \in X$ the inverse image $\pi^{-1}(x) = \{x\} \times E$, called the fibre at the point x, is canonically isomorphic to the Banach space E.

A morphism Φ acting from $(\mathcal{E}_1, X_1, \pi_1)$ into $(\mathcal{E}_2, X_2, \pi_2)$ is a total space morphism $\Phi : \mathcal{E}_1 \to \mathcal{E}_2$ which is a consistent with the additional structure. It is defined by a pair $\Phi = \{\varphi, F\}$ where $\varphi \in \text{Hom}\,(X_1, X_2)$, $F \in \text{Hom}\,(X, L\,(E_1, E_2))$ and acts in the following way

$$\Phi : (x_1, a_1) \mapsto (\varphi\,(x_1)\,, F\,(x_1)\,a_1).$$

Thus, a morphism of local vector bundles maps the fibre at the point x_1 into the fibre at the point $\varphi\,(x_1)$ and is a linear mapping on each fibre.

Local vector bundles constitute a category which will be denoted VB_{loc}.

Let $\mathcal{E}$ be a local manifold homeomorphic to the product $U \times E$, where U is a local manifold as well, and

$$S : \mathcal{E} \to U \times E$$

be an l.m. morphism.

Then one may introduce a structure of l.v.b. in $\mathcal{E}$ by identifying it with $U \times E$, by the homeomorphism S. Moreover, $X = \Phi^{-1}\,(U \times \{0\})$ will be the base of this l.v.b. and the submanifold $\mathcal{E}_x = \Phi^{-1}\,(\{x\} \times 0)$ will be the fibre at the point

$$x = \Phi^{-1}\,(\tilde{x} \times \{0\}), (\tilde{x} \in U) .$$

The fibre $\mathcal{E}_x$ inherits in a natural way the linear space structure of E.

Next, the mapping $F_\Phi\,(x) = \Phi|_{\mathcal{E}_x} : \mathcal{E}_x \to (x) \times E$ is, evidently, linear. Finally, the projection π is defined by the following commutative diagram

$$\begin{array}{ccc}
\mathcal{E} & \xrightarrow{\quad\Phi\quad} & U \times E \\
\Big\downarrow{\scriptstyle \pi} & & \Big\downarrow{\scriptstyle \tilde{\pi}} \\
X & \xrightarrow{\quad\varphi\quad} & U
\end{array}$$

where $\tilde{\pi}$ is the canonical projection and $\varphi = \Phi|_X$.

The homeomorphism Φ is called a trivializing mapping for $\mathcal{E}$ and the vector bundle constructed above is called a trivial bundle.

Let ψ be another trivialization of the l.m. $\mathcal{E}$, such that $\psi \circ \Phi^{-1}$ is a l.v.b. morphism. Then, by changing Φ into ψ, we do not disturb in any way the l.v.b. structure introduced in $\mathcal{E}$.

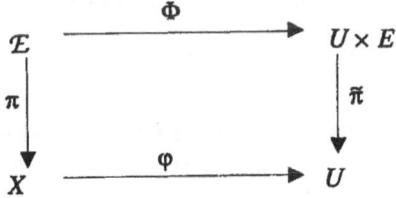

DEFINITION 1.2. A triple $(\mathcal{E}, X, \pi)$ is called a vector bundle (v.b.) if both $\mathcal{E}$ (the total space of the bundle) and X (the base of the bundle) are manifolds, while π (the projection) is a morphism $\pi : \mathcal{E} \to X$ such that $\pi(\mathcal{E}) = X$ and the following local triviality property holds: each point $x \in X$ possesses a neighborhood $U \subset X$ such that $\mathcal{E}_u = \pi^{-1}(U)$ has l.v.b. structure $\left(\mathcal{E}_u, U, \pi_u = \pi|_{\mathcal{E}_u}\right)$ and, moreover, for any pair of intersecting neighborhoods U_1, U_2, these structures may be identified on $U_1 \cap U_2$.

A Banach space E, which for $\mathcal{E}_u \sim U \times E$ is called a typical bundle fibre, and a set of pairs $(\pi^{-1}(U), \Phi_u)$, where $\Phi_u = (\varphi_u, F_u)$ are trivializations corresponding to different neighborhoods $U \supset X$ (local trivializations of the v.b.) constitute a system of charts of the total space $\mathcal{E}$ or an atlas of $\mathcal{E}$, while a set of pairs (U, φ_u) constitute an atlas of the base X.

We use the notation $(\mathcal{E}, X, \pi)$, or simply π, to denote a vector bundle.

To conclude the description of the vector bundle category, we must define its morphisms.

Let $\pi_1 : \mathcal{E}_1 \to X_1$, $\pi_2 : \mathcal{E}_2 \to X_2$ be a pair of vector bundles. Choose as a morphism $\Phi : \pi_1 \to \pi_2$ a mapping $\Phi : \mathcal{E}_1 \to \mathcal{E}_2$ such that its restriction to each neighborhood $\pi^{-1}(U_1)$ $(U_1 \subset X_1)$ which admits a trivialization maps it into a neighborhood $\pi_2^{-1}(\varphi(U_1))$ which admits a trivialization as well, and is a l.v.b. morphism. The morphism Φ is defined by a pair (φ, F_Φ), where $\varphi \in \mathrm{Hom}(X_1, X_2)$ and for each $x_1 \in X_1$ $F(x_1)$ is a linear map from the fibre $\mathcal{E}_{1x_1}$ into the fibre $\mathcal{E}_{2\varphi(x_1)}$.

Now we want to consider some operations over vector bundles. To construct them, we will use the following standard scheme.

Let S be a functor in the Banach space category $\mathcal{Z}$. By definition, it puts in correspondence a Banach space $S(B)$ to each $B \in \mathcal{Z}$ and a linear map

$$S(F) \in \mathrm{Mor}\big(S(B_1), S(B_2)\big) \quad \text{or} \quad \mathrm{Mor}\big(S(B_2), S(B_1)\big)$$

(respectively, for covariant or contravariant functors S), to each $F \in \mathrm{Mor}(B_1, B_2)$. We say that a functor S is smooth if for the map $F(x)$, smoothly depending on a certain parameter x, the mapping $S(F(x))$ possesses the same order of smoothness with respect to x as $F(x)$ does.

A smooth functor S may be extended in a natural way to the VB_{loc} category

$$S(X \times E) = X \times S(E), \qquad S(\Phi) = \big(\varphi, S(F(x))\big),$$

where $\Phi = (\varphi, F(x))$ and $S(\pi) : X \times S(E) \to X$ is the natural projection.

Given $\pi : \mathcal{E} \to X \in VB$ a bundle $S(\pi)$ may be constructed in the following way. For each chart $U \subset X$ and a trivialization Φ_u of the v.b. π construct a l.v.b. $S(\pi_u)$. Define the total space $S(\mathcal{E})$ of the new bundle as a free union of l.m. $U \times S(E)$, over all charts U, factorized with respect to an equivalence relation, which is generated by mappings $S(\Phi_v \circ \Phi_u^{-1})$ on intersections $U \cap V$ for each pair of charts. Since S is a smooth functor, it brings morphism compositions into morphism compositions and does not change the order of smoothness of the morphism. Therefore, we obtain a correctly defined bundle $S(\pi) : S(\mathcal{E}) \to X$.

By a similar argument, we may prove that in order to define in a correct way the morphisms $S(\Phi)$, one needs only to define their restrictions to each chart. Notice that multicomponent functors on $\mathcal{Z}$ may be extended to the VB category in the same way.

EXAMPLES 1. The functor $*: \pi \to \pi^*$ puts in correspondence to each v.b. its cobundle. In this way it couples $X \times E^*$ to $X \times E$ and $\Phi^* = (\varphi, F^*(x))$ with $F^*(x) \in L = (E_1^* E^*)$ to $\Phi = (\varphi, F(x))$ with $F(x) \in L(E, E_1)$.

2. The functor $L(B, \cdot): \pi \to L(B, \pi)$ which brings π into the bundle $L(B, \pi)$ with fibres consisting of linear maps from a fixed space B into corresponding fibres of π. Hence we have as l.v.b.

$$L(B, X \times E) = X \times L(B, E), \quad L(\Phi) = (\varphi, L(\cdot, F(x))),$$

where $L(\cdot, a): B \to E, a \in E$.

3. The Witney sum $\pi_1 \oplus \pi_2$ of two vector bundles π_1 and π_2 over the same base X may be constructed in the above manner by means of the 'Banach space direct sum' functor.

Below, we shall also need another approach to the construction of vector bundles.

Let $\pi: \mathcal{E} \to X$ be a vector bundle and $f: Y \to X$ be a manifold morphism. The vector bundle $f^*(\pi): f^*(\mathcal{E}) \to Y$ is called the inverse image of the bundle π with respect to the mapping f if the diagram

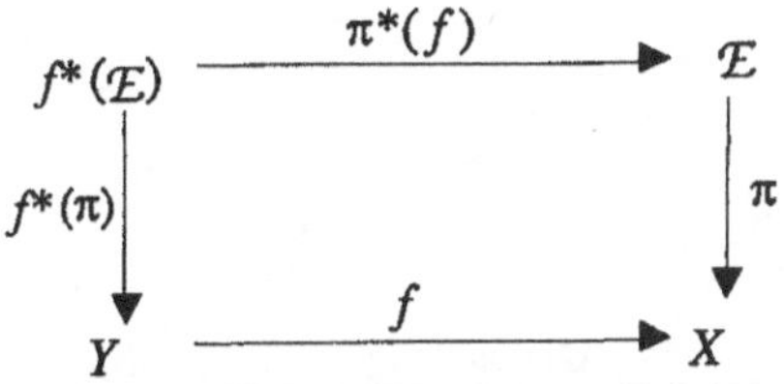

is commutative. The mapping $\pi^*(f)$ acts as an isomorphism on each fibre, identifying $f^*(\mathcal{E})_y$ with $\mathcal{E}_{f(y)}$ and the pair $(f, \pi^*(f))$ is vector bundle morphism.

For the local vector bundle $\pi: X \times E \to X$ the inverse image is the local v.b.

$$f^*(\pi): Y \times E \to Y$$

and

$$\pi^*(f): (y, a) \mapsto (f(y), a), \quad a \in E.$$

Assume that $Y = \mathcal{E}$ and $f = \pi$. In this case, in order to avoid confusion in dealing with $\pi^*(\pi)$, we shall use the following notation

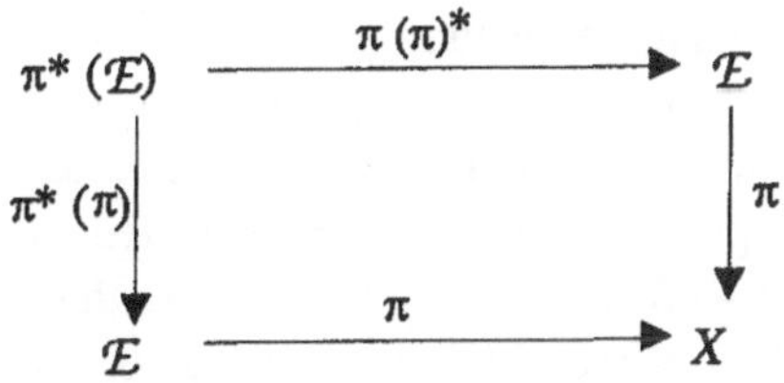

which implies that

$$\pi^* (\pi) : (x, a , b) \mapsto (x, a),$$

$$\pi (\pi)^* : (x, a, b) \mapsto (x, b), \quad a, b \in E.$$

1.3. TANGENT BUNDLE

One may put in correspondence to each manifold X with model space B its tangent bundle $\tau_X : T X \to X$ with typical fibre B. In addition T is a functor, acting from the Man category into the VB category since, by definition, it puts in correspondence the tangent map $Tf \in$ Mor (TX, TY) to each mapping

$$f \in \text{Mor}(X, Y) \text{ and } T(f \circ g) = Tf \circ Tg.$$

Write $TX = X \times B$ for a local manifold $X \subset B$ and define a morphism $Tf = (f, f'(x)) : TX_1 \to TX_2$ by a relation $Tf(x, y) = (f(x), f'(x)y)$ for a l.m. morphism $f : X_1 \to X_2$.

For a nonlocal manifold X let us construct first a local bundle $\tau_u : U \times B \to U$ for each chart U of an atlas of X. Define next a pair of consistent local manifold structures on U by

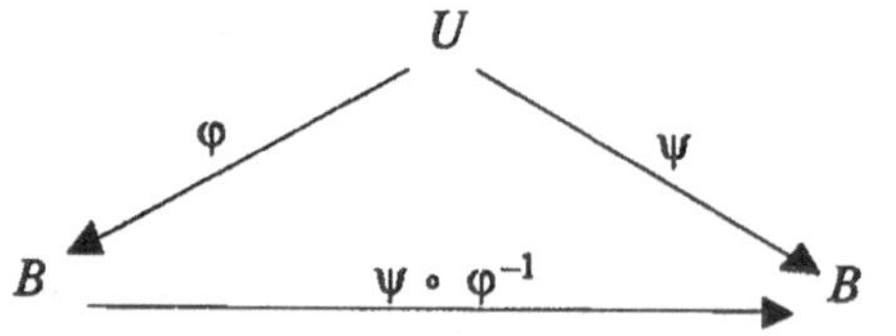

Notice that, due to derivative properties, the diagram

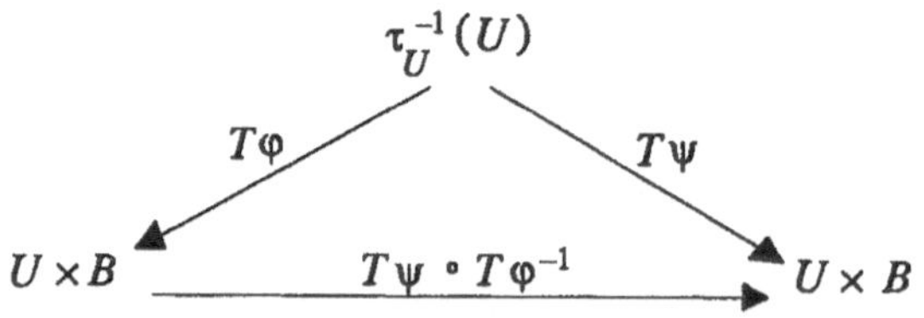

is commutative. Hence, the corresponding l.v.b. structures on the chart $U \times B$ are consistent. Since v.b. morphisms are defined in a local way, it suffices to describe mappings Tf for each trivialization, as has been pointed out above.

Assume that the considered manifold is equipped with a structure of vector bundle total space. Let us describe some important constructions connected with its tangent bundle.

Given a vector bundle $\pi : \mathcal{E} \to X$, denote $\tau_{\mathcal{E}} : T\mathcal{E} \to \mathcal{E}$ the tangent bundle over $\mathcal{E}$. If we apply the functor T to π, we obtain another vector bundle $T\pi : T\mathcal{E} \to TX$.

In this way, we determine two vector bundle structures on $T\mathcal{E}$ connected by the following commutative diagram

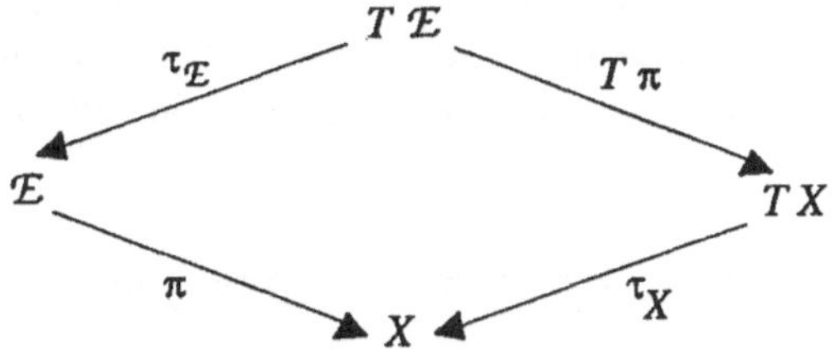

. (1.1)

To make specific features of those structure morphisms more evident, consider a trivialization of the bundle $\mathcal{E}$. Put $\mathcal{E} = X \times E$ and $TX = X \times B$, where E is a typical fibre of π and B is a model of X.

It is easy to see that a tangent bundle over a direct product of l.m. is a direct product as well, i.e.

$$\tau_{\mathcal{E}} : T\mathcal{E} = X \times E \times B \times E \to X \times E = \mathcal{E}.$$

The mappings in diagram (1.1) act in the following way:

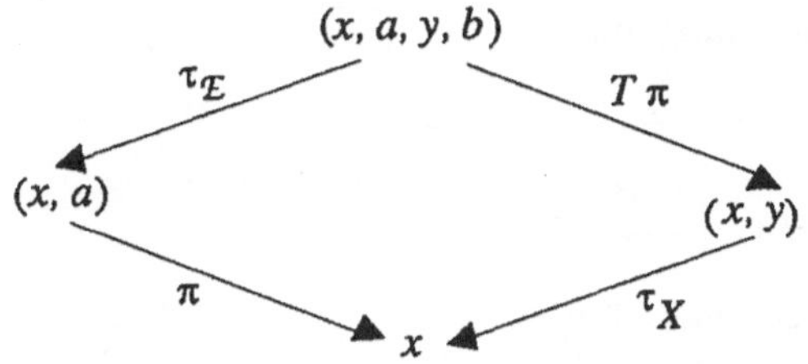

.

Hence, if $\Phi = (f, F) : X \times E \to X \times E$ is a v.b. morphism, then its tangent morphism $T\Phi$ is given by the formula

$$T\Phi : (x, a, y, b) \mapsto (f(x), F(x)\, a, f'(x)\, y,\; F(x)\, b + F'(x)\, (y, a)). \tag{1.2}$$

In particular, it follows from (1.2) that

$$T\Phi : (x, a, 0, b) \mapsto (f(x), F(x)\, a, 0, F(x)\, b) \tag{1.3}$$

and, therefore,

$$\tau_{\mathcal{E}} : X \times E \times 0 \times E \to X \times E$$

turns out to be subbundle of $T\mathcal{E}$, which is called the vertical tangent bundle $VT\mathcal{E}$. It also follows from (1.3) that $VT\mathcal{E}$ is isomorphic to the inverse image of the bundle π with respect to π

$$\pi^* (\pi) : X \times E \times E \to X \times E,$$

$$\pi^* (\pi) : (x, a, b) \to (x, a). \tag{1.4}$$

Denote this morphism

$$\upsilon : (x, a, 0, b) \to (x, a\ , b). \tag{1.5}$$

It is easy to see that v is a bundle morphism and that, hence, the above local trivial representations may be patched together to give global mappings of bundles over $\mathcal{E}$. Notice that the morphism $\upsilon^{-1} : \pi^* (\mathcal{E}) \to V\,T\,\mathcal{E} \subset T\,\mathcal{E}$ imbeds $\pi^* (\mathcal{E})$ into $T\,\mathcal{E}$

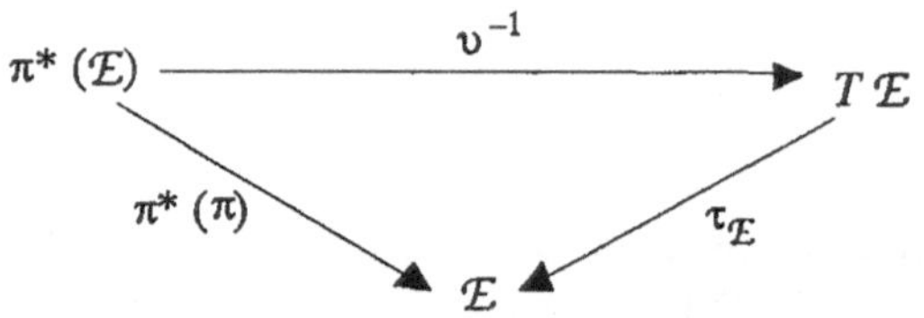

Consider elements of the form $(x, a\,y, 0)$. It follows from (1.0) that they do not form a subbundle over $\mathcal{E}$. To construct a horizontal bundle $H\,T\,\mathcal{E}$ over $\mathcal{E}$ such that $T\,\mathcal{E} = V\,T\,\mathcal{E} \oplus H\,T\,\mathcal{E}$, one needs a special construction (a connection), which will be described in the next section.

Using the inverse image mapping

$$\pi\,(\pi)^* : \pi^* (\mathcal{E}) \to \mathcal{E},$$

we obtain a vector bundle mapping υ_π

$$\upsilon_\pi = \pi\,(\pi)^* \circ \upsilon : V\,T\,\mathcal{E} \to \mathcal{E},$$

$$V\,T\,\mathcal{E} \xrightarrow{\ \upsilon\ } \pi^* (\mathcal{E}) \xrightarrow{\ \pi\,(\pi)^*\ } \mathcal{E}$$

which acts as an isomorphism on each fibre

$$\upsilon_\pi : (x, a, 0, b) \mapsto (x, b).$$

2. Bundle Section, Connections, Differential Operations

2.1. BUNDLE SECTIONS

A morphism $\xi : X \to \mathcal{E}$ such that $\pi \circ \xi = \mathrm{id}_X$ is called a section of a vector bundle $\pi : \mathcal{E} \to X$. By definition, $\xi_x \in \mathcal{E}_x$.

For a trivial vector bundle

$$\mathcal{E} = X \times E,\ \xi\,(x) = (x, \bar{\xi}\,(x)) \ \text{ where } \ \bar{\xi}\,(x) = \mathrm{Pr}_E\ \xi\,(x).$$

In this way we give rise to a mapping $\bar{\xi} : X \to E$ which is called the principal part of the section. If it will not lead to confusion, we shall sometimes omit the bar over ξ and denote by the same symbol both a section and its principal part.

Let $\Phi = (\varphi, F) : X \times E \to X \times E$ be a local v.b. isomorphism; then, the principal part $\xi\,(x)$ is transformed in accordance with

$$\bar{\xi}(x) \mapsto \xi^{\Phi}(x) = F(x)\,\xi(x). \tag{2.1}$$

Sections of the tangent (cotangent) bundle τ_X are called vector (covector) fields and, in general, sections of a tensor bundle are called tensor fields.

A section ξ of the v.b. π is said to be local if its support is entirely in the domain of a certain chart $U \subset X$ of the bundle base.

To construct a local section ξ, it suffices to describe its principal part $\bar{\xi}$ for some chosen trivialization. In fact, the relation (2.1) defines $\bar{\xi}$ for any trivialization and so defines the whole section in a correct way. Denote σ_{loc} or $\sigma_{\mathrm{loc}}(\mathcal{E})$ the set of local sections of the vector bundle $\mathcal{E}$.

To construct a nonlocal section, we may proceed in the following way. Choose a partition of unity that is a set of smooth positive local functions $\{\varphi_j(x)\}_{j=1}^{\infty}$ on X such that

$$\sum_{j-1}^{\infty} \varphi_j(x) = 1 \tag{2.2}$$

and suppose that for any point $\xi \in X$ only a finite number of functions $\varphi_j(x)$ are nonzero.

Notice that $\xi_j = \varphi_j \circ \xi$ is a local section of π for an arbitrary section ξ of π and, evidently, $\xi = \Sigma_{j=1}^{\infty} \xi_j$.

It is known that a partition of unity does exist for a manifold with a smooth Banach model (for example, with a Hilbert model).

Denote by $\sigma(\xi)$ the set of global sections of the bundle $\mathcal{E}$. When we need to stress that sections are C_k-smooth, we use the notation $\sigma_k(\mathcal{E})$.

A curve in X is a morphism $\alpha : J \to X$ from an open interval $J \subset R^1$ into a manifold X. For a given manifold morphism $g : X \to Y$, $g \circ \alpha$ is a curve in Y. For each curve $\alpha(t)$, one may consider the induced mapping of tangent bundles

$$\begin{array}{ccc} J \times R^1 & \xrightarrow{\;T\alpha\;} & TX \\ \Big\downarrow{\scriptstyle \tau_J} & & \Big\downarrow{\scriptstyle \tau_X} \\ J & \xrightarrow{\;\alpha\;} & X \end{array}$$

In what follows we shall sometimes use the notations $\alpha'(t)$ or $d\alpha/dt$ for $T\alpha$. Hence, $\alpha'(t)$ is a curve in TX if α is a curve in X.

Given $g : X \to Y$, we have

$$(g \circ \alpha)'(t) = T g\, \alpha'(t)$$

due to the properties of the functor T.

Let 0 belong to an interval J. Consider the curve α defined over J such that $\alpha(0) = x$. Two curves $\alpha_1(t)$ and $\alpha_2(t)$ are called tangent at

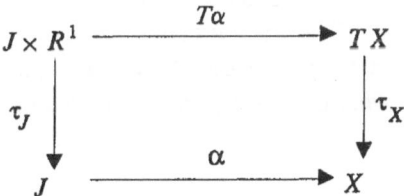

$t = 0$ if $\alpha'_1(0) = \alpha'_2(0)$.

The set of equivalence classes (under tangency at $t = 0$) of curves $\alpha(t)$ such that $\alpha(0) = x$ is isomorphic to the tangent space $T_x X$.

By an ordinary differential equation on X, we mean a vector field $Y(x)$ over X such that

$$\frac{\mathrm{d}\,x}{\mathrm{d}\,t} = Y(x(t)). \tag{2.3}$$

A curve $x(t)$ which maps an open interval $J \subset R^1$, such that $0 \in J$, into X and satisfies (2.3) for all $t \in J$ is called a solution of (2.3) or an integral path of the vector field Y. If $x(0) = x$, then one says that $x(t)$ is a solution of (2.3) with initial condition $x(0) = x$.

The vector field Y is said to possess a global flow if for any point x one may choose a $J = R^1$ and thus one may put in correspondence to each x a mapping $S_t : x \to x(t)$ from the manifold X into itself defined by the relation $x(t) = S_t \circ x$. In addition, evidently, this satisfies $S_{t+\tau} = S_t \circ S_\tau$.

2.2. CONNECTION MAPPING

To construct differential operations over v.b. sections, it is natural to use the tangent mapping T. In doing this, we find the following obstacle. Let $\xi : X \to \mathcal{E}$ be a section of the v.b. π. Then $T\xi : TX \to T\mathcal{E}$ is a section of the bundle $T\pi : T\mathcal{E} \to TX$ since the functor T brings $\pi \circ \xi = \mathrm{id}_X$ into $T\pi \circ T\xi = \mathrm{id}_{TX}$. Now, if $\eta : X \to TX$ is a vector field, then it would be natural to call the mapping $T\xi \circ \eta : X \to T\mathcal{E}$ the derivative of the section ξ along the vector field η.

Nevertheless, $T\xi \circ \eta$ is a section of the bundle $\tau_X \circ T\pi : T\mathcal{E} \to X$ rather than a section of the initial bundle π.

To make the construction correct, we need a v.b. mapping $T\mathcal{E} \to \mathcal{E}$ which does not coincide with a projection $\tau_{\mathcal{E}} : T\mathcal{E} \to \mathcal{E}$ since the latter forgets the information about the principal part of any section $\mathcal{E} \to T\mathcal{E}$. As far as a fibre of the bundle π is isomorphic to a fibre of the vertical tangent bundle $VT\mathcal{E}$, it suffices to construct a projection $T\mathcal{E} \to VT\mathcal{E}$ or, what is the same, to construct a horizontal tangent bundle $HT\mathcal{E}$ such that

$$T\mathcal{E} = VT\mathcal{E} \oplus HT\mathcal{E}.$$

For this purpose, consider first the inverse image $\pi^*(TX)$ of the tangent bundle τ_X given by the diagram

$$\begin{array}{ccc} \pi^*(TX) & \xrightarrow{\ \tau_X^*(\pi)\ } & TX \\ {\scriptstyle \pi^*(\tau_X)}\downarrow & & \downarrow{\scriptstyle \tau_X} \\ \mathcal{E} & \xrightarrow{\ \ \pi\ \ } & X \end{array} \qquad . \tag{2.4}$$

In the local case the diagram looks like

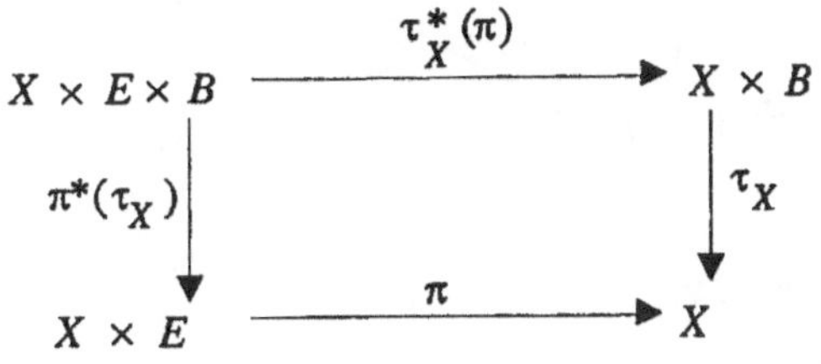

where the mappings act in the following way

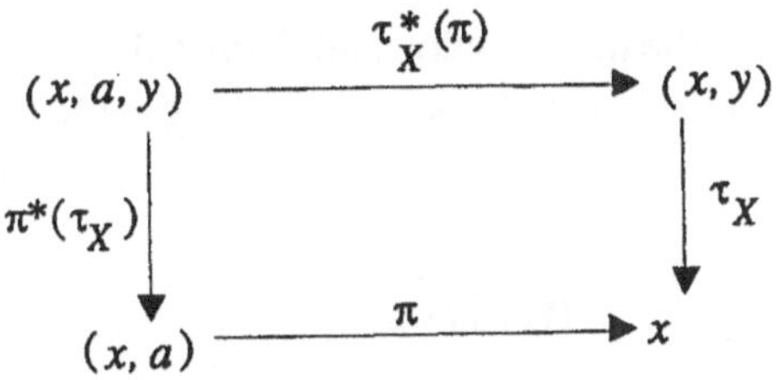

One may define morphisms of the bundle $\pi^*(TX)$ by

$$(x, a, y) \mapsto (f(x), F(x)\, a, f'(x)\, y) \, , \tag{2.5}$$

which shows that all the mappings in the above diagrams are in fact v.b. morphisms and, hence, may be extended to global mappings.

A horizontal tangent subbundle $HT\mathcal{E}$ of the v.b. $T\mathcal{E}$ (or a connection) is, by definition, the image of an embedding $\Gamma^\pi : \pi^*(TX) \to T\mathcal{E}$ possessing the following properties

(a) $T\pi \circ \Gamma^\pi = \tau_X^*(\pi)$, that is

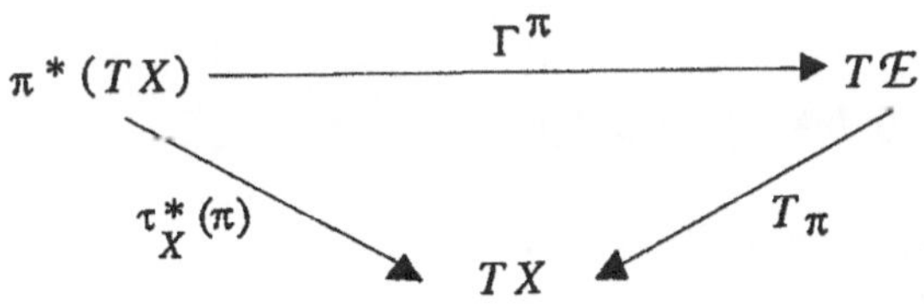

(2.6)

(b) Γ^π gives rise to a mapping from the v.b. $\pi^*(TX)$ into the v.b. $T\mathcal{E}$ which is linear over each fibre of the base $\mathcal{E}$ (having a v.b. structure).

For each trivialization, diagram (2.6) has the form

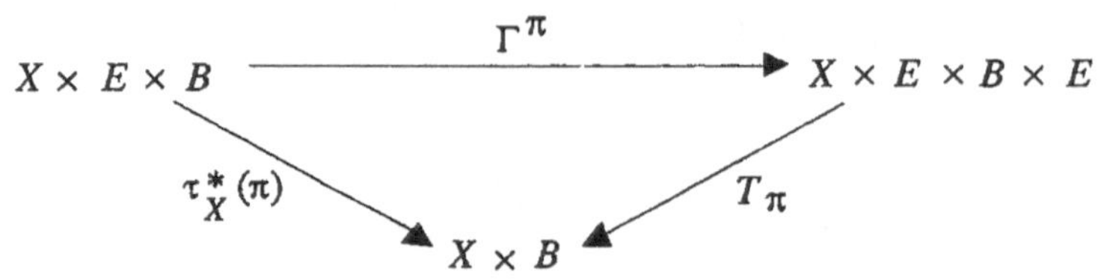

(2.7)

and the mapping Γ^{π} may be given by

$$\Gamma^{\pi}(x, a, y) = (x, a, y, b)$$

with $b \in E$ depending linearly both on $y \in B$ and $a \in E$. In this way, we obtain a bilinear mapping which will be convenient to denote (changing the sign)

$$b = -\Gamma^{\pi}_{x}(y, a), \quad \Gamma^{\pi}_{x} : B \times E \to E. \tag{2.8}$$

Here the mapping Γ^{π}_{x} depends on the trivialization and defines the principal part of the mapping Γ^{π}. It is called the connection coefficient corresponding to the given trivialization.

Thus

$$\Gamma^{\pi} : (x, a, y) \mapsto \big((x, a, y, -\Gamma^{\pi}_{x}(y, a)\big). \tag{2.9}$$

It follows from (1.2) that under the action of a morphism $\Phi : \mathcal{E} \to \mathcal{E}$ the connection coefficients Γ^{π}_{x} are transformed according to

$$\Gamma^{\pi\Phi}_{f(x)}(f'(x)\, y, F(x)\, a) = F(x)\, \Gamma^{\pi}_{x}(y, a) - F'(x)\,(y, a), \tag{2.10}$$

which corresponds to the fact that Γ^{π} plays the role of a section of the bundle

$$T\mathcal{E} \to \pi^{*}(TX).$$

Notice, moreover, that this bundle is in no case a vector bundle, since mapping (2.10) which governs the transformation of the fourth coordinate of $T\mathcal{E}$, is an affine rather than a linear mapping.

Nevertheless, we may construct a connection by the same procedure which had been suggested for the construction of a global v.b. section that is by patching together the local sections with the help of the partition of unity.

The expansion

$$(x, a, y, b) = \big(x, a, 0, b + \Gamma^{\pi}_{x}(y, a)\big) +$$
$$+ \big(x, a, y, -\Gamma^{\pi}_{x}(y, a)\big) \tag{2.11}$$

is invariant under the action of the $T\mathcal{E}$ morphisms and, therefore, $T\mathcal{E}$ may be written as the Whitney sum

$$T\mathcal{E} = HT\mathcal{E} \oplus VT\mathcal{E} \tag{2.12}$$

of the vertical and horizontal subbundles. In addition, (2.12) generates in an invariant way a projection

$$P^{\pi}_{\upsilon} : T\,\mathcal{E} \to T\,\mathcal{E}$$

which is in local coordinates

$$P^{\pi}_{\upsilon} : (x, a, y, b) \mapsto \big(x,\ a,\ 0,\ b + \Gamma^{\pi}_{x}\ (y, a)\big)\,. \tag{2.13}$$

The mapping

$$K^{\pi} = \upsilon_{\pi} \circ P^{\pi}_{\upsilon} : T\,\mathcal{E} \to T\,\mathcal{E},$$

$$K^{\pi} : (x, a, y, b) \mapsto \big(x,\ b + \Gamma^{\pi}_{x}\ (y, a)\big) \tag{2.14}$$

is called a connection map.

Evidently, the horizontal subbundle is the kernel of the connection map and, hence, those two objects are in one-to-one correspondence.

It is worth mentioning that, as follows from (2.10), the connection is defined in a unique way up to a summand which is a section of the v.b. $\mathcal{E}$. That is, for two different connections Γ^{π}_{1} and Γ^{π}_{2} we have

$$\Gamma^{\pi}_{1} - \Gamma^{\pi}_{2}\ : \pi^{*}\,(T\,X) \to V\,T\,\mathcal{E} \sim \mathcal{E}.$$

2.3. PARALLEL DISPLACEMENT

We now discuss those possibilities which arise after constructing a connection map K^{π}. Let η be a vector field on X. The inverse image map $\pi^{*}\,(\tau_X)$ gives rise to the map

$$\pi^{*}\,\eta : \mathcal{E} \to \pi^{*}\,(T\,X)$$

and, hence, generates a map

$$\Gamma^{\pi} \circ \pi^{*}\,\eta : \mathcal{E} \to H\,T\,\mathcal{E}$$

which is a horizontal vector field on $\mathcal{E}$.

In this way, a connection of the bundle $\mathcal{E}$ generates a map

$$\eta \mapsto \Gamma^{\pi} \circ \pi^{*}\,\eta$$

which puts in correspondence a horizontal vector field over $\mathcal{E}$ to each vector field over X. We call it a horizontal lift of η to $\mathcal{E}$.

Local considerations easily lead to the formula

$$\Gamma^{\pi} \circ \pi^{*}\,\eta : (x, a) \mapsto (x, a, \eta\,(x), -\,\Gamma^{\pi}_{x}\ (\eta\,(x), a))$$

for the principal part of this field. It shows, in particular, that

$$T\,\pi \circ \Gamma^{\pi} \circ \pi^{*}\,\eta = \eta.$$

Notice that the restriction of horizontal lift of a vector field to a certain subset of X does not depend upon field values outside this subset.

In particular, we may consider that horizontal lift of a vector field defined along a curve $x\,(t) \in X, t \in [0, 1]$, for example the horizontal lift of a tangent field $x'\,(t) \in T\,X$.

A curve $a\,(t) \in \mathcal{E}, t \in [0, 1]$ is called horizontal if the vector field $a'\,(t) \in T\,\mathcal{E}$ is horizontal and, thus, appears to be a horizontal lift of the tangent field $x'(t)$ for $x(t) = \pi a(t)$. We say in this case that two elements $a(t), a(\tau)$ for each pair $t, \tau \in [0, 1]$ are connected by parallel translation along the curve $t \mapsto x(t)$

$$a(t) = U(t, \tau; x(\cdot))\, a(\tau). \tag{2.15}$$

To examine the properties and methods of calculation of parallel transport, it suffices to obtain local results, then to prove them to be invariant and, finally, to patch them together.

Let $X \subset B$ be a local manifold, $\pi : \mathcal{E} = X \times E \to X$ be a local v.b. Consider a vector field $(x, \xi(x))$ over X. The principal part of its horizontal lift has the form

$$\left(\xi(x) - \Gamma_x^\pi(\xi(x), a(x))\right),$$

and we may consider a system of differential equations for the integral paths of this field

$$\frac{dx}{dt} = \xi(x), \qquad \frac{da}{dt} = -\Gamma_x^\pi(\xi(x), a). \tag{2.16}$$

Those integral paths are, by definition, horizontal.

If $\gamma = \{x(t), t \in [0, 1]\}$ is a curve in X and $x'(t)$ is the corresponding tangent vector field, then the integral paths $\{x(t), a(t)\}$ of its lift satisfy the following equation

$$\frac{da}{dt} + \Gamma_{x(t)}^\pi(x'(t), a(t)) = 0 \tag{2.17}$$

with $A(t) = \Gamma_{x(t)}^\pi(x'(t), \cdot) \in L(E)$ continuously depending on t. Such an equation defines an evolution family of linear operators $U(t, \tau) \in L(E)$

$$U(t, s)\, U(s, \tau) = U(t, \tau), \qquad U(t, t) = I,$$

which maps E into itself in a one-to-one way. In addition, $a(t) = U(t, \tau)\, a(\tau)$ so that $U = U(t, \tau; x(\cdot))$ is a parallel transport map.

Since the solution of Equation (2.17) is uniquely defined and the equation itself is invariant under the morphism which patch together local trivializations of π, it is easy to see that parallel transport is correctly defined along an arbitrary smooth curve $\gamma = \{x(t), t \in [0, T]\}$ in X. It follows immediately from the definition that the following properties hold:

(a) For any curve $\gamma = \{x(t)\}$ in X it is defined as an evolution family of continuous linear isomorphisms

$$U_\gamma(t, \tau) : \mathcal{E}_{\gamma(\tau)} \to \mathcal{E}_{\gamma(t)}$$

possessing the properties

$$U_\gamma(t, s)\, U_\gamma(s, \tau) = U_\gamma(t, \tau), \qquad U_\gamma(t, \tau) = U_\gamma^{-1}(\tau, t).$$

(b) For each point $b \in \mathcal{E}_{x(\tau)}$ the only horizontal path in $\mathcal{E}$, which defines the parallel transport of an element b along γ, is

$$b(t) = U_\gamma(t, \tau)\, b.$$

According to the properties mentioned above, to calculate the parallel displacement of an element $b = (x(t), a)$ one needs to consider a trivialization of the bundle π and to find $B(t) = (x(t), a(t))$ solving (2.17) for those t such that $\gamma(t)$ is inside a single chart containing $x(\tau)$ and satisfying the relation $a(\tau) = a$.

The path $b(t)$ obtained in this way is invariant under coordinate transformations and may be extended to another chart.

Thus, we have proved that a bundle connection defines a parallel transport in the bundle along a curve on the base.

The converse holds as well. Namely, given a parallel transport operation, one may define a bundle connection in a unique way.

In fact, consider trivializations of π and τ_X over X such that $b = (x, a) \in \mathcal{E}$, $\eta = (x, y) \in TX$ and a $\gamma = \{x(t)\}$ such that $x(0) = x, x'(0) = y$.

Then it follows from (2.17) that

$$\Gamma_x^\pi(y, a) = -\frac{d}{dt} U_\gamma(t, 0)\, a|_{t=0} , \tag{2.18}$$

and the result does not depend upon the path we have chosen, provided it satisfies the above properties.

2.4. INDUCED CONNECTIONS

Given a connection map on a vector bundle π, we would want to construct connection maps in the various bundles associated with π. Now we shall describe some ways to do it.

Let S be a smooth functor in the Banach space category. As has been shown in Section 1 of this chapter, its action may be extended to the Banach vector bundle category. Let Γ^π be a given connection in the vector bundle $\pi : \mathcal{E} \to X$. We want to construct a connection $\Gamma^{S(\pi)}$ in the bundle $S(\pi) : S(\mathcal{E}) \to X$. For this purpose, we consider an evolution family $U_\gamma(t, \tau)$ of parallel translations along the path γ and a family $S(U_\gamma(t, \tau))$ of parallel translations in the bundle $S(\pi)$ associated with π. In a local trivialization, as it follows from (2.18), onc has

$$-\Gamma_x^{S(\pi)}(y, \alpha) = \frac{d}{dt} S(U_\gamma(t, 0))\, \alpha|_{t=0}, \quad y \in B, \quad \alpha \in S(E). \tag{2.19}$$

In addition, the properties of the functor S grant that all connection properties hold.

EXAMPLE 1. Let $S = *$ be the operation which changes an object to an adjoint one. Then for $\alpha \in E$, $\beta \in E^*$, we have

$$\begin{aligned} \langle \alpha, \Gamma_x^{\pi *}(y, \beta)\rangle &= -\frac{d}{dt} \langle \alpha, U_\gamma^*(0, t)\, \beta\rangle \Big|_{t=0} \\ &= -\frac{d}{dt} \langle U_\gamma(t, 0)\, \alpha, \beta\rangle \Big|_{t=0} \\ &= -\langle \Gamma_x^\pi(y, \alpha), \beta\rangle . \end{aligned}$$

Here we have used the fact that $*$ is a contravariant functor which changes arrow directions. In accordance with it, while calculating $\Gamma^{\pi*}$, we have inverted the ordering of time moments 0 and t and then used the differentiation rule for evolution families, which leads to

$$\frac{\mathrm{d}\, U(t,\tau)}{\mathrm{d}\, t} = A(t)\, U(t,\tau) \mapsto \left.\frac{\mathrm{d}\, U(t,\tau)}{\mathrm{d}\, t}\right|_{t=\tau} = A(t).$$

EXAMPLE 2. Given a pair of vector bundles π and π_1 over the same base X, consider a functor L which puts in correspondence to them a bundle $L(\pi_1, \pi)$ whose fibre consists of linear operators acting from a fibre E_1 of the bundle π_1 into a fibre E of the bundle π. Given $\alpha \in E_1$ and $\beta \in L(E_1, E)$, put

$$\begin{aligned}\Gamma_x^{L(\pi_1,\pi)}(y, \beta)\,\alpha &= -\frac{\mathrm{d}}{\mathrm{d}\, t}\, U_\gamma^\pi(t,0)\,\beta\, U_\gamma^{\pi_1}(0,t)\,\alpha\,\Big|_{t=0} \\ &= \Gamma_x^\pi(y, \beta\,\alpha) - \beta\,\Gamma_x^{\pi_1}(y,\alpha)\,. \end{aligned} \tag{2.20}$$

Since

$$\Gamma_x^\pi(y,\alpha) = \Gamma_x^\pi(y)\,\alpha,$$

where

$$\Gamma_x^\pi(y) \in L(E) \quad \text{and} \quad \Gamma_x^{\pi_1}(y) \in L(E_1),$$

we may transform (2.20) to obtain

$$\Gamma_x^{L(\pi_1,\pi)}(y, \beta)\,\alpha = \left(\Gamma_x^\pi(y)\,\beta - \beta\,\Gamma_x^{\pi_1}(y)\right)\alpha\,.$$

In the case $\pi_1 = \pi$, the right-hand side of the latter expression turns out to be the commutator of the operators $\Gamma_x^\pi(y)$ and β

$$\Gamma_x^{L(\pi_1,\pi)}(y, \beta)\,\alpha = [\,\Gamma_x^\pi(y), \beta\,]\,\alpha.$$

This fact is consistent with (2.20) which shows that $\Gamma_x^{L(\pi_1,\pi)}$ acts as a differentiation when it is correctly defined.

Notice that given the above connections, parallel translations come to be defined in a reasonable way in all the bundles

$$U_\gamma^\pi(t,\tau)\,\beta\,\alpha = U_\gamma^{L(\pi_1,\pi)}(t,\tau)\,\beta\, U_\gamma^{\pi_1}(t,\tau)\,\alpha.$$

EXAMPLE 3. Using similar arguments, one may prove that for

$$\alpha_j \in E_j, \quad \beta \in L(E_1 \times \ldots \times E_n, E), \quad j = 1, \ldots, n,$$

$$\Gamma_x^{L(\pi_1 \times \ldots \times \pi_n,\pi)}(y, \beta)\,(\alpha_1, \ldots, \alpha_n)$$

$$= \Gamma_x^{\pi}\left(y, \beta\left(\alpha_1, \ldots, \alpha_n\right)\right) - \Sigma\, \beta\left(\alpha_1, \ldots, \Gamma_x^{\pi_j}\left(y, \alpha_j\right), \ldots, \alpha_n\right).$$

Let, in particular, $\mathcal{E} = X \times E$ be a trivial vector bundle with a trivial connection $\Gamma^{\mathcal{E}} = 0$. In this case $L(E_1 \times \ldots \times E_n, R^1)$ is the space of n-linear forms and

$$\Gamma_x^{L(\pi_1 \times \ldots \times \pi_n, \pi)}(y, \beta)\left(\alpha_1, \ldots, \alpha_n\right)$$

$$= \sum_{j=1}^{n} \beta\left(\alpha_1, \ldots, \Gamma_x^{\pi_j}\left(y, \alpha_j\right), \ldots, \alpha_n\right).$$

We now consider another way to construct a connection. Recall that a connection of the tangent bundle $\tau : TX \to X$ is called a linear connection of the manifold X and denoted either Γ^{τ} or Γ^{X}.

Let $\pi : \mathcal{E} \to X$ be a vector bundle. Given Γ^{π} and Γ^{τ}, construct a connection $\Gamma^{\mathcal{E}}$. To construct a connection map

$$K^{\mathcal{E}} : T^2\mathcal{E} \to T\mathcal{E} = VT\mathcal{E} \oplus HT\mathcal{E},$$

it suffices to construct its horizontal and vertical components

$$K_V^{\mathcal{E}} : T^2\mathcal{E} \to VT\mathcal{E}, \quad K_H^{\mathcal{E}} : T^2\mathcal{E} \to HT\mathcal{E}.$$

Since there exists the isomorphism $\upsilon : VT\mathcal{E} \to \pi^*(\mathcal{E})$, we may restrict ourselves to constructing a map from $T^2\mathcal{E}$ into $\pi^*(\mathcal{E})$.

For this purpose, consider the inverse image of a certain map from $T^2\mathcal{E}$ into $\mathcal{E}$, using the connection map $K^{\pi} : T\mathcal{E} \to \mathcal{E}$ and its tangent map $TK^{\pi} : T^2\mathcal{E} \to T\mathcal{E}$. So, put

$$K_V^{\mathcal{E}} = \upsilon^{-1} \circ \left(\pi\left(\pi^*\right)\right)^{-1}\left(K^{\pi} \circ TK^{\pi}\right)$$

and calculate the local representation of this map action. Let $TX = X \times B$, $\mathcal{E} = X \times E$ then

$$K^{\pi} : (x, a, y, b) \mapsto \left(x, b + \Gamma_x^{\pi}(y, a)\right),$$

$$TK^{\pi} : (x, a, y, b, z, c, u, d) \mapsto$$

$$\mapsto \left(x, b + \Gamma_x^{\pi}(y, a), z, d + \Gamma_x^{\pi}(u, a) + \right.$$

$$\left. + \Gamma_x^{\pi}(y, c) + \left(\Gamma_x^{\pi}\right)'(z, y, a)\right),$$

which leads to

$$K^{\pi} \circ TK^{\pi}(x, a, y, b, z, c, u, d)$$

$$= \left(x, d + \Gamma_x^{\pi}(u, a) + \Gamma_x^{\pi}(y, c) + \Gamma_x^{\pi}(z, b) + \right.$$

$$\left. + \left(\Gamma_x^{\pi}\right)'(z, y, a) + \Gamma_x^{\pi}\left(z, \Gamma_x^{\pi}(y, a)\right)\right).$$

While calculating the inverse image, we must change the base X to the base $\mathcal{E}$ or, correspondingly, change x to (x, a). Thus

$$K_V^{\mathcal{E}} : (x, a, y, b, z, c, u, d) \mapsto$$

$$\mapsto \big(x, a, 0, d + \Gamma_x^{\pi}\ (u, a) + \Gamma_x^{\pi}\ (y, c) +$$

$$+ \Gamma_x^{\pi}\ (z, b) + (\Gamma_x^{\pi})'\,(z, y, a) + \Gamma_x^{\pi}\ \big(z, \Gamma_x^{\pi}\ (y, a)\big)\big)\ . \tag{2.21}$$

In the same way, one may look for the horizontal component in the form $K_H^{\mathcal{E}} = \Gamma^{\pi} \circ\ (\pi^*)^{-1}\ (\kappa)$ where $\kappa : T^2\mathcal{E} \to T X$ is a map of the form $\kappa = K^X \circ T^2\ \pi$. Thus

$$K_H^{\mathcal{E}} = \Gamma^{\pi} \circ (\tau_X^*\ (\pi^*))^{-1}\ (K^X \circ T^2\ \pi).$$

Let us calculate the local action of this map. In the above notations, we have

$$\pi : (x, a) \mapsto x, \qquad T\,\pi : (x, a, y, b) \mapsto (x, y),$$

$$T^2\ \pi : (x, a, y, b, z, c, u, d) \mapsto (x, y, z, u),$$

$$K^X : (x, y, z, u) \mapsto \big(x, u + \Gamma_x^{\tau}\ (z, y)\big).$$

Therefore,

$$K_H^{\mathcal{E}}\ (x, a, y, b, z, c, u, d) = \Gamma^{\pi} \circ (\tau_X^*\ (\pi^*))^{-1}\ \big(x, u + \Gamma_x^{\tau}\ (z, y)\big)$$

$$= \big(x, a, u + \Gamma_x^{\tau}\, z, y),\ - \Gamma_x^{\tau}\ (u, a) - \Gamma_x^{\pi}\ (\Gamma_x^{\tau}\ (z, y)\ \ a)\big)\ . \tag{2.22}$$

Summing up $K_H^{\mathcal{E}}$ and $K_V^{\mathcal{E}}$, we obtain

$$K^{\mathcal{E}} = \upsilon^{-1} \circ (\pi\,(\pi^*))^{-1}\ (K^{\pi} \circ TK^{\pi}) \oplus \Gamma^{\pi} \circ (\tau_X^*\ (\pi^*))^{-1}\ (K^X \circ T^2\ \pi). \tag{2.23}$$

For ease of the reader, in describing the local action of this map, we give elements of the product $X \times E$ in the form of column $\binom{x}{a}$. Then it follows from (2.21) to (2.23) that

$$K^{\mathcal{E}} : \left(\binom{x}{a}, \binom{y}{b}, \binom{z}{c}, \binom{u}{d}\right) \mapsto$$

$$\mapsto \left(\binom{x}{a},\ \begin{pmatrix} u + \Gamma_x^{\tau}\ (z, y) \\ d + B_x^{\mathcal{E}}\ (a, y, b, z, c) \end{pmatrix}\right), \tag{2.24}$$

where

$$B_x^{\mathcal{E}}\ (a, y, b, z, c)$$

$$= \Gamma_x^{\pi}\ (y, c) + \Gamma_x^{\pi}\ (z, b) + \big(\Gamma_x^{\pi}\big)'\,(z, y, a) +$$

$$+ \Gamma_x^{\pi}\ \big(z, \Gamma_x^{\pi}\ (y, a)\big) - \Gamma_x^{\pi}\ \big(\Gamma_x^{\tau}\ (z, y), a\big)\ .$$

On the other hand, by definition

$$K^{\mathcal{E}} : \left(\binom{x}{a}, \binom{y}{b}, \binom{z}{c}, \binom{u}{d}\right) \mapsto$$

$$\mapsto \left(\binom{x}{a},\ \Gamma^{\mathcal{E}}_{\binom{x}{a}} \left(\binom{y}{c} \binom{z}{b} \right) \right). \tag{2.25}$$

Comparing the above expressions, we obtain an explicit form of connection coefficients

$$\Gamma^{\mathcal{E}}_{\binom{x}{a}} \left(\binom{y}{c}, \binom{z}{b} \right) = \begin{pmatrix} \Gamma^{\tau}_{x} (z, y) \\ \\ B^{\mathcal{E}}_{x} (a, y, b, z, c) \end{pmatrix}. \tag{2.26}$$

It follows from (2.24), (2.25) that $K^{\mathcal{E}}$ possesses all the properties which are required for a connection of the manifold $\mathcal{E}$ and it is, moreover, compatible with the structure of the vector bundle π in the sense of linear dependence on a on each fibre of this bundle.

2.5. COVARIANT DIFFERENTIATION OF SECTIONS

Now we are able to define covariant differentiation on sections of a given vector bundle $\pi : \mathcal{E} \to X$.

A section $\eta : X \to \mathcal{E}$ belongs to the class $\sigma_k (\mathcal{E})$ if for any point $x \in X$, its principal part η is continuously differentiable up to kth order for a given (and, hence, for any) trivialization of π. In the case where the section is smooth enough so that all operations are correct (but we are not interested in the exact degree of smoothness), the index k will be omitted. Thus, we shall speak about smooth sections and denote them $\eta \in \sigma (\mathcal{E})$.

If $\eta \in \sigma_k (\mathcal{E})$, then its tangent map $T_\eta : TX \to T\mathcal{E}$ is a smooth map of order k–1. Given a vector field $y : X \to TX$, define the map $T_\eta \circ y : X \to T\mathcal{E}$. Its vertical component may be identified in a canonic way with an element $z : X \to \mathcal{E}$ and denoted

$$z = \nabla^{\pi}_{y} \eta = K^{\pi} \circ (T_\eta \circ y). \tag{2.27}$$

Notice that $\nabla^{\pi}_{y} \eta$ is a section of the bundle π called the covariant derivative of the section η along the vector field y. For a trivial bundle, the principal part of the covariant derivative may be represented, due to (2.14), in the form

$$z (x) = \eta' (x)\, y (x) + \Gamma^{\pi}_{x} (y (x), \eta (x)). \tag{2.28}$$

It is easy to check that $\nabla_y \eta$ possesses the following properties:

(a) $\nabla_y \eta$ depends linearly on both y and η.

(b) Given the smooth functions $\alpha : X \to R^1$ and $\beta : X \to R^1$, we have

$$\nabla^{\pi}_{\alpha y} \eta = \alpha \nabla^{\pi}_{y} \eta,$$

$$\nabla^{\pi}_{y} (\beta \eta) = \beta \nabla^{\pi}_{y} \eta + \eta \nabla^{R^1}_{y} \beta,$$

where

$$\nabla^{R^1}_{y} \beta = \langle y (x), \beta' (x) \rangle = (y, D) \beta .$$

(c) Let η be a section of π, A be a section of $L (\pi, \pi_1)$ (see example 3) and let the connections of these bundles be compatible in the sense of that Example. Then

$$\nabla_y^{\pi^1}(A\,\eta) = \left(\nabla_y^{L(\pi,\pi_1)} A\right)\eta + A\,\nabla_y^{\pi}\eta,$$

which shows that the covariant derivative is a differentiation. In particular, if α is a section of the bundle π^*, then

$$\nabla_y^{R^1}\langle\eta,\alpha\rangle = \left\langle\nabla_y^{\pi}\eta,\alpha\right\rangle + \left\langle\eta,\nabla_y^{\pi^*}\alpha\right\rangle .$$

Similar relations may be proved for multilinear operations. Below, we shall omit the upper index in ∇_y^{π} if it does not lead to misunderstanding.

We now consider some important operations associated with covariant derivatives.

Given an affine connection on a manifold X, consider $\nabla_y z - \nabla_z y$; this is a bilinear operation for the pair of vector fields z and y. The principal part of this operation in a local trivialization looks like

$$z'(x)\,y(x) - y'(x)\,z(x) + \Gamma_x^{\tau}(y(x), z(x)) - \Gamma_x^{\tau}(y(x), z(x)). \tag{2.29}$$

It is easy to check that $z'(x)\,y(x) - y'(x)\,z(x)$ represents the principal part of a vector field which is usually denoted $[z, y]$ and called the vector field commutator. The vector field

$$\mathcal{K}_y^X(y, z) = \nabla_y z - \nabla_z y - [y, z] \tag{2.30}$$

gives rise to a $L(\tau\times\tau, \tau)$-type tensor field $\mathcal{K}_y^X$ with a principal part

$$\mathcal{K}_x^X(y, z) = \Gamma_x^{\tau}(y, z) - \Gamma_x^{\tau}(z, y). \tag{2.31}$$

This tensor is called the torsion of the given connection.

Now let $y(x), z(x)$ be vector fields and $u(x)$ be a scalar function on X. By definition, the derivative acts as $u'(x): T_x X \to R^1$ and, hence, $u'(x) \in T_x^* X$; that is, $u': X \to T^* X$, is a section of the cotangent bundle. In this case, the covariant derivative has the form

$$\nabla_y^{R^1} u(x) = \langle y(x), u'(x)\rangle_x = u'(x)\,y(x) \tag{2.32}$$

in agreement with the assumption that the connection of a trivial bundle $X \times R^1$ is trivial. Next, compute

$$\begin{aligned}
&\nabla_z^{R^1}\nabla_y^{R^1} u(x)\\
&\quad= \nabla_z^{R^1}\langle y(x), u'(x)\rangle_x\\
&\quad= \left\langle\nabla_x^{\tau} y(x), u'(x)\right\rangle + \left\langle y(x), \nabla_z^{\tau^*} u'(x)\right\rangle .
\end{aligned}$$

Given the cobundle connection coefficients, we obtain

$$\begin{aligned}
&\left\langle y(x), \nabla_z^{\tau^*} u'(x)\right\rangle\\
&\quad= \left\langle y(x), u''(x)\,z(x) + \Gamma_x^{\tau^*}(z(x), u'(x))\right\rangle
\end{aligned}$$

$$= u''(x)\,(y(x), z(x)) -$$

$$- \left\langle \Gamma_x^{\tau}(z(x), y(x)), u'(x) \right\rangle . \tag{2.33}$$

Therefore,

$$\nabla_z^{R^1} \nabla_y^{R^1} u(x) = u''(x)\,(y(x), z(x)) +$$

$$+ \langle y'(x)\, z(x), u'(x) \rangle . \tag{2.34}$$

Notice, moreover, that

$$\left(\nabla_z^{R^1} \nabla_y^{R^1} - \nabla_y^{R^1} \nabla_z^{R^1}\right) u = \langle [z, y], u'(x) \rangle = \nabla_{[z,y]}^{R^1} u(x). \tag{2.35}$$

Consider now the differential operator

$$\mathcal{D}(z, y)\, u = \left(\nabla_z \nabla_y - \nabla_{\nabla_z y}\right) u$$

$$= u''(z, y) - \left\langle \Gamma^{\tau}(z, y), u' \right\rangle.$$

The left-hand side of this equality shows that this expression is invariant while its right-hand side shows that it is symmetric (in case of zero torsion) with respect to z and y,

$$[\mathcal{D}(z, y) - \mathcal{D}(y, z)]\, u = \langle \mathcal{K}(y, z), u' \rangle . \tag{2.36}$$

Consider a section b of the vector bundle $\pi : \mathcal{E} \to X$. In a local trivialization,

$$\nabla_z^{\pi} \nabla_y^{\pi} b(x) = \nabla_z^{\pi}\left(b'(x)\, y(x) + \Gamma_x^{\pi}(y(x), b(x))\right)$$

$$= b''(x)\,(y(x), z(x)) + b'(x)\, y'(x)\, z(x) +$$

$$+ \Gamma_x^{\pi}(z(x), b'(x)\, y(x)) + \Gamma_x^{\pi}\left(z(x), \Gamma_x^{\pi}(y(x), b(x))\right) +$$

$$+ \left(\Gamma_x^{\pi}\right)'(z(x), y(x), b(x)) + \Gamma_x^{\pi}(y(x), b'(x)\, z(x)) +$$

$$+ \Gamma_x^{\pi}(y'(x)\, z(x), b(x)).$$

By alternating, we obtain

$$[\nabla_z \nabla_y - \nabla_y \nabla_z]\, b$$

$$= b'[z, y] + \Gamma^{\pi}([z, y], b) + R(z, y, b)$$

$$= \nabla_{[z,y]}\, b + R(z, y, b), \tag{2.37}$$

where

$$R(z, y, b) = \Gamma^{\pi}(z, \Gamma^{\pi}(y, b)) + (\Gamma^{\pi})'(z, y, b) -$$

$$- \Gamma^{\pi}(y, \Gamma^{\pi}(z, b)) + (\Gamma^{\pi})'(y, z, b). \tag{2.38}$$

It follows from (2.37) that

$$R(z, y, b) = \nabla_z \nabla_y b - \nabla_y \nabla_z b - \nabla_{[z,y]}\, b \tag{2.39}$$

is a section of the bundle π and, thus, R defines a tensor of the $L(\tau\times\tau\times\pi, \pi)$-type. This tensor is called the curvature tensor of the given connection. Equation (2.38) also implies that a trivial connection curvature is zero, which is consistent with (2.35) for a scalar function.

A simple transformation changes $\mathcal{D}(z, y)\, b$ to the following

$$\begin{aligned}
&\mathcal{D}(z(x), y(x))\, b(x) \\
&\quad = b''(x)\,(y(x), z(x)) + \Gamma^{\pi}_{x}\,(z(x), b'(x)\, y(x)) + \\
&\qquad + \Gamma^{\pi}_{x}\,(y(x), b'(x)\, z(x)) - b'(x)\,\Gamma^{\tau}_{x}\,(z(x), y(x)) + \\
&\qquad + \Gamma^{\pi}_{x}\left(z(x), \Gamma^{\pi}_{x}\,(y(x), b(x))\right) + (\Gamma^{\pi}_{x})'\,(z(x), y(x), b(x)) - \\
&\qquad - \Gamma^{\pi}_{x}\left(\Gamma^{\tau}_{x}\,(z(x), y(x)), b(x)\right)
\end{aligned}$$

and by alternation, we obtain

$$\begin{aligned}
&[\mathcal{D}(z, y) - \mathcal{D}(y, z)]\, b \\
&\quad = R\,(z, y, b) + b'\,\mathcal{K}(y, z) + \Gamma^{\pi}\,(\mathcal{K}(y, z), b) \\
&\quad = R\,(z, y, b) + \nabla_{\mathcal{K}(y,z)}\, b.
\end{aligned} \tag{2.40}$$

Notice that $\mathcal{D}(z, y)$ turns out to be bilinear with respect to y and z in the sense that

$$\mathcal{D}(\alpha_1 z_1 + \alpha_2 z_2, y) = \alpha_1\, \mathcal{D}(z_1, y) + \alpha_2\, \mathcal{D}(z_2, y),$$

$$\mathcal{D}(z, \beta_1 y_1 + \beta_2 y_2) = \beta_1\, \mathcal{D}(z, y_1) + \beta_2\, \mathcal{D}(z, y_2)$$

with $\alpha_j\, \beta_j$ functions on $X, j = 1, 2$.

Using (2.31), we may derive the following expression for the torsion of a linear connection on the tangent space $T\mathcal{E}$ of the bundle total space $\mathcal{E}$.

$$\mathcal{K}^{\mathcal{E}}_{\binom{x}{a}}\left(\binom{y}{a}, \binom{z}{c}\right) = \begin{pmatrix} \mathcal{K}^{X}_{x}\,(y, z) \\ R^{\pi}_{x}\,(y, z, a) - \Gamma^{\pi}_{x}\left(\mathcal{K}^{X}_{x}\,(y, z), a\right) \end{pmatrix}. \tag{2.41}$$

In particular, if the base torsion is trivial then the torsion tensor of the total space $\mathcal{E}$ connection coincides with the connection Γ^{π} curvature tensor. To be more precise, the bundle curvature is the vertical component of the total space torsion tensor.

We now introduce the symbolic notations for covariant derivatives similar to those we have used in Section 1 of chapter 1.

By ∇^{π} we mean a formal section of the bundle $L_2(\tau\times\pi, \pi)$, while given a vector field y, ∇^{π}_{y} stands for a section of $L(\pi, \pi)$. Finally, by ∇^{π}_{η} we mean a section of

$L(\tau, \pi)$ for a given section η of the bundle π. Thus $(\nabla^{\pi}_{\eta})\, y = \nabla^{\pi}_{y}\, \eta$ is the covariant derivative of η along y.

Given a local trivialization of a bundle π, we introduce the operator $\Gamma^{\pi}(\eta) \in L(B, E)$ by

$$\Gamma^{\pi}(y, \eta) = \hat{\Gamma}^{\pi}(\eta)\, y. \tag{2.42}$$

In the above, notations (2.28) reads

$$\nabla^{\pi}_{y}\eta = (D \otimes \eta)\, y + \hat{\Gamma}^{\pi}(\eta)\, y,$$

which implies

$$\nabla^{\pi}\eta = (D \otimes \eta) + \hat{\Gamma}^{\pi}(\eta)$$

and, finally,

$$\nabla^{\pi} = D \otimes I_E + \hat{\Gamma}^{\pi}. \tag{2.43}$$

In particular, given a trivial connection we have $\nabla^{\pi} = D \otimes I_E$. In the end, in the scalar case $E = R^1$, the covariant differentiation symbol coincides with the usual differentiation $\nabla = D$.

Let us calculate some differential expressions using the above notations.

Given a section b of the bundle $L(L(\tau, \pi), \pi_1)$ and sections ξ, η of bundles τ^* and π, we notice that the map

$$(\xi \otimes \eta) : y \mapsto \langle y, \xi \rangle\, \eta$$

generates a section of $L(\tau, \pi)$ and thus a section $b(\xi \otimes \eta)$ of the bundle π_1 is correctly defined.

Consider the trivial bundles

$$\tau : X \times B \to X, \quad \pi : X \times E \to X, \quad \pi_1 : X \times E_1 \to X,$$

then we obtain for the section principal parts

$$\xi \otimes \eta : B \to E, \quad b(\xi, \eta) = b(\xi \otimes \eta) \in E_1.$$

Define a first-order differential operation

$$b(D, \eta) = b(D \otimes \eta)$$

for a given function $\eta : X \to E$.

The covariant analogue of this operation will be denoted

$$b(\nabla, \eta) \overset{\text{def}}{=} b(\nabla^{\pi}\eta). \tag{2.44}$$

Its principal part is

$$b(\nabla, \eta) = b\ (\nabla^{\pi}\eta) = b(D \otimes \eta) + b(\hat{\Gamma}(\eta)). \tag{2.45}$$

Let, in particular, $\pi = \tau^*$ and $\eta = \nabla u = D u$ for $u : X \to R^1$.
Define a second-order differential operation

$$B(\nabla, \nabla) \overset{\text{def}}{=} b(\nabla^{\tau^*}\nabla) : u \to b(\nabla^{\tau^*}\nabla u). \tag{2.46}$$

Given the sections α and β of $L(\pi, \pi_1)$ and τ, respectively, consider

$$b = \beta \otimes \alpha : \tilde{\eta} \mapsto \alpha \, \tilde{\eta} \, \beta, \qquad \tilde{\eta} \in L(\tau, \eta). \tag{2.47}$$

Then

$$b\,(\nabla^{\pi} \eta) = (\beta \otimes \alpha)\,(\nabla^{\pi} \eta) = \alpha\left((\nabla^{\pi} \eta)\, \beta\right),$$

and

$$\begin{aligned}(\beta \otimes \alpha)\,&(\nabla^{\pi} \eta)\\ &= (\beta \otimes \alpha)\,(D \otimes \eta) + \alpha\,(\hat{\Gamma}\,(\eta))\,\beta\\ &= \alpha\,((\beta, D)\,\eta) + \alpha\,(\Gamma^{\pi}\,(\beta, \eta)).\end{aligned} \tag{2.48}$$

If $\pi = \tau^*$ and $\eta = \nabla u$, we obtain

$$(\beta \otimes \alpha)\,(\nabla^{\tau^*} \nabla)\, u = \alpha\left((\beta, D)\, D + \Gamma^{\tau^*}(\beta, D)\right) u. \tag{2.49}$$

In the sequel, putting $b = \Sigma_j \beta_j \otimes \alpha_j$, we shall use these formulas to compute differential operations.

2.6. TANGENT BUNDLE CONNECTION AND EXPONENTIAL MAPPING

Let X be a smooth Banach manifold, τ be its tangent bundle $\tau : TX \to X$, K^X the connection map, Γ^X its local coefficients.

Given a curve $\alpha : J \to X$ a curve $\beta : J \to TX$ is said to be a lift of the curve α if $\tau\,\beta = \alpha$. The curve $\alpha'(t) = T\alpha\,(t)$ is called the canonical lift of the curve $\alpha\,(t)$.

Recall that a first-order differential equation on X is simply a vector field over X. By a second-order differential equation on X, we mean a vector field η on TX such that for all $\upsilon \in TX$ the equality

$$T\tau \circ \eta\,(\upsilon) = \upsilon \tag{2.50}$$

holds.

It follows from the definition that η is a second-order differential equation on X if and only if each integral path $\beta\,(t)$ of the field η gives the canonical lift of the curve $\tau\,\beta\,(t)$ that is $(\tau\,\beta\,(t))' = \beta\,(t)$.

Consider a vector field η on TX as being the horizontal lift of a vector field ξ. Assume that ξ is, in turn, a canonical lift of a curve $x\,(t)$ in X. It follows from the result of Section 3 that the principal part of η may be written in a given trivialization in the form

$$\left(x'(t),\ -\Gamma^{\pi}_{x(t)}\,(x'(t), x'(t))\right), \tag{2.51}$$

and the integral curves of this field satisfy the relations

$$\frac{d\,x}{d\,t} = \xi, \qquad \frac{d\,\xi}{d\,t} = -\Gamma^{\pi}_{x(t)}\,(\xi, \xi).$$

By definition, integral curves $\beta\,(t) = (x\,(t), x'\,(t))$ are horizontal ones.

The above relations imply that an integral curve of a horizontal lift of $x'\,(t)$ satisfies the equation

$$\frac{d^2 x}{d t^2} + \Gamma^{\pi}_{x(t)} (x'(t), x'(t)) = 0. \tag{2.52}$$

The initial values

$$x(0) = x, \quad x'(0) - \upsilon \tag{2.53}$$

determine the solutions of this equation in a unique way.

The curve $x(t)$ satisfying (2.52) is called a geodesic curve in X.

Let us point out that a canonical geodesic lift satisfies the equation

$$\nabla_{x'(t)} x'(t) = 0, \tag{2.54}$$

which describes a parallel translation in TX.

Consider a mapping from TX into X

$$\upsilon \to x_\upsilon (1),$$

which associates to a vector field υ the value $x_\upsilon(1)$ of the solution of the Cauchy problem (2.52), (2.53) at the point 1. Due to the results of chapter 1 concerning solutions of first-order differential equations, this mapping is a morphism from an open set $\mathfrak{A}$ of TX into X.

Denote this morphism by

$$\exp : \mathfrak{A} \to X.$$

Hence,

$$\exp \upsilon = x_\upsilon (1). \tag{2.55}$$

Let $\exp_x$ be the restriction of $\exp$ to the tangent space $T_x X$ at the point $x \in X$. Notice that (2.28) implies that $\beta(t) = (x(t), x'(t))$ satisfies the relation

$$\exp (s\upsilon) = \tau \beta_{s\upsilon}(1) = \tau \beta_\upsilon (s) \tag{2.56}$$

and thus

$$\frac{d}{ds} \exp (s\upsilon)|_{s=0} = \frac{d}{ds} \tau \beta_\upsilon (s)|_{s=0} = T\tau \circ \frac{d \beta_\upsilon (s)}{ds}\Big|_{s=0}. \tag{2.57}$$

Let us differentiate (2.55) twice and use (2.56), (2.57). We obtain

$$\begin{aligned} &\frac{d^2}{ds^2} \exp (s\upsilon)|_{s=0} \\ &= \frac{d^2}{ds^2} \tau \beta_\upsilon (s))|_{s=0} \\ &= \frac{d}{ds}\left(T\tau \circ \frac{d \beta_\upsilon (s)}{ds}\right)\Big|_{s=0} = \frac{d}{ds}\Big|_{s=0}. \\ &= -\Gamma^X_x (\upsilon, \upsilon). \end{aligned} \tag{2.58}$$

Therefore the following assertion holds.

PROPOSITION 2.1. *The map* $\exp_x : T_x X \to X$ *induces a local isomorphism from a neighborhood of zero into itself and, moreover,* $T \exp_x$ *is the identity map at zero.*

Given a manifold morphism $f : X \to Y$, we say the connection maps K^X and K^Y are f-connected if their local coefficients satisfy the relation

$$\Gamma^Y_{f(x)} \; (f'(x)\, y, f'(x)\, z) \\ = f'(x)\, \Gamma^X_x \; (y\,, x) + f''(x)\,(y, z) \tag{2.59}$$

for $y, z \in T_x X$.

Exponent maps $\exp^X : T X \to X$ and $\exp^Y : T Y \to Y$ associated with f-connected connection maps are called f-connected as well. Given f-connected exponential maps, we obtain the following commutative diagram

$$\begin{array}{ccc} TX & \xrightarrow{\;Tf\;} & TY \\ {\scriptstyle \exp^X}\downarrow & & \downarrow{\scriptstyle \exp^Y} \\ X & \xrightarrow{\;f\;} & Y \end{array} \quad .$$

A connection map K^X of the manifold X, generates a connection of the tangent bundle TX total space, as has been shown in Section 2.4. In accordance with (2.26) the local coefficients of this connection may be given by the relation

$$\tilde{\Gamma}_{\binom{x}{a}} \left(\binom{y}{b}, \binom{z}{c} \right) \\ = \begin{pmatrix} \Gamma^X_x \, (z, y) \\ B^{TX}_x \, (a, y, b, z, c) \end{pmatrix},$$

where

$$B^{TX}_x \, (a, y, b, z, c) \\ = \Gamma^X_x (y, c) + \Gamma^X_x (z, b) + \\ + \left(\Gamma^X_x\right)' (z, y, a) + \Gamma^X_x \left(z, \Gamma^X_x (y, a)\right) - \Gamma^X_x \left(\Gamma^X_x (z, y), a\right) .$$

Let the connection map K^X be symmetric so that its torsion is zero. In this case, as is easy to see, one may construct another connection on $T X$ by choosing as its local coefficients

$$\Gamma^{TX}_{\binom{x}{a}} = \tilde{\Gamma}^{TX}_{\binom{x}{a}} - \binom{0}{R}_{\binom{x}{a}} ,$$

where R is the curvature tensor of the connection map K^X.

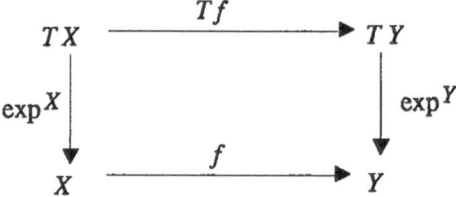

In this way we obtain the relation

$$\Gamma^{TX} = \begin{pmatrix} \Gamma^{X} \\ B^{TX} - R \end{pmatrix},$$

which leads, due to the equality,

$$\begin{aligned}
&\left(B^{TX} - R_x\right)(a, y, b, z, c) \\
&= \Gamma_x^X(y, c) + \Gamma_x^X(z, b) - \Gamma_x^X\left(\Gamma_x^X(z, y)\, a\right) + \\
&\quad + \left(\Gamma_x^X\right)'(z, y, a) + \Gamma_x^X\left(z, \Gamma_x^X(y, a)\right) - \left(\Gamma_x^X\right)'(z; a, y) + \\
&\quad + \left(\Gamma_x^X\right)'(a, z, y) - \Gamma_x^X\left(z, \Gamma_x^X(y, a)\right) + \\
&\quad + \Gamma_x^X\left(a, \Gamma_x^X(z, y)\right) \\
&= \left(\Gamma_x^X\right)'(a, z, y) + \Gamma_x^X(y, c) + \Gamma_x^X(z, b)
\end{aligned}$$

to the following expression for Γ^{TX}

$$\begin{aligned}
&\Gamma^{TX}_{\binom{x}{a}}\left(\binom{y}{b}, \binom{z}{c}\right) \\
&= \begin{pmatrix} \Gamma_x^X(z, y) \\ \left(\Gamma_x^X\right)'(a, z, y) + \Gamma_x^X(y, c) + \Gamma_x^X(z, b) \end{pmatrix}.
\end{aligned} \tag{2.60}$$

Notice that the connection map K^{TX} with local coefficients (2.60) possesses the following properties

$$\tau_1 \circ K^{TX} = \tau_1 \circ \tau_2, \quad \tau_k : T^{k+1} X \to T^k X, \quad k = 1, 2, \tag{2.61}$$

$$T\tau \circ K^{TX} = K^X \circ T^2 \tau, \tag{2.62}$$

$$K^X \circ K^{TX} = K^X \circ T K^X - R\left(T\tau \circ T^2\tau, \tau_1 \circ T\tau_1, \tau_1 \circ T^2\tau\right). \tag{2.63}$$

Denote ∇^{TX} as the covariant derivative corresponding to the connection map K^{TX}

$$\nabla^{TX}_{\alpha} z = K^{TX} \circ Tz \circ \alpha, \quad \alpha, z \in T^2 X. \tag{2.64}$$

PROPOSITION 2.2. *A curve* $\beta(t) \in TX$ *is a geodesic curve through the point* $\beta(0) = \kappa$, *that is*

$$\nabla^{TX}_{\beta'} \beta' = 0, \tag{2.65}$$

if and only if it is a lift of a geodesic curve $\gamma(t)$ *in* X *and satisfies the following relations*

$$\beta' = (0) = \xi,$$

$$\nabla_{\gamma'} \nabla_{\gamma'} \beta + R_\gamma (\beta, \gamma', \gamma') = 0, \tag{2.66}$$

$$\gamma'(0) = T\tau \circ \xi, \quad \beta(0) = \tau_1 \circ \xi, \quad \nabla_\beta (0) = K^X \circ \xi. \tag{2.67}$$

Proof. Let β satisfy (2.65). Then, given $\gamma - \tau\beta$, we have $\gamma' = T\tau \circ \beta'$ and, thus,

$$\gamma'(0) = T\tau \circ \xi, \quad \nabla_{\gamma'} = K^X \circ T_{\gamma'} = K^X \circ T^2 \tau \circ T\beta'$$

$$= T\tau \circ K^{TX} \circ T(\beta'_{\gamma'}) = T\tau \circ \nabla^{TX} \beta = 0.$$

Now, it follows from $\nabla \beta = K^X \circ T\beta$ that

$$\nabla^2 \beta = K^X \circ T(K^X \circ T\beta)$$

and, due to (2.63), we obtain

$$\nabla^2 \beta(z, y)$$

$$= K^X \circ K^{TX} + R(T\tau \circ T^2 \tau, \tau_1 \circ T\tau_1, \tau_1 \circ T^2 \tau) \circ T^2 \beta(z, y)$$

$$= R(z, \beta, y) = -R(\beta, z, y);$$

that is,

$$\nabla^2 \beta(z, y) + R(\beta, z, y) = 0$$

and, moreover,

$$\beta(0) = \tau_1 \circ T\beta(0) = \tau_1 \circ \xi,$$

$$\nabla \beta(0) = K^X \circ T\beta(0) = K^X \circ \xi,$$

$$T\tau \circ \beta'(0) = T(\tau \circ \beta)(0) = T\gamma,$$

$$\tau \circ \beta = \gamma.$$

To prove the converse, suppose that $\beta(t)$ solves (2.66) along the geodesic curve $\gamma(t)$ in X. Write

$$\xi = (\tau_1, T\tau, K^X)^{-1} (\beta(0), \gamma'(0), \nabla \beta(0)).$$

Then

$$K^X \circ \nabla^{TX} \beta' = K^X \circ K^{TX} \circ \beta'' = (K^X \circ T K^X - R) \circ \beta''$$

$$= \nabla^2 \beta - R = 0,$$

$$T\tau \; \nabla^{TX} \beta' = T\tau K^{TX} \circ \beta'' = K^X \circ T^2 \tau \circ \beta''$$

$$= K^X \circ T^2 (\tau \circ \beta) = K^X \circ T\gamma' = \nabla \gamma' = 0.$$

And, finally, since $\operatorname{Ker} K^X \cap \operatorname{Ker} T\tau = 0$, we have $\nabla^{TX} \beta' = 0$.

Now, we have that

$$K^X \beta'(0) = \nabla \beta(0) = K^X \circ \xi,$$

$$T\tau \circ \beta'(0) = T(\tau \circ \beta)(0) = \gamma'(0) = T\tau \circ \xi$$

so that $\beta'(0) = \xi$.

In the end, we shall see what are the links between exponential maps $\exp^X : TX \to X$ and $\exp^{TX} : T^2X \to TX$ associated with K^X and K^{TX}, respectively.

Let $S : T^2X \to T^2X$ be a morphism from T^2X into itself which has the form

$$S(x, a, y, b) = (x, y, a, b)$$

in a local trivialization. Then

$$\exp^{TX} = T\exp^X \circ S. \tag{2.68}$$

To prove that (2.68) holds, let us use the definition of the map $\exp^{TX}$, which yields

$$\exp^{TX}_{\binom{x}{a}} \begin{pmatrix} \alpha \\ \beta \end{pmatrix} = \begin{pmatrix} \gamma_\alpha(1) \\ \eta_\beta(1) \end{pmatrix}, \tag{2.69}$$

where $\begin{pmatrix} \gamma_\alpha(t) \\ \eta_\beta(t) \end{pmatrix}$ is a geodesic curve in TX with initial data

$$\gamma'_\alpha(0) = \alpha, \quad \eta'_\beta(0) = \beta, \quad \gamma_\alpha(0) = x, \quad \eta_\beta(0) = a.$$

Since $\exp^x_\alpha = \gamma_\alpha(1)$ and

$$(T\exp^X)_{\binom{x}{\alpha}} \begin{pmatrix} \alpha \\ \beta \end{pmatrix} = \begin{pmatrix} \gamma_\alpha(1) \\ \\ (T\gamma_\alpha)_{\binom{a}{b}}(1) \end{pmatrix} \tag{2.70}$$

by comparing (2.69) and (2.70), we obtain (2.68).

3. Hilbert and Hilbert–Schmidt Bundles

By a Hilbert bundle we mean a bundle γ whose typical fibre is a Hilbert space H. Here, by a Hilbert space in a wide sense, we mean a linear topological space which admits the introduction of a Hilbert structure generating a topology equivalent to the initial one.

Along with γ, we shall need below the bundle $L(\gamma, \gamma^*)$ such that the linear maps from $\mathcal{H}_x = \gamma^{-1}(x)$ into $\mathcal{H}^*_x = (\gamma^*)^{-1}(x)$ form its fibre through the point x. A section G_x of this bundle gives rise to a bilinear functional on $\mathcal{H}_x$:

$$(\xi_x, \eta_x)_x = \langle \xi_x, G_x \eta_x \rangle_{\mathcal{H}_x}. \tag{3.1}$$

This functional (as well as the section G itself) is called a Riemannian metric on γ if it is nondegenerate and nonnegative for any $x \in X$ and, thus, determines a Hilbert structure with inner product (3.1) on each $\mathcal{H}_x$. A Riemannian metric on a Hilbert bundle may be constructed in the usual manner of section construction.

A Hilbert bundle with a Riemannian metric is called a Riemannian bundle.

Given local trivializations of the Riemannian bundle over neighborhoods U_{x_0} of each point $x_0 \in X$, one may choose a set of local sections $\{e_k(x)\}_{k=1}^{\infty}$ forming an orthonormal basis in each $\mathcal{H}_x$, $x \in U_{x_0}$ and such that

$$(e_k(x), e_j(x))_x = \langle e_k(x), G_x e_j(x)\rangle_{\mathcal{H}_x} = \sigma_{jk}.$$

This set is called a local orthonormal basis.

Let Φ be a section of $L(\gamma, \gamma^*)$. Given the orthonormal basis $\{e_k(x)\}$, define the functional

$$\begin{aligned} \mathrm{Tr}_G \Phi(x) &= \sum_k \langle e_k(x), \Phi(x) e_k(x)\rangle_{\mathcal{H}_x} \\ &= \sum_k \left(e_k(x)\ G_x^{-1} \Phi(x) e_k(x)\right)_x = \mathrm{Tr}_{\mathcal{H}_x} G_x^{-1} \Phi(x) \end{aligned} \tag{3.2}$$

assuming that $G_x^{-1}\Phi(x)$ is a nuclear operator in a Hilbert space $\mathcal{H}_x$. Evidently, (3.2) does not depend upon the chosen orthonormal basis and is, therefore, defined in an invariant way.

Let now $\gamma : \mathcal{H} \to X$ be a Riemannian bundle, $\tau = \tau_X : TX \to X$ be a tangent bundle and, finally, A be a section of the bundle $L_{12}(\gamma, \tau)$.

Consider the map

$$b_A : \Phi \mapsto \mathrm{Tr}_G A^* \Phi A$$

acting on $L(\tau, \tau^*)$ sections.

Given trivializations of the above bundles

$$\gamma_u : U \times H \to U, \quad \tau_u : U \times B \to U,$$

principal parts

$$A_x : H \to B, \quad A_x^* : B^* \to H^*, \quad G_x : H \to H^*, \quad \Phi(x) : B \to B^*$$

and a local orthonormal basis $\{e_k(x)\}$, we obtain

$$\begin{aligned} &b_{A_x}(\Phi(x)) \\ &\quad = \sum_k \langle e_k(x),\ A_x^* \Phi(x) A_x e_k(x)\rangle_{\mathcal{H}_x} \\ &\quad = \sum_k \langle A_x e_k(x), \Phi(x) A_x e_k(x)\rangle_{\mathcal{H}_x} \\ &\quad = \Big(\sum_k A_x e_k(x) \otimes A_x e_k(x)\Big) \Phi(x), \qquad x \in U. \end{aligned} \tag{3.3}$$

This expansion converges, since A_x is a Hilbert–Schmidt map.

Remark 3.1. Given $B = H$ and $A = I$, it follows from (3.3) that

$$\mathrm{Tr}_G = \sum_{k=1}^{\infty} e_k(x) \otimes e_k(x), \tag{3.4}$$

but this expansion is valid only for those sections Φ which belong to $L_{11}(\tau, \tau^*)$.

Define a second order differential operation

$$b_A(\nabla, \nabla) \overset{\mathrm{def}}{=} b_A(\nabla^{\tau^*} \nabla).$$

By means of (2.49) and (3.3), we obtain for the principal parts in the considered trivialization

$$\begin{aligned} b_{A_x}&(\nabla, \nabla)\, u(x) \\ &= \Big(\sum_{k=1}^{\infty} A_x e_k(x) \otimes A_x e_k(x)\Big)(\nabla^{\tau^*} \nabla)\, u(x) \\ &= \sum_{k=1}^{\infty} \big\langle A_x e_k(x), \langle A_x e_k(x), D\rangle D\, u(x)\big\rangle - \\ &\qquad - \Big\langle \sum_{k=1}^{\infty} \Gamma_x^{\tau}(A_x e_{kk}(x), A_x e_k(x)), D\, u \Big\rangle \\ &= \sum_{k=1}^{\infty} u''(x)(A_x e_k(x), A_x e_k(x)) - \\ &\qquad - \Big\langle \sum_{k=1}^{\infty} \Gamma_x^{\tau}(A_x e_k(x), A_x e_k(x))\, u'(x) \Big\rangle, \end{aligned}$$

and thus

$$\begin{aligned} b_A(\nabla, \nabla)\, u(x) &= \mathrm{Tr}_G\, A_x^* u''(x) A_x - \\ &\quad - \big\langle \mathrm{Tr}_G\, A_x^* \Gamma_x^{\tau} A_x, u'(x) \big\rangle. \end{aligned} \tag{3.5}$$

It easily follows from those relations that

$$\begin{aligned} b_A(\nabla, \nabla) &= \sum_k \big(\nabla_{Ae_k} \nabla_{Ae_k} - \nabla_{\nabla_{Ae_k} Ae_k}\big) \\ &= \mathrm{Tr}\big(\nabla_A \nabla_A - \nabla_{\nabla_A A}\big). \end{aligned} \tag{3.6}$$

Let us investigate in detail a differential operator of (2.44) type acting on vector bundle sections. Given a section β of the bundle $L(L(\gamma, \pi), \pi_1)$ and a section Φ of $L(\pi, \pi)$, consider $b_A(\Phi) = B(\Phi A)$. It follows from (2.45) by putting $\beta(\xi, \eta) = \beta(\xi \otimes \eta)$, that

$$b(\nabla^{\pi} \eta) = \beta((D \otimes \eta)) A) + \beta\big(\hat{\Gamma}^{\pi}(\eta) A\big).$$

Given the basis $\{e_k(x)\}$ in H, consider the representation

$$\beta\big(\hat{\Gamma}^{\pi}(\eta) A\big) = \beta\Big(\sum_k e_k \otimes \Gamma^{\pi}(A e_k, \eta)\Big)$$

$$= \sum_k \beta\left(e_k, \Gamma^\pi\ (A\, e_k, \eta)\right),$$

which yields

$$b_A\,(\nabla^\pi \eta) = \beta\,(A^*\,D, \eta) + \sum_k\ \beta\left(e_k, \Gamma^\pi\ (A\, e_k, \eta)\right) \tag{3.7}$$

by taking into account $(\xi \otimes \eta)\,A = A^*\,\xi \otimes \eta$.

DEFINITION 3.2. We say that a manifold X is equipped with a Hilbert–Schmidt structure (τ_x, γ, i) if, given the Hilbert bundle γ, one may define a bundle embedding $i : \gamma \to \tau_x$, $\tau_x \circ i = \gamma$ possessing the following property: for each $x \in X$ the map $i_x : \mathcal{H}_x = \gamma^{-1}\ (x) \to T_x X = B_x$ belongs to $L_{12}\ (\mathcal{H}_x, B_x)$ and $i_x\, \mathcal{H}_x$ is dense in $T_x X$. We say this structure is nuclear is the map i_x belongs to $L_{11}\ (\mathcal{H}_x, B_x)$. In the case where γ is a Riemannian bundle with an inner product (3.1)

$$(\xi_x\,, \eta_x)_x = \langle \xi_x\,, G_x\ \ \eta\ x\,\rangle_{\mathcal{H}_x}\,,$$

we say that the H S structure is Riemannian.

We may put in correspondence to the bundle map $i : \gamma \to \tau$ its adjoint map $i^* : \tau^* \to \gamma^*$ which is an embedding as well. Moreover, for each point $x \in X$, we equip $T_x X$ with an H S structure

$$T_x^* X \xrightarrow{\ j_x = G_x^{-1}\, i_x^*\ } G_x^{-1}\ \mathcal{H}_x^* = \mathcal{H}_x \xrightarrow{\ i\ x\ } T_x X\,.$$

By identifying spaces with their images under embeddings, we obtain a rigged Hilbert space

$$\mathcal{H}_x^* = G_x^{-1}\ \ T_x^* X \subset \mathcal{H}_x \subset T_x X = \mathcal{H}_x^- \tag{3.8}$$

with the pairing

$$\langle h^+, h\rangle = (h^+, h)_x = \langle G_x\, h^+, h\rangle_{\mathcal{H}_x}$$

for $h^+ \in \mathcal{H}_x, h \in \mathcal{H}_x$ which introduce in $\mathcal{H}_x^+$ the structure of space adjoint to $\mathcal{H}_x^-$.

In local trivialization

$$\tau_u : U \times B \to U, \qquad \tau_x^* : U \times B^* \to U,$$

$$\gamma : U \times H \to U, \qquad \gamma^* : U \times H^* \to U, \tag{3.9}$$

we obtain the corresponding maps

$$i : H \to B, \quad i^* : B^* \to H^*, \quad G_x : H \to H^*$$

with a constant map i, $(x \in U)$ but with $j_x = G_x^- \ i^* : B^* \to H$ depending upon x. Therefore, after corresponding identifications, the space triple

$$H_x^+ = G_x^- B \subset H \subset H^-$$

does not generally give (for all x) a rigged space H. This must be kept in mind while dealing with manifolds equipped with H S structures.

We say that an affine connection on the manifold X equipped with a Hilbert–Schmidt structure is compatible with this structure if the local connection coefficients possess the property

$$\Gamma_x^\gamma = \Gamma_x^\tau \Big|_{B\times H} : B \times H \to H \quad (x \in U). \tag{3.10}$$

Introduce $\Gamma_x^{\gamma *} : B \times H^* \to H^*$ by

$$\left\langle z, \Gamma_x^{\gamma *}(y, \upsilon)\right\rangle_H = -\left\langle \Gamma^\gamma(y, z), \upsilon\right\rangle_H . \tag{3.11}$$

Denote by $\sigma_k(\gamma)$ the class of vector fields belonging to $\sigma_k(\tau)$ and valued in $\mathcal{H}_x \subset B_x$ for each $x \in X$.

Given $\eta \in \sigma_{k+1}(\gamma)$ and $\xi \in \sigma_k(\tau)$, put

$$\nabla_\xi^\gamma \eta \stackrel{\text{def}}{=} \nabla_\xi^\tau \eta \quad \text{and} \quad \nabla_{\xi_x}^\gamma \eta_x = \eta_x' \xi_x + \Gamma_x^\gamma(\xi_x, \eta_x),$$

which yields $\nabla_\xi^\gamma \eta \in \sigma_k(\tau)$ if Γ_x^γ is smooth enough.

Here η_x' and $\hat{\Gamma}_x^\gamma(\eta) : \xi \to \Gamma^\gamma(\xi, \eta)$ belong to $L(B, H)$ and, hence, possess $L_{12}(H)$ restrictions to H.

Due to this fact, the operation $\nabla^\gamma \eta : \xi \mapsto \nabla_\xi^\gamma \eta$ over each $\mathcal{H}_x$ possesses the same property.

Let X be a manifold equipped with a Riemannian structure (τ, γ, i). We say that an affine connection on the manifold is compatible with this structure if it satisfies (3.10), is symmetric (that is, $\mathcal{K}_x = 0$) and $\nabla^{L(\gamma,\gamma^*)} G = 0$, for the covariant differentiation generated by this connection.

It follows immediately from the definition that in this case the Riemannian metric G is invariant under the parallel displacement of tangent vectors and for vector fields $\varphi, \psi \in \sigma_1(\gamma)$, $z \in \sigma_1(\tau)$, one has

$$\nabla_z(\varphi, \psi) = \left(\nabla_z^\gamma \varphi, \psi\right) + \left(\varphi, \nabla_z^\gamma \psi\right).$$

A connection compatible with a given Riemannian metric is uniquely defined. In fact, given trivialization of considered bundles and fields φ, ψ, z belonging to $\sigma_1(\gamma)$, we have by the definition of the covariant derivative that

$$\begin{aligned}
&\langle \nabla_z G \varphi, \psi\rangle \\
&\quad = \left\langle (G' z)\varphi, \psi\right\rangle + \left\langle \Gamma^{L(\gamma,\gamma^*)}(z, G)\varphi, \psi\right\rangle \\
&\quad = \left\langle (G' z)\varphi, \psi\right\rangle + \left\langle \Gamma^{\gamma^*}(z, G\varphi) - G\,\Gamma^\gamma(z, \varphi), \psi\right\rangle .
\end{aligned}$$

Therefore, it follows from (3.11) and $\nabla_z G = 0$ that

$$\left\langle (G' z)\varphi, \psi\right\rangle = \langle \Gamma^\gamma(z, \varphi), G\psi\rangle + \langle \Gamma^\gamma(\psi, z), G\varphi\rangle .$$

By a cyclic permutation and summing, we get the formula

$$\langle \Gamma^\gamma(\varphi, \psi)\,,\; G\varphi\rangle$$

$$= \frac{1}{2} \left[\langle (G'\varphi)\psi, z\rangle + \langle (G'\varphi) z, \varphi\rangle - \langle (G'z)\varphi, \psi\rangle \right]. \tag{3.12}$$

Suppose G satisfies the relation

$$\langle (G'z)\varphi, \psi\rangle = G_1(z, i\varphi, i\psi), \quad z \in B, \varphi, \psi \in H, \tag{3.13}$$

where $G_1 \in L_3(B, R^1)$. We call this relation a nuclear condition as it implies the estimate

$$|\langle (G'z)\varphi, \psi\rangle| \le \text{const} \; \| i\varphi \|_B \; \| i\psi \|_B ,$$

and, as a result, we have convergence of the series

$$\sum_k |\langle (G'z) e_k, e_k\rangle| < \infty, \quad z \in B \tag{3.14}$$

for any local basis $\{e_k\}$.

Given a nuclear Riemannian metric, we may prove, thanks to (3.12), that a similar statement holds as well for local connection coefficients compatible with this metric. In particular, in the sequel, we shall need the estimate

$$\sum_{k=1}^{\infty} |\langle \Gamma^{\gamma}(e_k, \psi)\, Ge_k\rangle| < \infty. \tag{3.15}$$

We assume that (3.15) holds, even when Γ and G are not compatible.

Notice that there is another situation in which (3.15) is satisfied; namely a nuclear structure.

Remark 3.2. To assume that conditions (3.13) or (3.15) are invariant under coordinate transformations of a manifold, we have to put a restriction on the set of such allowed transformations.

Denote $\sigma_k^*(\gamma)$ the class of vector fields belonging to $\sigma_k(\tau)$ which satisfy the condition

$$G_x z_x \in I_x^* \; T_x^* \; X.$$

DEFINITION 3.3. We say that the divergence of the vector field $\eta \in \sigma_k^*(\gamma)$ exists if $\nabla^{\gamma} \eta|_H$ is a nuclear map, and define it by

$$\text{div}_G\, \eta(x) = \text{Tr}_G\, G\, \nabla^{\gamma} \eta.$$

THEOREM 3.1. *Let both the Riemannian metric and connection be nuclear. Then a vector field belonging to* $\sigma_1^*(\gamma)$ *possesses a finite divergence.*

Moreover,

$$\text{div}_G\, \eta(x) = (\nabla, \eta)_x, \tag{3.16}$$

where ∇ *is a differential operator with the principal part in the form*

$$\nabla = D + \Gamma$$

and where Γ *is defined by*

$$(\Gamma_x, \eta_x)_x = \text{Tr}_G\, \hat{\Gamma}_x^{\gamma}(\eta_x),$$

$$\Gamma_x = -G_x^{-1} \sum_{k=1}^{\infty} \Gamma_x^{\gamma *} (e_k(x), G_x e_k(x)). \tag{3.17}$$

Proof. Since the assertion is local, it suffices to consider trivial bundles. Let $\eta_x = G_x^{-1} i^* v_x$, where $v_x \in B^*$, $x \in B$. Then, given $h \in H \subset B$, we obtain

$$\begin{aligned} G\eta' h &= G_x (G_x^{-1})' i\, h (i^* v_x) + i^* v_x' i\, h \\ &= i^* v_x' i\, h - G_x' (i\, h\, G_x^{-1}\, i^* v_x) \end{aligned}$$

which implies, due to (3.13), that the map $h \mapsto G\eta' h$ is a nuclear operator. By similar arguments, we may prove the convergence of the series (3.17)

$$\begin{aligned} &\mathrm{Tr}_G G \hat{\Gamma}_x^{\gamma} (\eta_x) \\ &= -\left(G^{-1} \sum_{k=1}^{\infty} \Gamma_x^{\gamma *} (e_k(x), G_x e_k(x)), \eta\right) \\ &= \sum_{k=1}^{\infty} \langle e_k(x), G_x\, \Gamma_x^{\gamma} (e_k(x), \eta(x))\rangle . \end{aligned}$$

Furthermore, using (3.4) we have

$$\begin{aligned} \mathrm{div}_G\, \eta(x) &= \mathrm{Tr}_G G\, \nabla^{\gamma} \eta(x) = \left(\sum_k e_k(x) \otimes e_k(x)\right) \times \\ &\quad \times (G\, \nabla^{\gamma} \eta) = \sum_{k=1}^{\infty} \langle e_k(x), G\, \nabla^{\gamma} \eta(x)\, e_k(x)\rangle_H \\ &= \sum_{k=1}^{\infty} \left\langle (e_k(x)\ G\ \{D\eta\, e_k(x) + \Gamma_x^{\gamma} (e_k(x)\ \eta(x))\}\right\rangle \\ &= \mathrm{Tr}_H D\eta + (\Gamma_x, \eta_x)_x = (D + \Gamma_x, \eta)_x . \end{aligned}$$

Remark 3.3. By the same arguments which had been used in the proof of the theorem, we may prove the nuclearity of the map

$$y \mapsto R(\eta_1, y, \eta_2)$$

for the vector fields $\eta_1, \eta_2 \in \sigma_1(\gamma)$.

The map trace is the Ricci tensor of the considered connection. Its principal part is given by

$$R(\eta_1, \eta_2) = \sum_k (R(\eta_1, e_k, \eta_2), e_k)_G .$$

PROPOSITION 3.1. *Let the conditions of the theorem hold. Then given the vector fields* $z_1, z_2 \in \sigma_2(\gamma)$ *and a connection compatible with the Riemannian structure, we obtain the relations*

$$\delta(z_1, z_2) = R(z_1, z_2) - \mathrm{Tr}_G \nabla_{z_2} \nabla_{z_1}, \tag{3.18}$$

$$\delta(z_1, z_2) = \nabla_{z_1} \operatorname{div}_G z_2 - \operatorname{div}_G \nabla_{z_1} z_2 . \tag{3.19}$$

Proof. Choose φ and ψ to be parallel vector fields along the integral path of the vector field z_1. Then

$$\nabla_{z_1} \langle (\nabla_{z_2}) \varphi, G\psi\rangle - \langle \nabla (\nabla_{z_1} z_2) \varphi, G\psi\rangle$$

$$= \langle (\nabla_{z_1} \nabla_\varphi - \nabla_\varphi \nabla_{z_1} - \nabla_{[z_1,\varphi]}) z_2, G\psi\rangle +$$

$$+ \langle \nabla_{[z_1,\varphi]} z_2, G\psi\rangle + \langle \nabla_\varphi z_2, \nabla_{z_1} G\psi\rangle$$

$$= \langle R(z_1, \varphi, z_2), G\psi\rangle + \langle \nabla_{[z_1,\varphi]} z_2, G\psi\rangle . \tag{3.20}$$

If follows from the parallel translation condition that

$$[z_1, \varphi] = -\nabla_\varphi z_1 = -\nabla_{z_1}\varphi \quad \text{and} \quad \nabla_{[z_1,\varphi]} z_2 = -\nabla_{z_2} \nabla_{z_1} \varphi.$$

Now, just substitute it into (3.20) and compute the trace by putting $\varphi = \psi = e_k$, where $\{e_k\}$ is a local basis.

Remark 3.4. The second term in the right-hand side of (3.20) is correctly defined even if $z_1, z_2 \in \sigma_1(\gamma)$. The right-hand side of (3.20) is, therefore, correct as well for such fields, provided we know, a-priori, that the Ricci tensor $R(z_1, z_2)$ is well defined for them.

4. Measures on Smooth Manifolds

4.1. LOGARITHMIC DERIVATIVE

The main results concerning measure differentiation in linear spaces may be easily extended to measures on smooth manifolds.

Let X be a smooth manifold modeled on a Banach space B. Denote $V(X)$ as the class of smooth vector fields over X possessing integral flows

$$S_t^{(z)} : X \to X, \quad t \in R^1 ,$$

$$\frac{d}{dt} S_t^{(z)}(x) = z\left(S_t^{(z)}(x)\right), \quad z \in V(X). \tag{4.1}$$

Let μ be a Borel measure defined on X and

$$\mu_t = S_t^{(z)} \mu = \mu \circ \left(S_t^{(z)}\right)^{-1}, \quad t \in R^1.$$

As in the linear case, we set

$$\nabla_z \mu = -\left.\frac{d}{dt} \mu_t \right|_{t=0}$$

or, what is the same,

$$\int_X \varphi(x)(\nabla_z \mu)(dx) = -\int_X \nabla_z \varphi(x)\, \mu(dx)$$

for a large enough set of smooth functions $\varphi(x)$. Next, if the measures $\nabla_z \mu$ and μ are absolutely continuous, $\nabla_z \mu \prec \mu$, then there exists a logarithmic derivative

$$\rho_\mu(z; x) = \frac{\nabla_z \mu(dx)}{\mu(dx)}$$

of the measure μ along the vector field z. In this case, the following integration by parts formula

$$\int_X (\nabla_z \varphi)(x)\, \mu(dx) = -\int_X \varphi(x)\, \rho_\mu(z; x)\, \mu(dx) \tag{4.2}$$

holds. Thanks to this formula, one may easily derive the relation

$$\rho_\mu(\alpha z; x) = \alpha(x)\, \rho_\mu(z; x) + \nabla_{z(x)} \alpha(x) \tag{4.3}$$

which shows that the map $z \mapsto \rho_\mu(z)$ is a first-order differential operator.

Now let $f: X \to Y$ be a smooth mapping from X into Y and z^f be f-connected with z the vector field over Y.

$$z^f(f(x)) = (Tf)(x)\, z(x).$$

Then the smoothness of the measure μ along z yields the smoothness of the measure $\mu^f = \mu \circ f^{-1}$ along z^f and, moreover,

$$\nabla_{z^f} \mu^f = (\nabla_z \mu)^f. \tag{4.4}$$

If f is, moreover, an invertible mapping, then

$$\rho_{\mu^f}(z^f; f(x)) = \rho_\mu(z; x). \tag{4.5}$$

The proof of these facts may be obtained in the same way as in the case of linear spaces.

We say a measure μ is local if there exists a chart $U \subset X$ and an open subset V with closure $\overline{V} \subset U$ such that the support of μ is contained in V.

To each measure μ, one may associate a collection of local measures $\mu_{u,\upsilon} = \varphi_{u,\upsilon}\, \mu$ (localizations) where $\varphi_{u,\upsilon}$ is a smooth function on X equal to unity on V and to zero outside U. Here U stands for an arbitrary chart of the manifold. Notice that given a set $A \subset V$, we have $\mu_u(A) = \mu(A)$. The measure μ may be patched up from the local measures

$$\mu = \sum_k \mu_{u_k}$$

by assuming that functions $\varphi_{u_k}(x)$ give a (locally finite) partition of unity. This sum over small enough sets (which may be covered by a finite collection of charts) contains only a finite number of non-zero terms. Therefore, in order to investigate measure smoothness, it suffices to investigate the smoothness of its localization. We remark that if we assume that both the vector fields $z(x)$ and the functions $\varphi(x)$ in (4.1) and (4.2) are local (which means that their supports are contained in a single chart) then this investigation may be done to the moment of exit out of the chart. It results in that we may restrict ourselves to the investigation of a linear space situation, still verifying all formulas to be invariant under local manifold morphisms.

Consider a manifold X with an additional Riemannian H S structure (B, H, i, G) which gives rise to a Hilbert subbundle γ of the tangent bundle τ_X, equipped with the Riemannian metric $G_X : \gamma_x^{-1} \to (\gamma_x^*)^{-1}$. Recall that $\sigma_k(\gamma)$ is the class of vector fields belonging to $\sigma_k(\tau)$ which are sections of γ and $\sigma_k^*(\tau)$ are vector fields which are sections of $G^{-1}\tau^*$.

Assume that the given Riemannian metric is nuclear and, thus, due to Theorem 3.1, any vector field belonging to $\sigma_1(\gamma)$ possesses a finite divergence.

Let us try to obtain an invariant form of (1.3.8) by writing it down in the form

$$\rho_\mu(z; x) = (\Lambda(x), z(x))_X + \mathrm{div}_G\, z(x). \tag{4.6}$$

Define a new class of measures $\mathcal{M}_1^2(X)$. A measure μ belongs to $\mathcal{M}_1^2(X)$ if it possesses the following properties. Given a map ψ from the chart into the model space B and the localization $\mu_{u,\upsilon}$ of the measure μ, consider its image $\mu_{u,\upsilon}^{\psi}$ under the map ψ. $\mu \in \mathcal{M}_1^2(X)$ if $\mu_{u,\upsilon}^{\psi}$ has a logarithmic derivative $\lambda_{u,\upsilon}^{\psi}$ which is differentiable at $x \in \psi(V)$ along H and satisfies the estimate

$$\sup_{x \in V} \| \lambda'(x) \|_{L(\mathcal{H}_x)} < \infty .$$

Evidently, this condition is invariant under a change of chart. It results from Theorem 1.3.2 that $\mu_{u,\upsilon}^{\psi}$ is differentiable over subsets V along H and has a square integrable logarithmic derivative for H-valued vector fields.

To obtain (4.6), it suffices to consider a local situation. Hence, we identify the domain $U \subset X$ with its image in B and consider the trivial bundles τ and γ

$$\tau : U \times B \to U, \quad \gamma : U \times H \to U.$$

In addition, consider the dual bundles

$$\tau^* : U \times B^* \to U, \quad \gamma^* : U \times H^* \to U$$

and the Riemannian metric $G_X : H^* \to H$. In this way, we obtain a collection of rigged spaces

$$H_{+,x} = G_x^{-1} B^* \subset H \subset H_- = B$$

with a pairing

$$\langle h, h_+ \rangle_X = (h, h_+)_X = \langle h, G_X h_+ \rangle_B , \tag{4.7}$$

where $h_+ \in H_{+,x}$ $h \in H \subset B$, and $(\cdot, \cdot)_X$ is the inner product in H corresponding to the above Riemannian structure.

Assume that the following condition is satisfied: there exists a linear set $\mathcal{D}$ which is dense in H and such that $G_X \mathcal{D} \subset B^*$ for all $x \in U$, that is $\mathcal{D} \subset H_{+,x}$.

First consider a vector field $z(x) \equiv h \in H$. Due to Theorem 1.3.2, given $\mu \in \mathcal{M}_1(U)$, one may prove the existence of the logarithmic derivative

$$\rho_\mu (h; x) \overset{\text{def}}{=} (\lambda (x), h)_x . \tag{4.8}$$

Here $\lambda (x) \in B$. Given $h \in \mathcal{D}$, the right-hand side of (4.8), due to (4.7), may be written in the form

$$(\lambda (x), h)_x = \langle \lambda (x), G_x h \rangle_B .$$

As has been mentioned in Remark 3.1 to Theorem 1.3.2, given $h \in H$, relation (4.8) determines $\lambda (x)$ to be an extension of the linear functional defined a.e. with respect to x. By assumption $\lambda \in C_1 (U, H, B)$.

Let, now, $z (x) = \alpha (x) h$, where $\alpha (x)$ is a functional belonging to $C_1 (U, H, R^1)$. Then it results from (4.3) that

$$\begin{aligned} \rho_\mu (z; x) &= \alpha (x) \rho_\mu (h; x) + \nabla_h \alpha (x) \\ &= (\lambda (x), \alpha (x) h)_x + \nabla_h \alpha (x). \end{aligned}$$

Notice that, given a local orthonormal basis $\{e_k (x)\}$ in H, we have the relation

$$D_h \alpha (x) = \sum_k (h, e_k (x))_x D_{e_k (x)} \alpha (x) = \mathrm{Tr}_H D \alpha(x) h,$$

which leads to

$$\rho_\mu (z; x) = (\lambda (x), z (x))_x + \mathrm{Tr}_H D \alpha z (x).$$

Let us now use the formula

$$\mathrm{div}_G z (x) = \mathrm{Tr}_H D z (x) + (\Gamma_x, z (x))_x ,$$

which had been derived in the proof of the theorem and introduce a measurable function on U

$$\Lambda (x) = \Lambda (x) - \Gamma_x = \lambda (x) + G_x^{-1} \sum_{k=1}^{\infty} \Gamma^{\gamma^*} (e_k, Ge_k) \qquad (\mu - \text{a.e.}). \tag{4.9}$$

Then,

$$\rho_\mu (z; x) = (\Lambda (x), z (x))_x + \mathrm{div}_G z (x).$$

Since ρ_μ and $\mathrm{div}_G z (x)$ are invariant, it results that

$$(z; x) = (\Lambda (x), z (x))_x = \langle \Lambda (x), G_x h \rangle_B$$

is invariant as well.

One may extend (4.6) first to linear combinations of the form $\Sigma \alpha_k (x) h_k$ and then, by passing to the limit, to $H_{+,x}$-valued vector fields z for which each term in the right-hand side makes sense.

Coming back to the manifold X, we may state the following assertion.

THEOREM 4.1. *Given $\mu \in \mathcal{M}_2 (X)$ for μ-almost all $x \in X$ there is defined a measurable section $x \mapsto \Lambda (x)$ of the tangent bundle (the vector logarithmic derivative of μ) such that*

$$\rho_\mu (z; x) = (\Lambda (x), z (x))_x + \mathrm{div}_G z (x) \tag{4.10}$$

for $z \in \sigma^* (\gamma)$.

Remark 4.1. Given a smooth metric and connection, we may prove that $\Lambda(x)$ is differentiable along the vector fields belonging to $\sigma^*(\tau)$. Moreover,

$$\nabla \Lambda(x) : H_x \to T_x X, \qquad \sup_x \| \nabla \Lambda(x) \|_x < \infty . \tag{4.11}$$

The above arguments permit us to state an invariant definition. We say that μ belongs to the class $\mathcal{M}_1(X, G)$ if its logarithmic derivative exists and (4.10), (4.11) hold.

Given a measure $\mu \in \mathcal{M}_1(X, G)$, we calculate the derivative

$$\begin{aligned}
&\nabla_{z_1} \rho_\mu(z_2; x) \\
&\quad = \langle \nabla_{z_1} \Lambda(x), G_x z_2 \rangle + \\
&\qquad + \langle \Lambda(x), G_x \nabla_{z_1} z_2 \rangle + \nabla_{z_1} \operatorname{div} z_2 \\
&\quad = \rho_\mu\left(\nabla_{z_1} z_2; x\right) + \langle \nabla_{z_1} \Lambda(x), G_x z_2 \rangle + \delta(z_1, z_2),
\end{aligned} \tag{4.12}$$

where

$$\delta(z_1, z_2) = \langle R(x) z_1, z_2 \rangle - \operatorname{Tr}_H \nabla_{z_1} \nabla_{z_2}$$

(see (3.18)).

Substituting the above expressions into the relation

$$\begin{aligned}
&\int_X \rho_\mu(z_1; x)\, \rho_\mu(z_2; x)\, \mu(d x) \\
&\quad = \int_X \rho_\mu(z_2; x) \left(\nabla_{z_1} \mu\right)(d x) \\
&\quad = -\int_X \nabla_{z_1} \rho_\mu(z_2; x)\, \mu(d x),
\end{aligned}$$

we obtain the relation

$$\begin{aligned}
&\int_X \rho_\mu(z_1; x)\, \rho_\mu(z_2; x)\, \mu(d x) \\
&\quad = \int_X \left\{ \operatorname{Tr} \nabla_{z_1} \nabla_{z_2}(x) - \langle R(x) z_1, z_2 \rangle - \right. \\
&\qquad \left. - \langle \nabla \Lambda(x) z_1(x), \ z_2(x) \rangle \right\} \mu(d x) .
\end{aligned} \tag{4.13}$$

Its square variant looks like

$$\begin{aligned}
&\int_X |\rho_\mu(z; x)|^2 \, \mu(d x) \\
&\quad = \int_X \left\{ \operatorname{Tr}_G (\nabla z(x))^2 - \langle \nabla \Lambda(x) z(x), z(x) \rangle - \right. \\
&\qquad \left. - \langle R(x) z(x), z(x) \rangle \right\} \mu(d x)
\end{aligned} \tag{4.14}$$

and differs from (1.3.13) by an additional summand depending on the Riemannian curvature tensor.

Assuming, in addition, that

$$\langle R(x)\, z(x), z(x)\rangle \geq \text{const}\ \| z(x) \|_H^2, \tag{4.15}$$

we obtain the following estimate

$$\int_X |\rho_\mu(z; x)|^2 \mu(dx) \leq \text{const}\ \sigma_{2,\mu}^2(z), \tag{4.16}$$

where

$$\sigma_{2,\mu}^2(z) = \int_X \{\sigma_2^2(\nabla z) + \| z(x) \|_H^2\}\ \mu(dx).$$

Let us mention that (4.15) holds, for example, if the Ricci tensor is nonnegative at every point.

The same argument as in the linear case lead to the following result.

THEOREM 4.2. *Let* $\mu \in \mathcal{M}_1(X, G)$ *and* $z \in \sigma_{12}(\gamma)$ *be such that* (4.15) *holds. Then* μ *is differentiable along* z *and the density* $\rho_\mu(z; x)$ *satisfies the estimate* (4.16).

Remark 4.2. Given a Riemannian nuclear structure, the Ricci tensor does exist for vector fields belonging to $\sigma_1(\gamma)$ and satisfies the estimate (4.15).

4.2. HIGHER-ORDER DIFFERENTIAL OPERATIONS. MEASURE MAPPING

Given a measure $\mu \in \mathcal{M}_1(X)$ and a pair of vector fields z_1 and z_2, we may calculate

$$\begin{aligned}
\nabla_{z_2} z_1 \mu &= \nabla_{z_2} \rho_\mu(z_1, x)\, \mu \\
&= \rho_\mu(z_2, x)\, \rho_\mu(z_1, x)\, \mu + \\
&\quad + [\nabla_{z_2} \rho_\mu(z_1, x)]\, \mu \\
&= \{[(\nabla + \Lambda(x), z_2(x))\, (\nabla + \Lambda(x), z_1(x))] + \\
&\quad + \nabla_{z_2}(\nabla + \Lambda(x), z_1(x))\}\, \mu \\
&= [(\nabla + \Lambda(x), z_2(x))\, (\nabla + \Lambda(x), z_1(x))]\, \mu.
\end{aligned} \tag{4.17}$$

It must be explained that the operator ∇ in (4.17) acts only on those fields which are located as factors to its right.

On the other hand, it follows from (4.12) that

$$\begin{aligned}
\nabla_{z_2} \nabla_{z_1} \mu &= \{\rho_\mu(z_2; x)\, \rho_\mu(z_1; x) + \\
&\quad + \rho_\mu(\nabla_{z_1} z_2; x) - \mathrm{Tr}_G(\nabla_{z_2})(\nabla_{z_1}) + \\
&\quad + ((\nabla \Lambda + R)\, z_2, z_1)\}\, \mu = \rho_2(z_2, z_1; x)\, \mu.
\end{aligned} \tag{4.18}$$

Under the conditions of the theorem, $\rho_2(z_2, z_1; x)$ is a μ-integrable function

It is worth remarking that in the proof of Theorem 1.3.3 nothing had been used but (1.3.15), general arguments based on Holder inequality, and integration by parts.

Changing (1.3.15) to (4.14) or (4.16), one may easily state the smoothness conditions which must be added to the Theorem 4.2 assumptions to guarantee that

$$\int_X |\rho_2(z_2, z_1; x)|^2 \mu(dx) < \infty. \tag{4.19}$$

Let us calculate the operator $b_A^*(\nabla, \nabla)$, acting on measures, which is adjoint to the differential operator (3.5). Given a local basis $\{e_k(x)\}_{k=1}^{\infty}$ in a certain trivialization, we have

$$b_A(\nabla, \nabla)u = \sum_{k=1}^{\infty} \left(\nabla_{Ae_k}\nabla_{Ae_k}u - \nabla_{\nabla_{Ae_k}Ae_k}u\right),$$

and (4.18) results in

$$\begin{aligned} &b_A^*(\nabla, \nabla)u \\ &= \sum_k \{\rho_2(Ae_k, Ae_k; x) + \rho_\mu((\nabla_{Ae_k}Ae_k; x)\}\mu \\ &= \sum_k \{\rho_\mu(Ae_k; x)\rho_\mu(Ae_k; x) + \\ &\qquad + 2\rho(\nabla_{Ae_k}Ae_k; x) - \mathrm{Tr}(\nabla Ae_k)^2 + ((\nabla\Lambda + R)Ae_k, Ae_k)\}\mu \\ &= \rho_{2,A}(x)\mu. \end{aligned} \tag{4.20}$$

Notice that $\rho_{2,A}(x)$ does not depend on the local basis. Its estimate for $\mu \in \mathcal{M}_1(X, G)$ may be derived by arguments similar to those used to prove (1.3.25). If, moreover, (4.16) is valid, then the result looks similar: $\rho_{2,A} \in \mathcal{L}_2(X, \mu)$ if the tensor field A belongs to $\sigma_{22}(\tau)$, namely has derivatives up to second order which are Hilbert–Schmidt mappings.

Formula (4.17) may be extended to higher-order operations

$$\nabla_{z_n}\nabla_{z_{n-1}}\cdots\nabla_{z_1}\mu = \rho_n(z_1, \ldots, z_n; x)\mu \tag{4.21}$$

under corresponding assumptions about vector fields $z_j(x)$ and $\Lambda(x)$ order of smoothness. Nevertheless, the expansion of type (4.18) appears to be rather complicated. Still, it results from (4.21) that there exist sufficient conditions, stated in terms of Hilbert–Schmidt norms of the vector fields considered, which grants ρ_n to be a square integrable function.

Logarithmic derivative $\mathcal{L}_2$-estimates give the possibility to formulate sufficient conditions which guarantee that the image of a smooth measure under smooth mapping is a smooth measure as well. Let $f: X \to Y$ be a smooth manifold mapping. We say that the differentiable operations A_x and A_y acting on functions defined, respectively, on X and Y are f-connected if

$$A_y\varphi(y)|_{y=f(x)} = A_x(\varphi \circ f)(x). \tag{4.22}$$

Given the measure $A_x^*\mu \prec \mu$,

$$\int_X A_x\varphi(x)\mu(dx) = \int_X \varphi(x)\rho_{A_x}(x)\mu(dx),$$

where

$$\rho_{A_x}(x) = \frac{(A_x^* \mu)(d\,x)}{\mu(d\,x)} \in \mathcal{L}_2(X, \mu).$$

Then the estimate

$$\left| \int_Y A_y \varphi(y) \mu^f(d\,y) \right|^2$$

$$= \left| \int_Y A_x (\varphi \circ f)(x) \mu(d\,x) \right|^2$$

$$= \left| \int_Y \varphi(f(x)) \rho_{A_x}(x) \mu(d\,x) \right|^2$$

$$\leq \int_Y |\varphi(y)|^2 \mu^f(d\,y) \int_X |\rho_{A_x}(x)|^2 \mu(d\,x)$$

holds. Moreover, there exists a function $\rho_{A_y} \in \mathcal{L}_2(Y, \mu^f)$ such that

$$\int_Y A_y \varphi(y) \mu^f(d\,y) = \int_Y \varphi(y) \rho_{A_y}(y) \mu^f(d\,y).$$

This means that there exists a density

$$\rho_{A_y}(y) = \frac{(A_y^* \mu^f)(d\,y)}{\mu(d\,y)}.$$

In particular, if ξ and $\eta = \xi^f$ are f-connected vector fields over X and Y, respectively, and $\rho_\mu(\xi, x) \in \mathcal{L}_2(X, \mu)$, then μ^f has a logarithmic derivative $\rho_{\mu^f}(\eta, y) \in \mathcal{L}_2(Y, \mu^f)$.

Moreover, if the vector fields $\eta_1, \ldots, \eta_m$ are f-connected with the vector fields $\xi_1, \ldots, \xi_m$, then the differential operations

$$A_y = \nabla_{\eta_m} \ldots \nabla_{\eta_1} \quad \text{and} \quad A_x = \nabla_{\xi_m} \ldots \nabla_{\xi_1}$$

are f-connected as well.

Finally, the fact $\rho_m(\xi_1, \ldots, \xi_m) \in \mathcal{L}_2(X, \mu)$ implies the validity of the corresponding fact for the measure image, that is

$$\rho_m(\eta_1, \ldots, \eta_m) \in \mathcal{L}_2(Y, \mu^f).$$

CHAPTER 3

Stochastic Equations in Banach Spaces

In this chapter, we set down a probability framework which will be needed in the sequel. We describe probabilistic machinery and, especially, Ito stochastic analysis, or stochastic calculus, in Banach spaces with smooth norms. We have tried to make the exposition detailed enough and adjusted to our future needs while dealing with smooth Banach manifolds.

1. Basic Notions

This first section gives a brief sketch of the main notions of probability theory which are necessary below, and is intended for use by a reader not expert in the field.

1.1. RANDOM VARIABLES. INDEPENDENCE

A triple $(\Omega, \mathcal{F}, P)$ is called a probability space if P is a σ-additive nonnegative measure on the measurable space $(\Omega, \mathcal{F})$ such that

$$P : \Omega \to [0, 1]$$

and $P(\Omega) = 1$. Points $\omega \in \Omega$ are called elementary events, sets $F \in \mathcal{F}$ ($\mathcal{F}$-measurable sets) are called events and P is called a probability.
Let $(X, \mathcal{Z})$ be another measurable space. A measurable mapping

$$\xi : (\Omega, \mathcal{F}) \to (X, \mathcal{Z})$$

is called an X-valued random variable defined on the given probability space. We shall omit, as a rule, the indication of σ-algebras if it is clear from the context which ones are meant.
A random variable ξ generates a measure

$$\mu_\xi = P^{(\xi)} = P \circ \xi^{-1}$$

on the σ-algebra $\mathcal{Z}$, which is called the random variable probability distribution. Notice that in this way a new probability space $(X, \mathcal{Z}, \mu_\xi)$ is determined.

Let ξ_λ be a collection of random variables defined on $(\Omega, \mathcal{F}, P)$. Denote $\mathcal{F}_\Lambda = \sigma\{\xi_\lambda, \lambda \in \Lambda\}$ the minimal σ-algebra with respect to which those variables are measurable. We say $\mathcal{F}_\Lambda$ is the σ-algebra generated by the collection of random variables or reference σ-algebra.

In this case, the initial probability space $(\Omega, \mathcal{F}, P)$ may be changed to

$(\Omega, \mathcal{F}_\Lambda, P)$.

The crucial notion of probability theory is independence. Events $A_1, \dots, A_n \in \mathcal{F}$ are called collectively independent if for any $k = 1, \dots, n$, $t_i \in \{1, \dots, n\}$

$$P\left(\prod_{i=1}^{k} A_{t_i}\right) = \prod_{i=1}^{k} P(A_{t_i}).$$

The set of σ-algebras $\mathcal{Z}_1, \dots, \mathcal{Z}_n$ is called collectively independent if each set of events $A_1, \dots, A_n$ $(A_k \in \mathcal{Z}_k, k = 1, \dots n)$ is independent.

The notion of independence may be defined for random variables as well. The random variables $\xi_1, \dots, \xi_n$ are independent if the σ-algebras $\mathcal{F}_{\xi_1}, \dots, \mathcal{F}_{\xi_n}$ generated by them are collectively independent.

Let $\xi_j : (\Omega, \mathcal{F}) \to (X_j, \mathcal{Z}_j)$. Consider the mapping

$$\xi = (\xi_1, \dots, \xi_n) : (\Omega, \mathcal{F}) \to (X, \mathcal{Z}),$$

where $X = \Pi_{j=1}^{n} X_j$ is a Cartesian product space and $\mathcal{Z} = \Pi_{j=1}^{n} \mathcal{Z}_j$ is the σ-algebra generated by sets of the form

$$\left(\prod_{j=1}^{n} A_j \subset X, \ A_j \in \mathcal{Z}_j\right).$$

It is easy to verify that the collective independence of the collection random variables $\xi_1, \dots, \xi_n$ is equivalent to the following distribution property

$$\mu_\xi = \prod_{j=1}^{n} \mu_{\xi_j},$$

that is

$$\mu_\xi\left(\prod_{j=1}^{n} A_j\right) = \prod_{j=1}^{n} \mu_{\xi_j}(A_j),$$

which means that the distribution of a set of independent random variables is equal to the product of their distributions. The measure μ_ξ in X space is called the mutual distribution of the random variables $\xi_1, \dots, \xi_n$.

1.2. MOMENTS, CHARACTERISTIC FUNCTIONS, CONDITIONAL EXPECTATION

In what follows, while dealing with random variables valued in a separable Banach space B, we take as the corresponding σ-algebra $\mathcal{Z}_B$ the Borel σ-algebra $\mathcal{Z}$, omitting the indication of the space if it will not lead to confusion.

The mean value (or the expectation) of a scalar random variable $\xi : (\Omega, \mathcal{F}) \to (R^1, \mathcal{Z}_{R_1})$ is the integral

$$E\xi = \int_\Omega \xi(\omega) P(d\omega). \tag{1.1}$$

For a Banach space B-valued random variable ξ, the mean value

$$E\xi = \int_\Omega \xi(\omega) P(d\omega)$$

is an element of the space B such that

$$\langle E\,\xi, \varphi\rangle = E\,\langle \xi, \varphi\rangle$$

holds for any $\varphi \in B^*$.

If $\|\xi(\omega)\|$ is a measurable function on $(\Omega, \mathcal{F})$ and

$$\int_\Omega \|\xi(\omega)\|\; P(\mathrm{d}\,\omega) < \infty,$$

then the mean value (1.1) does exist and may be represented as a Bochner integral. In this case ξ is said to have a strong first order.

By changing variables under the integral sign, one may easily deduce that

$$E\,\xi = \int_B x\;\mu_\xi(\mathrm{d}\,x),$$

which shows that in order to calculate the expectation of a random variable, one only needs to know its distribution.

In general, if $\eta = f \circ \xi$,

$$(\Omega, \mathcal{F}) \xrightarrow{\;\xi\;} (X, Z_X) \xrightarrow{\;f\;} (B, Z_B),$$

then

$$E\,\eta = \int_B y\;\mu_\eta(\mathrm{d}\,y) = \int_X f(x)\;\mu_\xi(\mathrm{d}\,x) = E\,f(\xi),$$

assuming that all integrals do exist.

Various scalar characteristics of the probability distribution μ_ξ are called by definition the corresponding characteristics of the Banach space B-valued random variable ξ.

For example, the function

$$\chi_\xi(\theta) = E\,\exp\{i\,\langle \xi, \theta\rangle\} = \int_B e^{i\langle x,\theta\rangle}\,\mu_\xi(\mathrm{d}\,x) \tag{1.2}$$

is called the characteristic function of ξ, while the multilinear functionals

$$m_k(\theta_1, \ldots, \theta_k) = E\,\langle \xi, \theta_1\rangle \ldots \langle \xi, \theta_k\rangle$$
$$= \int_B \langle x, \theta_1\rangle \ldots \langle x, \theta_k\rangle\,\mu_\xi(\mathrm{d}\,x)$$

are called its kth order moments, $\theta, \theta_1, \ldots, \theta_k \in B^*$.

The condition

$$E\,\|\xi\|^k < \infty$$

is sufficient to guarantee the existence of the kth order moment m_k of the random variable ξ.

The bilinear form defined on B^*

$$m_2(\theta_1, \theta_2) = E\,\langle \xi, \theta_1\rangle\;\langle \xi, \theta_2\rangle$$

is called the variance of the random variable ξ, while the form

$$\tau(\theta_1, \theta_2) = E\,\langle \xi, \theta_1\rangle\,\langle \xi, \theta_2\rangle - E\,\langle \xi, \theta_1\rangle\,E\,\langle \xi, \theta_2\rangle$$

(which coincides with m_2 if $E\xi = 0$) is called its correlation.

Now we shall list some simple but important properties of the above random variable characteristics.

(a) If $\Phi \in L(B, B_1)$ and $\xi_1 = \Phi\xi$, then

$$\chi_{\xi_1}(\theta) = E\, e^{i\langle \xi_1, \theta\rangle} = E\, e^{i\langle \Phi\xi, \theta\rangle} = \chi_\xi(\Phi^*\theta).$$

(b) If

$$\xi = (\xi_1, \dots, \xi_n) : (\Omega, \mathcal{F}) \to \prod_{j=1}^{n} (X_j, Z_j)$$

is a collection of mutually independent random variables

$$f_j : (X_j, Z_j) \to (R^1, Z_{R^1}) \text{ and } f(x) = (f_1(x_1), \dots, f_n(x_n)),$$

then

$$Ef(\xi) = \prod_{j=1}^{n} \int_{X_j} f_j(x_j)\, \mu_{\xi_j}(\mathrm{d}x_j) = \prod_{j=1}^{n} Ef_j(\xi_j).$$

In particular, given $\theta = (\theta_1, \dots, \theta_n)$, we obtain

$$\chi_\xi(\theta) = \prod_{j=1}^{n} \chi_{\xi_j}(\theta_j).$$

(c) Given a pair of random variables $\xi_j : (\Omega, \mathcal{F}) \to (X_j, Z_j)$, $j = 1, 2$, we may calculate the correlation

$$\tau(\theta_1, \theta_2) = E\,\langle \xi_1, \theta_1\rangle \langle \xi_2, \theta_2\rangle - E\,\langle \xi_1, \theta_1\rangle\, E\,\langle \xi_2, \theta_2\rangle,$$

$$\theta_1 \in B_1^*, \theta_2 \in B_2^*.$$

It is easy to see that this form may be obtained as a restriction of the correlation of the random variable

$$\xi = (\xi_1, \xi_2) : (\Omega, \mathcal{F}) \to (X_1 \times X_2, Z_1 \times Z_2).$$

Random variables ξ_1 and ξ_2 are said to be uncorrelated if $\tau_{\xi_1\xi_2} = 0$. It is obvious that the independence of ξ_1 and ξ_2 implies that ξ_1 and ξ_2 are uncorrelated.

(d) A Banach space B valued random variable ξ is called a centered Gaussian random variable if its distribution μ_ξ is a Gaussian measure on B, that is

$$\chi_\xi(\theta) = \exp\left\{-\frac{1}{2} b(\theta, \theta)\right\}, \tag{1.3}$$

where b is a nonnegative bilinear functional on B.

It follows from (a) that the linear image of a Gaussian random variable is a Gaussian random variable again and

$$\chi_{\Phi\xi}(\theta) = \exp\left\{-\frac{1}{2} b(\Phi^*\theta, \Phi^*\theta)\right\}.$$

(e) Given a characteristic functional of a random variable ξ, we may compute its moments in the following way

$$m_k(\theta_1, \ldots, \theta_k) = \frac{1}{i^k} \frac{\partial^k}{\partial \tau_1 \ldots \partial \tau_k} \chi_\xi \left(\prod_{j=1}^{k} \tau_j \theta_j \right) \Bigg|_{\tau_1=0, \ldots, \tau_k=0} .$$

In particular, given a Gaussian random variable ξ with characteristic functional (1.3), we obtain

$$E \langle \xi, \theta \rangle = 0, \quad \tau(\theta_1, \theta_2) = b(\theta_1, \theta_2).$$

(f) Let $\xi = (\xi_1, \ldots, \xi_n)$ be a Gaussian random variable valued in $B_1 \times \ldots \times B_n$ whose components ξ_j, $(j = 1, \ldots, n)$ are mutually uncorrelated. Then given $\theta = (\theta_1, \ldots, \theta_n)$, we obtain

$$b(\theta, \theta) = \tau(\theta, \theta) = \sum_{j=1}^{n} \tau_j(\theta_j, \theta_j) = \sum_{j=1}^{n} b_j(\theta_j, \theta_j)$$

and

$$\chi_\xi(\theta) = \sum_{j=1}^{n} \chi_{\xi_j}(\theta_j), \quad \chi_{\xi_j}(\theta_j) = \exp\left\{ -\frac{1}{2} b_j(\theta_j, \theta_j) \right\} .$$

The factorization of the characteristic function yields the factorization of the probability distribution $\mu_\xi = \Pi_{j=1}^{n} \mu_{\xi_j}$, which means that the Gaussian random variables $\xi_1, \ldots, \xi_n$ are mutually independent.

Thus, for Gaussian random variables to be independent is the same as to be uncorrelated.

Denote $\mathcal{L}_p(\mathcal{F}, B)$ as the totality of (equivalence classes of) $\mathcal{F}$-measurable random variables valued in B and possessing a strong pth order. If $\mathcal{F}_1$ is a σ-subalgebra of the σ-algebra $\mathcal{F}$ and P_1 is a restriction of P to $\mathcal{F}_1$, then $\mathcal{L}_p(\mathcal{F}_1, B)$ is a closed subspace of the space $\mathcal{L}_1(\mathcal{F}, B)$.

Given a separable Banach space B, construct a continuous linear operator $E_{\mathcal{F}_1} : \mathcal{L}_1(\mathcal{F}, B) \to \mathcal{L}_1(\mathcal{F}_1, B)$ such that for each $\xi \in \mathcal{L}_1(\mathcal{F}, B)$ and $A_1 \in \mathcal{F}_1$,

$$\int_{A_1} E_{\mathcal{F}_1} \xi \, dP_1 = \int_{A_1} \xi \, dP . \tag{1.4}$$

The operator $E_{\mathcal{F}_1}$ may be constructed in the following way. Introduce a mapping $P_{\mathcal{F}_1} : \mathcal{F} \to \mathcal{L}_1(\mathcal{F}, R^1)$ such that for each $A \in \mathcal{F}$ and $A_1 \in \mathcal{F}_1$

$$\int_{A_1} P_{\mathcal{F}_1}(A) \, dP_1 = P(A \cap A_1).$$

Notice that the existence of $P_{\mathcal{F}_1}(A)$, which is called a conditional probability distribution, is the consequence of the Radon–Nykodim theorem.

Given $\xi = \Sigma_{k=1}^{n} x_k j_k$, where j_k is the characteristic function of the set

$$F_k, F_1, \ldots, F_n \in \mathcal{F}, \quad F_j \cap F_k = \emptyset , \quad k \neq j,$$

put

$$E_{\mathcal{F}_1} \xi = \sum_{k=1}^{n} x_k P_{\mathcal{F}_1}(F_k) .$$

In this way we define an operator $E_{\mathcal{F}_1}$ on a dense set in $\mathcal{L}_1(\mathcal{F}, B)$ which possesses the property (1.4). Moreover, $E_{\mathcal{F}_1}$ is a linear map on this set and

$$\| E_{\mathcal{F}_1} \xi \|_{\mathcal{L}_1} = \| \xi \|_{\mathcal{L}_1} \tag{1.5}$$

holds. This permits us to extend $E_{\mathcal{F}_1}$ to a linear operator on the whole $\mathcal{L}_1(\mathcal{F}, B)$ preserving the estimate (1.5).

The operator $E_{\mathcal{F}_1}: \mathcal{L}_1(\mathcal{F}, B) \to \mathcal{L}_1(\mathcal{F}_1, B)$ is called a conditional expectation operator with respect to the σ-algegra $\mathcal{F}_1$. A random variable $E_{\mathcal{F}_1}\xi$ is called a conditional expectation of ξ given $\mathcal{F}_1$.

Let $(X, \mathfrak{A})$ be a measurable space and $\eta : (\Omega, \mathcal{F}) \to (X, \mathfrak{A})$ be an X-valued random variable. Denote $\mathcal{F}_\eta = \{\eta^{-1}(A), A \in \mathfrak{A}\}$. The random variable $E_{\mathcal{F}_\eta}\xi$ is called the conditional expectation of ξ given the random variable η and sometimes is denoted $E_\eta \xi$.

The following properties of conditional expectations may be easily verified.

(a) $E_\xi \xi = \xi$ and given $\mathcal{F} = \{\emptyset, \Omega\}$ we obtain $E_{\mathcal{F}}\xi = E_\xi$.

(b) If $\xi \in \mathcal{L}_1(\mathcal{F}_1, B)$ then $E_{\mathcal{F}_1}\xi = \xi$, that is $E_{\mathcal{F}_1}$ is a projection operator from $\mathcal{L}_1(\mathcal{F}, B)$ to $\mathcal{L}_1(\mathcal{F}_1, B)$ and $\| E_{\mathcal{F}_1} \| = 1$.

(c) If $\mathcal{F}_1 \subset \mathcal{F}_2$, $\mathcal{F}_2$ is a σ-subalgebra of $\mathcal{F}$, then

$$E_{\mathcal{F}_1} E_{\mathcal{F}_2} \xi = E_{\mathcal{F}_1} \xi.$$

(d) If $f \in \mathcal{L}_1(\mathcal{F}_1, R^1)$ and $E \| f\xi \| < \infty$ then $E_{\mathcal{F}_1}(f\xi) = f E_{\mathcal{F}_1}\xi$ and, in particular,

$$E f E_{\mathcal{F}_1} \xi = E(f\xi).$$

1.3. RANDOM FUNCTIONS, MARKOV PROCESSES

A random function ξ_t defined on a set T and valued in a space $(X, \mathcal{Z})$ is a set of random variables

$$\xi_t : (\Omega, \mathcal{F}) \to (X, \mathcal{Z}), \quad t \in T$$

defined on the same probability space. If t is a real-valued parameter, then the random function is usually called a random (stochastic) process, while t is interpreted as the time parameter. In what follows, we shall deal only with this situation.

A stochastic process ξ_t generates a mapping

$$\omega \to \xi_t(\omega), \quad (t \in T),$$

which associates a curve in the space X, called a path of the process, to each elementary event ω. Notice that in the space $\mathfrak{X} = \Phi(T, X)$ of all X-valued functions defined on T, there is a natural σ-algebra $\tilde{\mathfrak{A}} = \tilde{\mathfrak{A}}(T, X)$ which is generated by the sets

$$C(t_1, \dots, t_n, \Delta_1, \dots, \Delta_n) = \left\{ x \in \Phi(T, X) : x(t_k) \in \Delta_k, \Delta_k \in \mathcal{Z}_k, k = 1, \dots, n \right\}$$

(cylindrical sets) for arbitrary sets $(t_1, \dots, t_n, \Delta_1, \dots, \Delta_n)$.

Given a fixed set $\mathbf{t} = (t_1, \dots, t_n)$, the cylindrical sets generate an algebra $\mathfrak{A}_{\mathbf{t}} \subset \mathfrak{A}$ and $\tilde{\mathfrak{A}}$ is the minimal σ-algebra of sets containing all $\mathfrak{A}_{\mathbf{t}}$.

Consider the mapping

$$\pi_t : x(\cdot) \mapsto (x(t_1), \dots, x(t_n)),$$

$$\pi_t : (\mathfrak{X}, \tilde{\mathfrak{A}}) \to (X^n, Z^n)$$

and notice that $\mathfrak{A}_{\mathbf{t}} = \pi_{\mathbf{t}}^{-1}(Z^n)$.

Let μ_ξ be the probability distribution of the random function $\xi(t)$ in the space $\mathcal{X}$, $\mu_t = \mu_\xi|_{\mathfrak{A}_{\mathbf{t}}}$ be the restriction to the algebra $\mathfrak{A}_{\mathbf{t}}$ and

$$\nu_{\mathbf{t}} = \mu_\xi^{\pi_{\mathbf{t}}} = \mu_{\mathbf{t}}^{\pi_{\mathbf{t}}} = \mu_\xi \circ \pi_{\mathbf{t}}^{-1}$$

be the image of μ_ξ under $\pi_{\mathbf{t}}$, called the mutual probability distribution of random variables $\xi_{\mathbf{t}} = (\xi(t_1), \dots, \xi(t_n))$. The collection of measures $\nu_{\mathbf{t}}$ on (X^n, Z^n) is called the collection of 'finite-dimensional' distributions of the stochastic process. Here we use inverted commas because the space X itself may have an infinite dimension.

The collection of $\nu_{\mathbf{t}}$ satisfies a natural consistency condition: given $\mathbf{t}' \subset \mathbf{t}$ and $\pi_{\mathbf{t}'\mathbf{t}} : x_{\mathbf{t}} \to x_{\mathbf{t}'}$, we have $\pi_{\mathbf{t}'\mathbf{t}} : \nu_{\mathbf{t}} \to \nu_{\mathbf{t}'}$.

Under some additional conditions concerning the space X, the converse assertion holds as well; namely, each consistent set of probability measures on (X^n, Z^n) generates the unique measure μ on $(\mathfrak{X}, \mathfrak{A})$. This assertion is the well known Kolmogorov theorem which is valid at any rate in the case X is a separable Banach space.

Thus, while considering a stochastic process defined on T and valued in X we may take $(\Phi(T, X), \tilde{\mathfrak{A}}, \mu)$ as probability space. In this case, by elementary events $\omega = \xi(\cdot)$ we mean process paths and random variables $\xi(t)$ are defined as evaluation mapping values

$$\xi(t) : \omega \mapsto e\upsilon_t(\omega).$$

Notice that in no case ought $\tilde{\mathfrak{A}}$ coincide with the minimal σ-algebra $\tilde{\mathfrak{A}}(T, X)$ which contains all cyclindrical sets. For example, it may differ from $\tilde{\mathfrak{A}}(T, X)$ by a certain collection of zero probability sets. Changing the values of the stochastic process on these sets, we do not change the finite-dimensional distributions of the stochastic process. This opens the way to construct different modifications $\eta(t)$ of the process $\xi(t)$ which are stochastically equivalent to it in the sense that

$$P\{\xi(t) = \eta(t)\} = 1.$$

Finite-dimensional distributions of stochastically equivalent processes do coincide. Nevertheless, while dealing with process modifications, we may be interested in those properties of the process which are not determined by 'a finite number of experiments'. For example, it may be important to know to what functional space process paths belong with probability one, having to hand only the information about the finite-dimensional distributions of the process. In particular, thanks to Kolmogorov, we know the following.

If there exist $C > 0, \alpha > 0, \mathcal{E} > 0$, such that the stochastic process $\xi(t)$-valued in the Banach space B, satisfies the estimate

$$E \| \xi(t + \Delta t) - \xi(t) \|^{\alpha} \leq C (\Delta t)^{1+\mathcal{E}}, \tag{1.6}$$

then there exists a modification $\eta(t)$ for which almost all paths are continuous (that is, the distribution μ_η is supported on $C(T, X)$).

Let $\mathcal{F}_t$ be a flow of σ-subalgebras of the σ-algebra $\mathcal{F}$ and $\mathcal{F}_{t_2} \subset \mathcal{F}_{t_1}$ if $t_2 < t_1$. A stochastic process is said to be adapted to the flow $\mathcal{F}_t$ if for each $t \in [0, T]$, the function $\xi(t)$ is $\mathcal{F}_t$-measurable.

The stochastic process $\xi(t)$ valued in the Banach space $(B, \mathcal{Z})$ is called a Markov process with respect to the σ-algebra flow $\mathcal{F}_t$, $(t \in [0, T])$ if it is adapted to $\mathcal{F}_t$ and for all $s, t, 0 \leq s \leq t < T$ and $A \in \mathcal{Z}$

$$P\{\xi(t) \in A/\mathcal{F}_s\} = P\{\xi(t) \in A/\xi_s\} \quad (\text{mod } P). \tag{1.7}$$

We now introduce the notations

$$\mathcal{F}^t = \sigma\{\xi(u) : u \in [0, T], u > t\},$$

$$\mathcal{F}_s^t = \sigma\{\xi(u) : u \in [0, T], s < u \leq t\},$$

$$\mathcal{F}_s = \sigma\{\xi(u) : u \in [0, T], u \leq s\}.$$

Some relations equivalent to (1.7) will be useful in the sequel; namely,

(a) For arbitrary bounded $\mathcal{Z}$-measurable function f

$$E\{f(\xi(t))/\mathcal{F}_s\} = E\{f(\xi(t))/\xi(s)\} \quad (\text{mod } P). \tag{1.8}$$

(b) For any bounded $\mathcal{F}^s$-measurable random variable η

$$E\{\eta/\mathcal{F}_s\} = E\{\eta/\xi(s)\} \quad (\text{mod } P). \tag{1.9}$$

The function $P(s, x, t, \Gamma)$ is called the transition probability of a Markov process if it is defined for all $x \in B$ $s, t \in [0, T]$ and Borel sets $\Gamma \in \mathcal{Z}$ and the equality

$$P\{\xi(t) \in \Gamma/\mathcal{F}_s\} = P(s, \xi(s), t, \Gamma) = P\{\xi(t) \in \Gamma/\xi(s)\} \tag{1.10}$$

holds.

By definition a transition probability possesses the following properties

(a) Given a fixed s, t, x, the function $P(s, x, t, \Gamma)$ is a measure on the σ-algebra $\mathcal{Z}$.

(b) Given a fixed $s < \tau < t$, $P(s, x, t, \Gamma)$ is a measurable function with respect to x.

(c) Given $s < \tau < t$, $x \in B, \Gamma \in \mathcal{Z}_B$

$$P(s, x, t, \Gamma) = \int_B P(s, x, t, dy)\, P(\tau, y, t, \Gamma). \tag{1.11}$$

The last equation is called the Chapman–Kolmogorov equation.

The transition probability enables us to compute all finite-dimensional distributions of the given Markov process where we know its initial distribution.

In fact, let $q : t_0 < t_1 < \dots t_n = t$ and $A_0, A_1, \dots, A_n \in \mathcal{Z}$. Then

$$P\left\{\xi(t_0) \in A_0,\ \xi(t_1) \in A_1, \dots, \xi(t_n) \in A_n\right\}$$

$$= \int_{A_0} \mu(\mathrm{d}\,x_0) \int_{A_1} P(t_0, x_0, t_1, \mathrm{d}\,x_1) \dots \int_{A_n} P(t_{n-1}, x_{n-1}, t_n, \mathrm{d}\,x_n), \tag{1.12}$$

where

$$\mu(A) = P\left\{\xi(t_0) \in A\right\}.$$

The proof of this relation may be obtained as an easy consequence of the relation

$$P\left\{\xi(t_0) \in A_0, \dots, \xi(t_n) \in A_n\right\}$$

$$= EP\left\{\xi(t_n) \in A_n / \xi(t_{n-1})\right\} \Pi\, j_{A_k}\ (\xi(t_k)),$$

where

$$j_A(x) = \begin{cases} 1, & \text{if } x \in A, \\ 0, & \text{if } x \,\overline{\in}\, A, \end{cases}$$

is the characteristic function of the set A.

It follows from (1.12) that for any Borel function $f(x_0, \dots, x_n)$, the equality

$$E f(\xi(t_0), \dots, \xi(t_n))$$

$$= \int \dots \int f(x_0, \dots, x_n) \times$$

$$\times P(t_0, x_0, t_1, \mathrm{d}\,x_1) \dots P(t_{n-1}, x_{n-1}, t_n, \mathrm{d}\,x_n), \tag{1.13}$$

holds if its right-hand side is correctly defined.

Let $W(B)$ and $\mathcal{M}(B)$ be, especially, the Banach spaces of measurable bounded functions on B and bounded Borel measures on $\mathcal{Z}$.

To each Markov process, we may assign the following families of linear bounded maps

$$U(\tau, t) f(x) = \int_B f(y) P(\tau, x, t, \mathrm{d}\,y) \tag{1.14}$$

and

$$V(t, \tau) \mu(\Gamma) = \int_B \mu(\mathrm{d}\,y) P(\tau, x, t, \Gamma)$$

acting, respectively, on $W(B)$ and $\mathcal{M}(B)$.

It follows from the Chapman–Kolmogorov equation that both $U(\tau, t)$ and $V(\tau, t)$ are evolution families.

By generator of the Markov process $\xi(t)$ we mean an infinitesimal operator of the evolution family $U(\tau, t)$ generated in accordance with (1.13) by the process transition probability, that is the operator

$$\mathfrak{A}(t) f(x) = \lim_{\Delta t \to 0} \frac{U(t - \Delta t, t) - U(t, t)}{\Delta t} f(x),$$

defined on those f for which the limit does exist.

Let $\langle \cdot , \cdot \rangle$ be a pairing of dual spaces $W(B)$ and $\mathcal{M}(B)$

$$\langle f, \mu \rangle = \int_B f(y)\, \mu(dy).$$

Evolution families $U(\tau, t)$ and $V(t, \tau)$ are connected by the relation

$$\langle U(\tau, t) f, \mu \rangle = \langle f, V(\tau, t)\mu \rangle \tag{1.15}$$

which shows that they give a pair of dual families. It may be easily deduced from (1.15) that generators $\mathfrak{A}_U$ and $\mathfrak{A}_V$ corresponding to dual families U and V are dual as well

$$\langle \mathfrak{A}_U f, \mu \rangle = \langle f, \mathfrak{A}_V \mu \rangle.$$

1.4. REAL WIENER PROCESSES AND STOCHASTIC WIENER INTEGRALS

A real-valued stochastic process $w_t(\omega) = w(t)$ defined on $[0, \infty] \times \Omega$, $(W/0) = 0$ is called a Wiener process if it possesses the following properties.

(a) For an arbitrary finite number of disjoint intervals $[t_i, t_i + \delta]$ the random variables $\Delta_i w = w(t_i + \delta) - w(t_i)$, $(i = 1, \dots, n)$ are independent.

(b) The random variable $\Delta w = w(t + \delta) - w(t)$ has Gaussian distribution with parameters

$$E\,\Delta w = 0, \qquad E(\Delta w)^2 = \sigma^2 \delta$$

and, thus, its distribution density is

$$p(x) = \left(\delta \sqrt{2\pi\delta}\right) \times \exp\left\{-\frac{x^2}{2\sigma^2\delta}\right\}.$$

It may be deduced immediately from the definition that for each collection $0 < t_1 < t_2 < \dots < t_n$, a set of random variables

$$w_t = \left(w(t_1), \dots w(t_n)\right)$$

has a distribution μ_t in R^n with the density

$$p_t(x) = \frac{\exp\left\{-\dfrac{1}{2\sigma^2} \Sigma \dfrac{(x_k - x_{k-1})^2}{t_k - t_{k-1}}\right\}}{\sqrt{(2\pi)^n \sigma^{2n} t_1 (t_2 - t_1) \dots (t_n - t_{n-1})}}.$$

It may be easily checked that this system of finite-dimensional distributions is consistent and, thus, there exists a correspondent probability measure on $\Phi([0, T], R^1)$ (for any $T > 0$).

Next, one may prove that

$$E(\Delta w)^4 = 3\sigma^4 \delta^2 \tag{1.16}$$

which shows, due to (1.6), that this measure is supported on continuous functions

$C([0, T], R^1)$. It means that a Wiener process has a modification with continuous paths. This modification will be taken as the Wiener process in what follows.

Notice that all Wiener process paths are Hölder continuous

$$|\Delta w| \leq C_{\mathcal{E}} \, |\delta|^{1/2-\mathcal{E}}$$

with an exponent smaller than 1/2.

Given a function $\varphi \in \mathcal{L}_2([0, T], \mathrm{d}\,t)$ define a stochastic Wiener integral

$$I(\varphi) = \int_0^T \varphi(s) \, \mathrm{d}\, w(s).$$

This integral cannot stand for a Stieltjes integral for each path, since those paths are functions of unbounded total variation.

Suppose φ is a step-wise function, that is for a collection $0 \leq t_1 < \ldots \leq t_n = t$,

$$\varphi(t) = \varphi(t_{k-1}), \; t_k < t < t_{k-1}, \quad k = 1, \ldots, n.$$

Then define

$$w'(\varphi) = \int_0^T \varphi(s) \, \mathrm{d}\, w(s) = \sum_{k=1}^{n} \varphi_k \left[w(t_{k+1})) - w(t_k) \right. . \tag{1.17}$$

As it follows from the definition of Wiener process, the random variable $w'(\varphi)$ possesses the following properties:

$$E\, w' = 0,$$

$$E\, |w'(\varphi)|^2 = \sigma^2 \sum_{k=1}^{n} |\varphi_k|^2 (t_{k+1} - t_k)$$

$$= \sigma^2 \int_0^T |\varphi(\tau)|^2 \, \mathrm{d}\,\tau.$$

The linear set S of step-wise functions is dense in $\mathcal{L}_2[0, T] = \mathcal{L}_2([0, T]\, \mathrm{d}\, t)$ with $\mathrm{d}\,t$ being a Lebesgue measure on $[0, T]$.

The correspondence $\varphi \mapsto w'(\varphi)$ enables us to obtain a linear map $w' : S \to \mathcal{H}$, where $\mathcal{H}$ is the Hilbert space of random variables with norm $\| \xi \|_{\mathcal{H}} = \{E \, \| \xi \|^2\}^{1/2}$. This map is continuous due to the estimate

$$\| w'(\varphi) \|_{\mathcal{H}} \leq \sigma \, \| \varphi \|_{\mathcal{L}_2 [0,T]}$$

and, thus, may be extended to the whole $\mathcal{L}_2[0, T]$. We preserve the above notation for the extended map as well

$$w'(\varphi) = \int_0^T \varphi(s) \; \mathrm{d}\, w(s), \qquad \varphi \in \mathcal{L}_2[0, T].$$

Obviously, the limit procedure does not violate the relations

$$E\, w'(\varphi) = 0, \quad E\,(w'(\varphi))^2 = \sigma^2 \int_0^T |\varphi(s)|^2 \; \mathrm{d}\, s.$$

Moreover, we may prove that

$$E\, w'(\varphi)\, w'(\varphi) = \sigma^2 \int_0^T \varphi(s)\, \psi(s)\, \mathrm{d}s.$$

Notice that random variables $w'(\varphi)$ and $w'(\varphi)$ have a mutual Gaussian distribution and, thus, are independent for orthogonal φ and ψ, $(\int_0^T \varphi(s)\, \psi(s)\, \mathrm{d}s = 0)$.

Let us point out another approach to the definition of the Wiener process. This approach is connected with the so-called 'white noise', that is with a generalized stochastic process.

Let φ be a function over $[0, T]$, $(\varphi(0) = 0,\ \varphi(T) = 0)$ with a square integrable derivative $\varphi'(t)$. Consider a sequence of step-wise functions

$$\varphi_n(t) = \varphi(t_{k-1}), \quad t_{k-1} \le t < t_k,\ t_k = \frac{kT}{n}.$$

This sequence converges uniformly to φ and, thus,

$$w'(\varphi) = \lim_{n\to\infty} w'(\varphi_n) = -\lim_{n\to\infty} \sum_{k=1}^{n} w(t_k)\left[\varphi(t_k) - \varphi(t_{k-1})\right].$$

On the other hand, as it is easy to see, the limit in the right-hand side is equal to $\int_0^T w(s)\, \varphi'(s)\, \mathrm{d}s$ for each continuous path $w(t)$. Therefore,

$$w'(\varphi) = -\int_0^T w(s)\, \varphi'(s)\, \mathrm{d}s = -(w, \varphi')_{\mathcal{L}_2[0,T]}$$

and almost all Wiener process paths generate a linear continuous functional w' on the Sobolev space $H_+ = W_2^1[0, T]$. Next, a path derivative in the sense of distributions is an element of the dual space $H_- = W_2^1[0, T]$ and, thus, coincides with w' (which justifies the notation we are using).

It is not difficult to prove, using (1.17), that an H-valued random variable generated in this way has a canonical Gaussian distribution associated with the triple

$$\mathcal{H}_+ \subset \mathcal{H}_0 = \mathcal{L}_2[0, T] \subset \mathcal{H}_-$$

of Hilbert spaces.

The generalized stochastic process $\kappa(t) = w'(t)$, which is a Wiener process derivative, has the name of white noise. In this way, we may set forth the following construction.

Consider a rigged Hilbert space

$$Z_+ \subset Z_0 \subset Z_-$$

and make the measurable space $(Z_-, \mathcal{Z}_-)$ to be a probability space by choosing a canonical Gaussian measure μ with characteristic functional

$$\chi_\mu(\varphi) = \exp\left\{-\frac{1}{2}\, \|\varphi\|_0^2\right\}$$

as the probability P.

Each element $\varphi \in Z_0$ gives rise to a measurable linear functional ℓ_φ on Z_-; that is, to a real random variable which will be denoted $\tilde{\varphi}$. Since the map $\varphi \mapsto \tilde{\varphi}$ is linear, this random variable has a Gaussian distribution and

$$E\,\tilde{\varphi} = 0, \qquad E\,|\,\tilde{\varphi}\,|^2 = \|\,\varphi\,\|^2\,.$$

Moreover,

$$E\,\tilde{\varphi}\,\tilde{\psi} = (\varphi, \psi)_0,$$

and, hence, random variables $\tilde{\varphi}$ and $\tilde{\psi}$ are independent if $(\varphi, \psi) = 0$.

Let now $Z_0 = \mathcal{L}_2\,[0, T]$. The probability space constructed above is called a Wiener space and the Z-valued random variable κ generated by the identity map is called white noise (corresponding to the rigged space Z_0).

If $\varphi\,(\Delta) = j\,(\Delta)$ is the characteristic function of the Borel set $\Delta \subset [0, T]$, then $\tilde{\varphi}\,(\Delta)$ is a real random measure defined on Borel sets of $[0, T]$ with Gaussian distribution and independent values on disjoint sets. Setting

$$w\,(t) = \tilde{\varphi}\,([0, T])$$

we obtain a Wiener process.

Moreover, for a step-wise function $\varphi\,(t) = \Sigma\, c_k\, j\,(\Delta_k)$, we have

$$\tilde{\psi} = \Sigma\, c_k\, \tilde{\varphi}\,(\Delta_k) = \int_0^T \psi\,(s)\,\varphi\,(s)\,\mathrm{d}\,s = \int_0^T \psi\,(s)\,\mathrm{d}\,w; \tag{1.18}$$

that is, $\tilde{\varphi}$ is a stochastic Wiener integral possessing the following properties

$$E\,\tilde{\psi} = 0, \qquad E\,|\tilde{\psi}|^2 = \int_0^T |\psi\,(s)|^2\,\mathrm{d}\,s.$$

By a limiting procedure, the above relations may be extended to every $\psi \in Z_0 = \mathcal{L}_2\,[0, T]$.

As has been shown above, a white noise is a generalized stochastic process $\kappa\,(t) = w'\,(t)$ and, in this sense,

$$\tilde{\psi} = \int_0^T \psi\,(s)\,\kappa\,(s)\;\mathrm{d}\,s = \langle \psi, \kappa \rangle$$

for smooth functions $\psi\,(t)$.

2. Stochastic Integrals for Vector and Operator Functions

2.1. VECTOR WIENER PROCESS

Consider a pair of Hilbert triples

$$Z_+ \subset Z_0 \subset Z_-\,, \qquad H_+ \subset H_0 \subset H_-$$

and a rigged Hilbert space

$$\mathcal{H}_+ \subset \mathcal{H}_0 \subset \mathcal{H}_- \tag{2.1}$$

formed by the tensor products

$$\mathcal{H}_+ = Z_+ \otimes H_+ , \quad \mathcal{H}_0 = Z_0 \otimes H_0 , \quad \mathcal{H}_- = Z_- \otimes H_- .$$

We shall deal with random variables defined on the probability space $(\mathcal{H}_-, \mathcal{Z}_-, \mu)$ with a canonical Gaussian measure μ as probability.

Denote by κ the random variable generated by the identity map $i\,d_{\mathcal{H}_-}$.

Consider the map

$$S_\alpha : \mathcal{H}_0 \to H_0, \quad (\alpha \in Z_0)$$

defined by the relation

$$(\varphi, S_\alpha h)_{H_0} = (\alpha \otimes \varphi, h)_{\mathcal{H}_0}, \quad \varphi \in H_0. \tag{2.2}$$

It follows from the obvious estimate

$$\| S_\alpha \| \le \| \alpha \|_{Z_0} \tag{2.3}$$

that $S_\alpha(\kappa)$ is an H_--valued random variable for any $\alpha \in Z_0$. Indeed, the composition of S_α with the embedding $J : \mathcal{H}_0 \to H_-$ is an operator $J S_0 \in L_{12}(\mathcal{H}_0, H_-)$. As has been shown in Chapter 1, this operator may be extended to the whole space $\mathcal{H}_-$ as a measurable linear operator. The probability distribution of the resulting set of random variables may be determined by the formulas

$$E \langle \varphi, S_\alpha(\kappa) \rangle = \int_{\mathcal{H}_-} \langle \varphi, S_\alpha(\kappa) \rangle \, \mu(d\kappa) = 0, \tag{2.4}$$

$$\begin{aligned} &E \langle \varphi, S_\alpha(\kappa) \rangle \langle \psi, S_\beta(\kappa) \rangle \\ &= \int_{\mathcal{H}_-} \langle \varphi, S_\alpha(\kappa) \rangle \langle \psi, S_\beta(\kappa) \rangle \, \mu(d\kappa) \\ &= \langle \alpha \otimes \varphi, \beta \otimes \psi \rangle_{\mathcal{H}_0} = (\alpha, \beta)_{Z_0} \, (\varphi, \psi)_{H_0} . \end{aligned} \tag{2.5}$$

As a result, we obtain the following proposition.

PROPOSITION 2.1. *The linear continuous map* $\alpha \mapsto S_\alpha$ *from* Z_0 *into the set of Gaussian random variables which are canonical with respect to*

$$H_+ \subset H_0 \subset H_- ,$$

having the parameter $\| \alpha \|^2_{Z_0}$ *and the mutual correlations* (2.5) *correspond to the rigged space* (2.1).

Let $Z_0 = \mathcal{L}_2[0, T]$ and $Z_+ = W_2^1[0, T]$ consist of functions with square integrable derivatives. Then $\mathcal{H}_0 = \mathcal{L}_2([0, T], H_0)$ and $\mathcal{H}_+ = W_2^1([0, T], H_0)$ consists of functions defined on $[0, T]$ and valued, respectively, in H_0 and H_+.

In this case, we say the probability space $(\mathcal{H}_-, \mathcal{Z}_-, \mu)$ is a Wiener space and the random variable κ – a white noise corresponding to the Hilbert space H_0. We now construct the H-valued vector Wiener process $w(t)$ and show that, given $h \in \mathcal{H}_0$, the measurable linear functional (h, κ) may be represented by a stochastic integral

$$(h, \kappa) = \int_0^T (h(t), \, d\,w(t))_{H_0} . \tag{2.6}$$

THEOREM 2.1. *Random variables* $\tilde{S}_{j[0,t]} = w(t)$ *form a Wiener random process corresponding to* H_0; *that is, an* H_-*-valued process whose increments* $\Delta w(t+\Delta t) - w(t)$ *are independent random variables with canonical Gaussian distribution* μ *and parameter* $\sigma^2 = |\Delta t|$.

The equality (2.6) holds and the stochastic integral in its right-hand side is determined for step-wise functions as a sum

$$\int_0^T (h(t), \mathrm{d}w(t)) = \sum_{k=0}^{n} (h_k, w(t_k) - w(t_{k-1}))_{H_0} \tag{2.7}$$

given

$$h_k = h(t_{k-1}) = h(t), \ t \in [t_{k-1}, t_k]$$

and extended to the whole H_0 by continuity.

The stochastic integral defined in this way has the following properties.

$$E \int_0^T (h(t), \mathrm{d}w(t)) = 0,$$

$$E \left| \int_0^T (h(t), \mathrm{d}w(t)) \right|^2 = \int_0^T \| h(t) \|^2 \, \mathrm{d}t. \tag{2.8}$$

For any $\varphi \in H_0$ random variables $w_\varphi(t) = \langle \varphi, w(t) \rangle$ constitute a scalar Wiener process. If $\{e_k\}$ is an orthonormal basis in H_0 and $w_k(t) = \langle e_k, w(t) \rangle$, then w_k are independent and

$$w(t) = \sum_{k=1}^{\infty} w_k(t)\, e_k, \tag{2.9}$$

where the series converges in a mean square sense

$$E \left\| w(t) - \sum_{k=1}^{n} w_k(t)\, e_k \right\|_{H_-}^2 \to 0, \quad \text{for } n \to \infty.$$

Proof. It follows from Proposition 2.1. that $w(t)$ possesses the above properties. So we have to derive (2.7). Given h and $\kappa \in \mathcal{H}_0$, we have

$$\begin{aligned}(h, \kappa) &= \sum \left(j(\Delta_k) \otimes h_k, \kappa \right) = \sum \left(h_k, S_{j(\Delta_k)} \kappa \right) \\ &= \sum \left(h_k, S_{[0,t_k]} \kappa - S_{[0,t_{k-1}]} \kappa \right)\end{aligned}$$

and, hence,

$$(h, \kappa) = \sum (h_k, w(t_k) - w(t_{k-1})).$$

The estimate (2.8) follows immediately from the properties of the measurable linear functional (h, κ). The series (2.9) converges since

$$E \left\| \sum_{k=n}^{n+m} w_k(t)\, e_k \right\|_{H_-}^2 = \sum_{k=n}^{n+m} \| e_k \|_{H_-} \to 0, \quad n \to \infty.$$

Remark 2.1. As in the scalar case, given $h \in \mathcal{H}_+$, we have

$$\int_0^T (h(t),\ \mathrm{d}w(t)) = -\int_0^T (h'(t), w(t))\,\mathrm{d}t = \int_0^T (h(t), \kappa(t)).$$

Here $\kappa(t) = w'(t)$ and for the Wiener process derivative, we take the derivative in the sense of distributions. Thus,

$$(h, \kappa) = \int_0^T (h(t), \kappa(t))\ \mathrm{d}t$$

we shall use this notation in the case $h \in \mathcal{H}_0$ as well.

2.2. RANDOM FUNCTION STOCHASTIC INTEGRAL

Our aim here is to construct more complicated white noise functionals, which arise if in the relation

$$(f, \kappa) = \int_0^T (f(t), \kappa(t))\ \mathrm{d}t = \int_0^T (f(t), \mathrm{d}w(t))$$

a determined element $f \in \mathcal{H}_0$ becomes a random variable $f(\kappa)$.

Given $f: \mathcal{H}_- \to \mathcal{H}_+$, by $(f(\kappa), \kappa)$ we mean a pairing of $\kappa \in \mathcal{H}_-$ and $f(\kappa) \in \mathcal{H}_+$. We still conserve the notation

$$(f(\kappa), \kappa) = \int_0^T (f(t), \mathrm{d}w(t))$$

for it and call it a stochastic integral. This integral may be defined for instance by integration by parts formula

$$\int_0^T (f(t, \kappa), \mathrm{d}w(t)) = -\int_0^T \left(\frac{\partial}{\partial t} f(t, \kappa),\ w\ (t)\right) \mathrm{d}t,$$

since $f(0, \kappa) = f(T, \kappa) = 0$.

As the domain of the definition of the above integral is too small, we would like to extend it in some way to $h: \mathcal{H}_- \to \mathcal{H}_0$.

Let $f: \mathcal{H}_- \to \mathcal{H}_0$ be a smooth mapping. Then the expression $\operatorname{div} f(\kappa) = \operatorname{Tr} f'(\kappa)$ is correctly defined.

We say

$$(f(\kappa), \kappa) - \operatorname{Tr} f'(\kappa) = -\rho_\mu(f; \kappa) \tag{2.10}$$

to be a regularized stochastic integral (it is usually called an extended stochastic integral) and denote

$$\int_0^{T\sim} (f(t, \kappa), \mathrm{d}w(t)).$$

In addition, the simple relation

$$\int_0^{T\sim} (f(t, \kappa), \mathrm{d}w(t)) = \int_0^T (f(t, \kappa)\,\mathrm{d}w)\ \operatorname{Tr} f'(\kappa). \tag{2.11}$$

holds.

Recall that $\rho_\mu(f;\kappa)$ (see Chapter 1) gives a logarithmic derivative of the canonical (with respect to the triple $\mathcal{H}_+ \subset \mathcal{H}_0 \subset \mathcal{H}_-$) Gaussian measure μ along the vector field f.

The random variable

$$\rho(f) : \kappa \mapsto \rho(f;\kappa)$$

satisfies the equality

$$E\,|\rho(f)|^2 = E\left\{\mathrm{Tr}_{\mathcal{H}_0}[f'] + \|f\|^2_{\mathcal{H}_0}\right\} \tag{2.12}$$

and, thus, may be extended in mean square sense to those random variables for which the right-hand side of (2.12) is well defined; that is, to

$$f \in C_1(\mathcal{H}_-, \mathcal{H}_0, \mathcal{H}_0), \quad f' \in L_{12}(\mathcal{H}_0).$$

The expansion (2.11) in this case does not make any sense, since its summands do not exist. We single out below a class of random variables for which $\mathrm{Tr}\, f'(\kappa) = 0$ and, hence the stochastic integral does coincide with the regularized integral, satisfies simple estimated and, therefore, may be extended to non-smooth functions.

DEFINITION 2.1. A measurable map $h : \mathcal{H}_- \to \mathcal{K}$, where $\mathcal{K}$ is a normed function space over $[0, T]$, is called nonanticipating if for any $y \in \mathcal{H}_-$ with support $\mathrm{supp}\,(y) \subset [t, T]$, $t \in [0, T]$,

$$h(\kappa + y, t) = h(\kappa, t) \qquad (\mu - \text{a.e.})\,.$$

In other words, $h(\kappa, t)$ does not depend upon values $\kappa(\tau)$ for $\tau > t$. Denote $\mathcal{P}(\mathcal{H}_-, \mathcal{K})$ as a collection of nonanticipated mappings. It is obvious that $\mathcal{P}(\mathcal{H}_-, \mathcal{K})$ is a linear space closed with respect to convergence in measure μ (and, therefore, in a mean square sense if the corresponding integrals do exist and the estimates which grant the possibility of passing to the limit under the integral sign hold).

Remark 2.2. Let $\mathcal{F}_t$ be a minimal σ-subalgebra of the Borel σ-algebra $\mathcal{Z}$ generated by characteristic functions j_Δ of the measurable sets $\Delta \subset [0, t]$. Evidently, $\mathcal{P}(\mathcal{H}_-, \mathcal{K})$ is the collection of $\mathcal{F}_t$-measurable maps for each t.

Consider a smooth mapping $f : \mathcal{H}_- \to \mathcal{H}_0$. Its derivative $f'(\kappa)$ is a linear belonging to $L_{12}(\mathcal{H}_0)$. Each operator Q of this class may be given in the form of an integral operator

$$(Q\varphi)(t) = \int_0^T q(t, \tau)\, \varphi(\tau)\, d\tau,$$

where $q(t, \tau)$ is a kernel with values in $L_{12}(\mathcal{H}_0)$. Moreover,

$$\sigma_2^2(Q) = \int_0^T \int_0^T \sigma_2^2\,(q(t, \tau))\; dt\, d\tau.$$

The kernel corresponding to the operator $f'(\kappa)$ is denoted $\tilde{f}'(\kappa)(t, \tau)$.

It follows immediately from the definition that given a nonanticipating mapping

$f \in \mathcal{P}(\mathcal{H}_-, \mathcal{H}_0)$, this kernel is triangular, $\tilde{f}'(\kappa)(t,\tau) = 0$, $(t < \tau)$, and

$$f'(\kappa)\,\varphi(t) = \int_0^T \tilde{f}'(\kappa)(t,\tau)\,\varphi(\tau)\,d\tau.$$

The square of the operator $f'(\kappa)$ may be represented as

$$[f'(\kappa)]^2\,\varphi(t) = \int_0^T b(t,\tau)\,\varphi(\tau)\,d\tau,$$

where the kernel

$$b(t,\tau) = \int_\tau^t \tilde{f}'(\kappa)(t,s)\,\tilde{f}'(\kappa)(s,\tau)\,ds$$

is equal to zero over the diagonal, $b(s,s) = 0$ and, thus,

$$\operatorname{Tr}[f'(\kappa)]^2 = 0.$$

Then, given a nonanticipating mapping $f \in \mathcal{P}(\mathcal{H}_-, \mathcal{H}_0)$, we easily deduce that the triangular kernel $\tilde{f}'(\kappa)(t,\tau)$ is continuous and, thus, at diagonal points

$$\tilde{f}'(\kappa)(t,\tau) = 0 \quad \text{and} \quad \operatorname{Tr} f'(\kappa) = 0.$$

This leads to the following assertion.

PROPOSITION 2.2. *Given a nonanticipating random variable $f \in \mathcal{P}(\mathcal{H}_-, \mathcal{H}_0)$ with the derivative $f' \in L_{12}(\mathcal{H}_0)$, there exists a stochastic integral which coincides with the regularized one*

$$\int_0^T (f(t,\kappa)\,d w(t)) = \int_0^{T\sim} (f(t,\kappa)\,d w(t))$$

and satisfies the relation

$$E\,\Big\| \int_0^T (f(t,\kappa),\ d w(t)) \Big\|^2 = \int_0^T E\,\|f(t)\|^2_{H_0}\,dt. \tag{2.13}$$

The equality (2.13) permits us to justify the desired extension of the stochastic integral to a wider class of random variables.

Denote $\mathcal{P}(\mathcal{H}_-, \mathcal{H}_0) \cup L_2(\mathcal{H}_-, \mathcal{H}_0, \mu) = N$.

PROPOSITION 2.3. *Let $f \in \mathcal{P}(\mathcal{H}_-, \mathcal{H}_0)$ and $E\,\|f\|^2 < \infty$. Consider a sequence $f_n \in \mathcal{P}(\mathcal{H}_-, \mathcal{H}_+)$ of functions $f_n \in C_1(\mathcal{H}_-, \mathcal{H}_0, \mathcal{H}_+)$ such that*

$$E\,\|f_n - f\|^2 \to 0, \quad n \to \infty.$$

Then the limit

$$\int_0^T (f(t,\kappa),\ d w) = \lim_{n\to\infty} \int_0^T (f_n(t,\kappa),\ d w)$$

exists in a mean square sense, and is called the stochastic integral of the random function f.

This limit does not depend upon the choice of the sequence and satisfies (2.13).

We shall now list some simple properties of the stochastic integral, which follow immediately from its definition given in Proposition 2.3.

(a)

$$\int_0^T (\alpha_1 f_1 + \alpha_2 f_2,\ \mathrm{d}\,w) = \alpha_1 \int_0^T (f_1, \mathrm{d}\,w) + \alpha_2 \int_0^T (f_2, \mathrm{d}\,w).$$

(b) Let $f_n \in N$ and $E\ \|f_n - f\|^2 \to 0$, $n \to \infty$, then

$$E \left| \int_0^T (f_n, \mathrm{d}\,w) - \int_0^T (f, \mathrm{d}\,w) \right|^2 \to 0.$$

(c) Let $j_\Delta(t)$ be the characteristic function of the set $\Delta \subset [0, T]$. Then

$$\int_\Delta (f(t),\ \mathrm{d}\,w\ (t)) = \int_0^T\ (f(t)\, j_\Delta(t),\ \mathrm{d}\,w\ (t)).$$

(d) Let $\varphi \in H_0$. Then

$$\int_\tau^T (\varphi,\ \mathrm{d}\,w\ (s)).$$

This relation is still valid in the case where φ does not depend on $w(s)$ values for $s \in [\tau, T]$. Thus, for a nonanticipating step random function

$$\varphi(t) = \varphi_k, \quad t_{k-1} \le t \le t_k,$$

where $\varphi_k(\kappa)$ is $\mathcal{F}_{t_{k-1}}$-measurable, we have

$$\int_\tau^T \varphi(s),\ \mathrm{d}\,w\ (s)) = \Sigma\left(\varphi_k, w(t_k) - w(t_{k-1})\right). \tag{2.14}$$

Remark 2.3. We may take (2.14) as the definition of stochastic integral for a nonanticipating step random function φ. It easily leads to the important relation

$$E \left| \int_0^T (\varphi(s), \mathrm{d}\,w\,(s)) \right|^2 = \int_0^T E\ \|\varphi(s)\|^2\ \mathrm{d}\,s.$$

which justifies the extension of the stochastic integral to nonanticipating functions which may be given as limits of stepwise functions.

This one is the usual way to construct the stochastic integral. But in this way, one shades the connection between stochastic integrals and white noise functionals.

Let now $F : \mathcal{H}_- \to L(\mathcal{H}_-, \mathcal{K})$ be an operator function whose values are bounded operators acting from $\mathcal{H}_-$ into a Banach space $\mathcal{K}$ for each $\kappa \in \mathcal{H}_-$. Then

$$F^*(\kappa) \in L(\mathcal{K}^*, \mathcal{H}_+).$$

Let $f_\varphi(t, \kappa) = F^*(\kappa)\,\varphi$ be an H_+-valued function. The mapping F is called nonanticipating if all $f_\varphi(t, \kappa)$, $(\varphi \in H_+)$ possess this property. For those F, stochastic integrals $\int_0^T (F_\varphi(t, \kappa),\ \mathrm{d}\,w\,(t))$ are correctly defined and

$$\langle F(\kappa)\,\kappa, \varphi\rangle = \langle \kappa, F^*(\kappa)\,\varphi\rangle = \int_0^T (f_\varphi(t,\kappa),\ \mathrm{d}w).$$

Hence,

$$E\,|\langle F(\varphi)\,\kappa, \varphi\rangle|^2$$

$$= E\,\left|\int_0^T (f_\varphi(t,\kappa),\ \mathrm{d}w(t))\right|^2$$

$$= \int_0^T E\,\|F^*(\kappa)\,\varphi\|^2_{H_0}\ \mathrm{d}t = E\,\|F^*(\kappa)\,\varphi\|^2_{\mathcal{H}_0}.$$

Assume now that K is a Hilbert space with a basis $\{\varphi_k\}_{j=1}^{\infty}$. In this case, we obtain the equality

$$E\,\|F(\kappa)\,\kappa\|^2 = \left|\sum E\,\langle F(\kappa)\,\kappa, \varphi_j\rangle\right|^2$$

$$= \sum_j E\,\|F^*(\kappa)\,\varphi_j\|^2_{\mathcal{H}_0}.$$

The right-hand side of this relation is correctly defined as well for $F^*(\kappa) \in L_{12}(\mathcal{K}_0, \mathcal{H}_0)$, that is for $F(\kappa) \in L_{12}(\mathcal{H}_0, \mathcal{K}_0)$. Therefore, by a standard limit procedure, we may extend $F(\kappa)\,\kappa$ to those mappings preserving the estimate

$$E\,\|F(\kappa)\,\kappa\|^2 = E\,\sigma_2^2(F(\kappa)).$$

Consider, in particular, the mapping

$$F(\kappa)\,h = \int_0^T A(t,\kappa)\,h(t)\ \mathrm{d}t,$$

where

$$A : [0, T] \times \mathcal{H}_- \to L_{12}(\mathcal{H}_0, \mathcal{K}) \tag{2.15}$$

is a nonanticipating mapping. In this case, denote

$$F(\kappa)\,\kappa = \int_0^T A(t,\kappa)\ \mathrm{d}w(t), \tag{2.16}$$

and call this expression a stochastic integral of an operator valued function.

PROPOSITION 2.4. *Let $A(t,\kappa)$ be a nonanticipating map of* (2.15) *type. Then the stochastic integral* (2.16) *does exist and satisfies the relation*

$$E\,\left\|\int_0^T A(t,\kappa)\ \mathrm{d}w(t)\right\|^2_{\mathcal{K}} = \int_0^T E\,\sigma_2^2(A(t,\kappa))\ \mathrm{d}t$$

if the right-hand side of this relation is correct.

To make the proof complete, we must mention that in the case considered

$$\sigma_2^2(F(\kappa)) = \int_0^T \sigma_2^2(A(t,\kappa))\ \mathrm{d}t.$$

Let us now find which conditions may guarantee the smoothness of those functions on $\mathcal{H}_-$ represented by stochastic integrals.

We say that a function $f: \mathcal{H}_- \to \mathcal{K}$ is differentiable along $\mathcal{H}_0$ f for each

$$f(\kappa + h) - f(\kappa) = \int_0^T q(\kappa + \tau h) h \, d\tau ,$$

where the derivative $q(\kappa) = f'(\kappa) \in L_{12}(\mathcal{H}_0, \mathcal{K})$ is defined for μ almost all $\kappa \in \mathcal{H}_-$.

In the case where f, q are square integrable functions, this definition is equivalent to the definition of the mean square smoothness of the function f

$$\lim_{\mathcal{E} \to 0} E \left\| \frac{f(\kappa - \mathcal{E} h) - f(\kappa)}{\mathcal{E}} - f'(\kappa) h \right\|^2 = 0.$$

Let $F(\kappa)$ be a smooth function in the above sense. Then for $F(\kappa) \in L(\mathcal{H}_-, \mathcal{K})$

$$\frac{d}{d\mathcal{E}} F(\kappa + \mathcal{E} h)(\kappa + \mathcal{E} h) \Big|_{\mathcal{E}=0} = (F'(\kappa) h) \kappa + F(\kappa) h .$$

By a limiting procedure, we may extend this relation to the smooth functions $F(\kappa) \in L_{12}(\mathcal{H}_0, \mathcal{K})$.

These arguments lead to the following result.

PROPOSITION 2.5. *Let* $A(t, \kappa) \in L_{12}(\mathcal{H}_0, \mathcal{K})$, $(\kappa \in \mathcal{H}_-, t \in [0, T])$ *be non-anticipating smooth function with mean square derivatives along* $\mathcal{H}_0$.

Then the functional $f(\kappa) = \int_0^T A(t, \kappa) \, d w(t)$ *satisfies the relation*

$$f'(\kappa) h = \int_0^T [A'_\kappa(t, \kappa) h(t)] \, d w(t) + \int_0^T A(t, \kappa) h(t) \, dt . \tag{2.17}$$

2.3. ESTIMATES OF BANACH SPACE-VALUED GAUSSIAN RANDOM VARIABLES

In the previous section, we have constructed Ito stochastic integrals of functions depending smoothly on $\kappa \in \mathcal{H}_-$. In the next section, we shall give a construction of stochastic integrals based on (2.14) which will not require $A(\kappa)$ to be a smooth function. We shall instead restrict ourselves to a smaller class of admissible Banach spaces, namely the class of Banach spaces with smooth norms. In this section, we obtain some preliminary estimates we will need later in order to construct a stochastic integral using this approach.

Let ξ be a canonical Gaussian random variable corresponding to the Hilbert triple $H_+ \subset H_0 \subset H_-$ and $A : H_0 \to B$ be a map belonging to $L_{12}(H, B)$. Then $\eta = A \xi$ is a Banach space B-valued random variable. Later, speculations will be essentially based on the estimate

$$E \| x + A \xi \|_B^2 \le \| x \|_B^2 + C \sigma_2^2 (A). \tag{2.18}$$

DEFINITION 2.2. A Banach space B belongs to the class τ_2 if (2.18) holds for each $x \in B$.

Evidently, Hilbert spaces belong to the τ_2 class.

Let us now show that Banach spaces with smooth enough norms belong to this class as well. To be more precise, τ_2 class contains Banach spaces B such that the functions $f(x) = \| x \|_B^2$ have Lipschitz continuous derivatives

$$\| f'(x) - f'(y) \| \le L \| x - y \|_B .$$

While speaking about this property of Banach space B we shall briefly say that it possesses a C_L^1 property.

Let us first reduce the estimate (2.18) to a simpler one.

PROPOSITION 2.6. *The estimate* (2.18) *is valid if for any* $x, y \in B$ *and the one-dimensional Gaussian random variable* ξ *the estimate*

$$E \| x + \xi y \|_B^2 \le \| x \|_B^2 + C \| y \|_B^2 \tag{2.19}$$

holds.

Proof. Substituting x in (2.19) by a random variable $\eta \in B$ which does not depend on ξ but satisfies the estimate $E \| \eta \|_B^2 < \infty$, we obtain, after averaging,

$$E \| \eta + \xi y \|_B^2 \le E \| \eta \|_B^2 + C \| y \|_B^2 .$$

Now let $\{e_k\}$ be an orthonormal basis in H_0; then ξ may be given in the form

$$\xi = \sum_{k=1}^{\infty} \xi_k e_k ,$$

where $\xi_k = (\xi, e_k)$ and the series converges in mean square sense in H_-. It follows from (2.19) that

$$E \left\| x + \sum_{k=m}^{\ell-1} \xi_k Ae_k \xi_\ell Ae_\ell \right\|^2$$

$$\le E \left\| x + \sum_{k=m}^{\ell-1} \xi_k Ae_k \right\|^2 + C \left\| Ae_\ell \right\|_B^2$$

and, by iteration, we obtain the estimate

$$E \left\| x + \sum_{k=m}^{\ell-1} \xi_k Ae_k \right\|_B^2 \le \| x \|^2 + C \sum_{k=m}^{n} \left\| Ae_k \right\|_B^2 .$$

First, for $x = 0$ it implies that the series $A\xi = \Sigma_k \xi_k Ae_k$ converges in B in the means square sense and, next, by letting n go to infinity for $m = 1, \dots, n$, we obtain the desired estimate (2.18) as well.

THEOREM 2.2. *If the spaces* B *possesses the property* C_L^1, *then it belongs to the* τ_2 *class and* $C = 1/2\, L$.

Proof. Consider the identity

$$\varphi(x+\xi y)-\varphi(x) = \langle \varphi'(x), y\rangle + \int_0^1 \langle \varphi'(x+\theta\xi, y)\rangle - \langle \varphi'(x), y\rangle \, d\theta.$$

By averaging it with respect to ξ, we obtain, due to C_L^1, property, that

$$E\,\varphi(x+\xi(y)) \le \varphi(x) + L\,\|y\|_B^2\, E\xi^2 = \varphi(x) + \frac{L}{2}\,\|y\|_B^2 \tag{2.20}$$

which leads to (2.19) for $\varphi(x)=\|x\|^2$.

Remark. We may extend (2.20) by a limiting procedure to functions given as

$$\varphi(x) = \lim_{n\to\infty} \varphi_n(x), \tag{2.21}$$

where φ_n is a nondecreasing sequence of functions satisfying (2.20).

This proves that the τ_2 class contains those spaces for which the square of the norm, that is, the function $\varphi(x)=\|x\|^2$, may be represented in the form (2.21).

COROLLARY. *Let $\xi_1, \dots, \xi_n$ be canonical Gaussian variables in $H_-, A_1, \dots, A_n$ be random operators belonging to the $L_{12}(H_-, B)$ class and ξ_k does not depend on the collection $\{\xi_1, \dots, \xi_{k-1}, A_1, \dots, A_{k-1}, A_k\}$ for each $k=1,2,\dots,n$. Arguments similar to those used in the proof of Proposition* 2.6 *show, due to* (2.8), *that the estimate*

$$E\left\|\sum_{k=1}^n A_k \xi_k\right\|^2 \le C\sum_{k=1}^n E\sigma_2^2(A_k) \tag{2.22}$$

holds.

2.4. STOCHASTIC INTEGRAL PROPERTIES. STOCHASTIC DIFFERENTIALS

Let the Banach space B belong to τ_2. In this section, we shall give a construction of the B-valued Ito stochastic integral based on its approximation by stochastic integrals of simple functions, and describe stochastic integral properties.

Let $w(t)$ be a vector Wiener process corresponding to H_0, $\mathcal{F}_t=\sigma\{w(s), s\le t\}$ and $A(t)\in L_{12}(H,B)$ be an $\mathcal{F}_t$-measurable operator-valued random process.

If $A(t)$ is a step-wise function, that is, for a partition $q: 0=t_0<t_1\le \dots \le t_n = T\; A(t)=A(t_i)=A_i$, for $t_i\le t\le t_{i+1}$, then the sum

$$I(A)=\int_0^T A(t)\,dw(t)=\sum_{i=0}^{n-1} A_i\Delta_i w,$$

where $\Delta_i w = w(t_{i+1})-w(t_i)$ is called a stochastic integral $I(A)$.

The stochastic integral defined in this way possesses the following properties.

(1) $E_{\mathcal{F}_0} I(A) = 0$.

(2) It follows from (2.22) that the estimate

$$E_{\mathcal{F}_0} \| I(A) \|^2 \le C \int_0^T E\, \sigma_2^2 (A(t))\, dt \tag{2.23}$$

holds.

(3) For any $N \ge 0$ and $K > 0$,

$$P\{ \| I(A) \| > K\} \le \frac{N C}{K^2} + P\left\{ \int_0^T \sigma_2^2 (A(t))\, dt > N \right\}. \tag{2.24}$$

To prove (2.24), introduce a function $A_N(t)$ such that $A_N(t) = A(t)$, for $t_k \le t \le t_{k+1}$ if

$$\sum_{j=0}^{k} \sigma_2^2 (A(t_j))(t_{j+1} - t_j) \le N$$

and $A_N(t) = 0$, if

$$\sum_{j=0}^{k} \sigma_2^2 (A(t_j))(t_{j+1} - t_j) > N, \quad k = 1, \dots, n$$

and notice that

$$\int_0^T \sigma_2^2 (A_N(t))\, dt = \sum_{j=0}^{\ell} \sigma_2^2 (A(t_j))(t_{j+1} - t_j) = \gamma_\ell,$$

where ℓ is the maximal number for which $\gamma_\ell \le N$. Hence,

$$E \int_0^T \sigma_2^2 (A_N(t))\, dt \le N.$$

Next, we see that $A(t) - A_N(t) = 0$ for all $t \in [0, t]$ if $\int_0^T \sigma_2^2\ (A(t))\, dt \le N$.

We now deduce that

$$\begin{aligned} P\{ \| I(A) \| \ge K\} &\le P\{ \| I(A_N) \| > K\} + \\ &\quad + P\{ I(A) \ne I(A_N)\} \\ &\le \frac{E \| I_N(A) \|^2}{K^2} + P\left\{ \int_0^T \sigma_2^2 (A(t))\, dt > N \right\} \\ &\le \frac{N C}{K^2} + P\left\{ \int_0^T \sigma_2^2 (A(t))\, dt > N \right\}. \end{aligned}$$

(4) Let $\Phi \in L_2\ (B \times B, R^1)$. Then for any $A, B \in L_{12}\ (H, B)$,

$$E\,\Phi\,(I\,(A), I\,(B)) = \int_0^T E\ \operatorname{Tr} B^*\,(t)\,\hat{\Phi}\,A\,(t)\ \mathrm{d}\,t \tag{2.25}$$

where

$$\operatorname{Tr}\ B^*\,\hat{\Phi}\,A = \sum_{k=1}^{\infty} \Phi\ \ (Ae_k, Be_k)\,.$$

Notice that (2.25) results immediately from the stochastic integral definition and the Wiener process properties, since

$$\begin{aligned} E\,\Phi\,(I\,(A), I\,(B)) &= E \sum_{i,\,j} \ \sum_{k,\,m} \ \Phi\,(A_i\,e_k\,\Delta_i\,w_k\,,\ B_j\,e_m\ \Delta_j\,w_m\,) \\ &= \sum_i \sum_m \Phi\,(A_i\,e_m,\ A_i\,e_m)\,\Delta_i\,t\,. \end{aligned}$$

Notice, in particular, that if $\Phi\,(x, y) = \langle f, x\rangle\,\langle \varphi, y\rangle$ with $f, \varphi \in B^*$, then

$$E\,\langle f, I\,(A)\rangle\,\langle \varphi, I\,(B)\rangle = \int_0^T E\ \ \langle f, A\,(t)\,B^*\,(t)\,\varphi\rangle\ \ \mathrm{d}\,t.$$

Denote $\mathcal{K}$ the set of functions $A\,(t) \in L_{12}\ \ (H, B)$ satisfying the estimate

$$\int_0^T E\ \sigma_2^2\,(A\,(t))\ \ \mathrm{d}\,t < \infty\,.$$

Let $A_n\,(t)$ be a sequence of step-wise functions belonging to $L_{12}\ \ (H, B)$ for which

$$\lim_{n\to\infty}\ \int_0^T E\ \sigma_2^2\,(A\,(t)) - A_n\,(t))\ \ \mathrm{d}\,t = 0. \tag{2.26}$$

Consider a sequence of B-valued random variables of the form

$$\xi_n\,(T) = \int_0^T A_n\,(t)\ \ \mathrm{d}\,w\,(t).$$

We call the limit of this sequence in the sense of convergence in probability (if it does exist) a stochastic integral and denote

$$I\,(A) = \int_0^T A\,(t)\ \ \mathrm{d}\,w\,(t) = P - \lim\ \int_0^T A_n\,(t)\ \ \mathrm{d}\,w\,(t). \tag{2.27}$$

THEOREM 2.3. *Let* $A \in \mathcal{K}$, *then the integral* (2.27) *does exist and possesses properties* (1)–(4).

Proof. It follows from property (2) that $\xi_n\,(T)$ constitutes a Cauchy sequence in the sense of convergence in probability

$$\begin{aligned} &\lim_{n\to\infty}\ P\left\{ \left\| \int_0^T A\,(s)\ \ \mathrm{d}\,w\,(s)\ -\ \int_0^T A_n\,(s)\ \ \mathrm{d}\,w\,(s) \right\| > \mathcal{E} \right\} \\ &\quad \le \lim_{n\to\infty}\ \int_0^T E\ \sigma_2^2\,(A\,(s)) - A_n\,(s))\ \ \mathrm{d}\,s = 0 \end{aligned}$$

and, hence, the limit does exist. Moreover, (2.26) justified the possibility of going to the limit in the above estimates.

A stochastic integral with a varying upper limit, may be defined by the relation

$$I_t(A) = I(A j_t) = \int_0^T A(s) j_t(s) \, \mathrm{d} w(s),$$

where

$$j_t(s) = j_t(s) = \begin{cases} 1, & \text{if } s \le t \\ 0, & \text{if } s > t \end{cases} .$$

Consider the stochastic process

$$\xi(t) = \xi_0 + \int_0^t a(s) \, \mathrm{d} s + \int_0^t A(s) \, \mathrm{d} w(s), \tag{2.28}$$

where $\xi_0 \in B$ is an $\mathcal{F}_0$-measurable random variable.

We say that the process $\xi(t)$, in this case, possesses the stochastic differential

$$\mathrm{d}\xi(t) = a(t) \, \mathrm{d} t + A(t) \, \mathrm{d} w(t). \tag{2.29}$$

With this definition, the stochastic differential is nothing more than a formal symbol and (2.29) is nothing but another equivalent way of writing down the integral relation (2.28).

For $A \equiv 0$, the stochastic differential (2.29) comes to be the ordinary differential

$$\mathrm{d}\xi = a(t) \, \mathrm{d} t$$

of the smooth random function $\xi(t) = \int_0^t a(s) \, \mathrm{d} s$.

This differential may be treated as follows: each $t \in [0, T]$ $\mathrm{d}\xi(t)$ is defined as the class of smooth function with coinciding first derivatives at the point t. Extending this approach, in a similar way we may treat relation (2.29), and this will, indeed, be done in the next chapter.

Here, while deriving some formal relations of stochastic differential calculus, we shall be satisfied with the above formal definition.

The so-called Ito's formula, which gives one of the main relations of stochastic differential calculus, is merely the way to differentiate a complicate random function.

THEOREM 2.4. *For a pair of* τ_2 *class spaces* B *and* B_1, *let* $f: [0, T] \times B \to B_1$ *be a mapping which is* C_1*-smooth with respect to* $t \in [0, T]$ *and* C_2*-smooth with respect to* $x \in B$. *Let* $\xi(t)$ *have a stochastic differential of the form* (2.29). *Then the stochastic process*

$$\eta(t) = f(t, \xi t))$$

has the stochastic differential

$$\mathrm{d}\eta(t) = \left\{ \frac{\partial f(t, \xi(t))}{\partial t} + f'_x(t, \xi(t)) \, a(t) + \right.$$

$$\left. + \frac{1}{2} \operatorname{Tr} A^*(t) f''_x(t, \xi(t)) A(t) \right\} \mathrm{d} t +$$

$$+f'_x(t,\xi(t))\,A(t)\;\mathrm{d}w(t)$$

$$=\frac{\partial f(t,\xi(t))}{\partial t}\,\mathrm{d}t+(D,f)\;\mathrm{d}\xi+\frac{1}{2}\,(A\,A^*\,D,D)f\;\mathrm{d}t. \qquad (2.30)$$

The fact that (2.30) includes a term with a second derivative is a crucial point of stochastic differential calculus. This peculiarity arises because of the necessity of using terms up to second order in Taylor expansion, while calculating $\mathrm{d}\eta$, since $(\mathrm{d}w)^2$ is of the same order as $\mathrm{d}t$.

We shall give here a sketch of Ito's formula derivation, omitting some of the technical details which may be found in other books. We intend here to explain this formula rather than to prove it rigorously.

First, let us mention that the precise sense of Ito's formula is defined by the integral relation

$$\eta(t)=\eta(t_0)+\int_{t_0}^{t}\Big[\frac{\partial f(t,\xi(s))}{\partial s}+f'_x(s,\xi(s))\,a(s)+$$

$$+\frac{1}{2}\,\mathrm{Tr}\,A^*(s)f''_x(s,\xi(s))\,A(s)\Big]\;\mathrm{d}s+$$

$$+\int_{t_0}^{t}f'_x(s,\xi(s))\,A(s)\;\mathrm{d}w(s)\,. \qquad (2.31)$$

While changing A and a in (2.31) to $A_n\to A$ and $a_n\to a$, which satisfy necessary conditions connected with estimates of the corresponding integrals, we may pass to the limit in (2.31). Therefore, it suffices to derive (2.31) for step-wise function a and A. Since both right and left sides of (2.31) present additive functions of the interval $[t_0,t]$, we may reduce the problem to investigating the case

$$a(t)=a_0(\omega),\quad A(t)=A_0(\omega),$$

where a_0, A_0 are $\mathcal{F}_{t_0}$-measurable variables.

Thus, let

$$\xi(t)=\xi_1(t)+\xi_2(t),$$

where

$$\xi_1(t)=\xi_0+a_0(\omega)\,(t-t_0)\,,$$

$$\xi_2(t)=A_0(\omega)\,\big[w(t)-w(t-t_0)\big]\,.$$

We may treat $\eta(t)$ as a function of the pair $(\xi_1(t),\ \xi_2(t))$ and, due to properties of the process $\xi_1(t)$ which does not depend on stochastic differentials, we can reduce the problem one again, considering the case $\xi(t)=\xi_2(t)$.

At last, using the approximation of the smooth function $f(t,x)$ by linear combinations of products $\psi(t)\,\psi(x)$, we reduce the investigation to the case $f(t,x)\equiv f(x)$ with the function f not depending on t.

Let $\xi(t)=A_0(\omega)\,[w(t)-w(t_0)]$ and $\eta(t)=f(\xi(t))$. Consider a partition

$t_0 < t_1 \ldots \le t_n = T$ of the interval $[t_0, T]$ and apply the Taylor formula. We obtain the relation

$$\begin{aligned}\eta(t) - \eta(t_0) &= \sum_{k=0}^{n-1} \left[f(\xi(t_{k+1})) - f(\xi(t_k)) \right] \\ &= \sum_{k=0}^{n-1} \Big[\langle f'(\xi(t_k)), A_0(\omega)\Delta_k w \rangle + \\ &\quad + \frac{1}{2} f''(\xi(t_k)) \, (A_0(\omega)\Delta_k w, \, A_0(\omega)\Delta_k w) \Big] + \\ &\quad + \sum_{k=0}^{n-1} \int_0^1 (1-\theta) \left[f''(\xi(t_k) + \theta A_0(\omega)\Delta_k w)) - \right. \\ &\quad \left. - f''(\xi(t_k)) \right] (A_0(\omega)\Delta_k w, \, A_0(\omega)\Delta_k w) \, d\theta. \end{aligned} \tag{2.32}$$

PROPOSITION 2.7. *Let* $f \in C_2(B, R^1)$; *then*

$$\lim_{n\to\infty} \sum_{k=1}^{n} f''(\xi(t_k)(A_0(\omega)\Delta_k w, \, A_0(\omega)\Delta_k w)$$

$$= \int_0^T \mathrm{Tr}\; A_0^*(\omega) f''(\xi(\theta)) \, A_0(\omega) \, d\theta. \tag{2.33}$$

Proof. Let $\mathcal{P}(H)$ denote the set of all finite orthogonal projections in a Hilbert space H equipped with a orthonormal base $\{e_k\}_{k=1}^{\infty}$. Consider

$$\sigma_{P_m} = \sum_{k=0}^{n} \Big[f''(\xi(t_k)) \, (A_0(\omega) P_m \Delta_k w, \, A_0(\omega) P_m \Delta_k w) -$$

$$- \sum_{k=0}^{m} f''(\xi(t_k)) \, (A_0(\omega) e_i, A_0(\omega) e_i) \, \Delta_k t \Big], \; P_m \in \mathcal{P}(H). \tag{2.34}$$

Since

$$P\left\{ \| \sigma_{P_m} \| > \varepsilon \right\} \le \varepsilon^{-2} E \| \sigma_{P_m} \|^2 \to 0, \quad \Delta_k \to 0,$$

it results that σ_{P_m} tends to zero with probability one and (2.33) is deduced from (2.34) passing to the limit $P_m \to I$. Notice that the Lebesgue theorem on the limit under integral sign justifies the limiting procedure.

PROPOSITION 2.8. *The stochastic integral* $I(A)$ *satisfies the estimate*

$$E \| I(A) \|^{2m} \le (m(2m-1))^m (T - t_0)^{m-1} C^m \int_0^T E \, \sigma_2^2 \, (A(s)) \, ds. \tag{2.35}$$

Proof. Apply Ito's formula to the process $\xi(t) = \int_0^t A(s) \, dw(s)$ and the function

$\varphi(x) = \| x \|^{2m}$. Due to the easily verified estimate

$$\begin{aligned} \langle \varphi''(x) h, h \rangle &= 2m \| x \|^{2(m-1)} \| h \|^2 + \\ &\quad + 2m(m-1) \| x \|^{2m-3} \| h \|^2 \\ &\le 2m(2m-1) \| x \|^{2(m-1)} \| h \|^2, \quad h \in B, \end{aligned} \tag{2.36}$$

we obtain the inequality

$$\begin{aligned} \operatorname{Tr} A^* \varphi'' A &= \sum_{k=1}^{\infty} \langle \varphi''(x) A e_k, A e_k \rangle \\ &\le 2m(2m-1) \| x \|^{2(m-1)} \sigma_2^2(A). \end{aligned}$$

Applying Ito's formula, we have

$$\begin{aligned} E \| \xi(t) \|^2 &= \frac{1}{2} E \left\{ \int_0^t \operatorname{Tr} A^*(s) \varphi''(\xi(s)) A(s) \, ds + \right. \\ &\quad \left. + \int_0^t (A^*(s) \varphi'(\xi(s)), dw(s)) \right\} \\ &= \frac{1}{2} \int_0^t E \operatorname{Tr} A^*(s) \varphi''(\xi(s)) A(s) \, ds. \end{aligned}$$

Formula (2.36) implies that φ'' is a positive operator and, thus, $E \| \xi(t) \|^{2m}$ is a monotone function of t.

Finally, Hölder inequality yields

$$\begin{aligned} & E \| \xi(t) \|^{2m} \\ &\quad \le m(2m-1) \int_0^t E \| \xi(s) \|^{2(m-1)} \sigma_2^2(A s)) \, ds \le \\ &\quad \le m(2m-1) \left\{ \int_0^t E \| \xi(s) \|^{2m} \right\}^{(m-1)/m} \times \\ &\quad\quad \times \left\{ \int_0^t \sigma_2^{2m}(A(s)) \, ds \right\}^{1/m} \le \\ &\quad \le m(2m-1)(T - t_0)^{(m-1)/m} \{ E \| \xi(T) \|^{2m} \}^{(m-1)/m} \times \\ &\quad\quad \times \left\{ \int_0^t E \sigma_2^{2m}(A(s)) \, ds \right\}^{1/m}, \end{aligned}$$

and the desired estimate may be obtained by simple transformations.

Assume that the spaces considered are equipped with norms such that the function $f(x) = \| x \|^2$ possesses two bounded continuous derivatives. In this case, Ito's formula enables us to state a number of important properties of stochastic integrals.

PROPOSITION 2.9. *Let $A(s) \in L_{12}(H,B)$. Then*

$$P\left\{\sup_{0\le t\le T}\left\|\int_0^t A(s)\,\mathrm{d}w(s)\right\|^2 > a\right\} \le \frac{1}{a^2}\int_0^T \sigma_2^2(A(s))\,\mathrm{d}s\,, \tag{2.37}$$

$$E\left\{\sup_{0\le t\le T}\left\|\int_0^t A(s)\,\mathrm{d}w(s)\right\|^2\right\}^\alpha \le \frac{\alpha}{\alpha-1}E\left\|\int_0^T A(s)\,\mathrm{d}w(s)\right\|^{2\alpha}. \tag{2.38}$$

Proof. Let $0 = t_{n0} < \ldots \le t_{nn} = T$ be a partition of the interval $[0, T]$ and let $A(t)$ be constant over $[t_{nk}, t_{nk+1}]$. Since

$$\xi(t_{nk}) = \int_0^{t_{nk}} A(s)\,\mathrm{d}w(s)$$

is $\mathcal{F}_{t_{nj}}$ measurable for $k \le j$, then

$$E\left(\int_0^{t_{nn}} A(s)\,\mathrm{d}w(s) - \int_0^{t_{nj}} A(s)\,\mathrm{d}w(s)/\xi(t_{n1})\ldots\xi(t_{nj})\right) = 0\,.$$

Denote

$$\mu(t) = \left\|\int_0^t A(s)\,\mathrm{d}w(s))\right\|^2, \quad \lambda_n = \sup_k \left\|\int_0^{t_{nk}} A(s)\,\mathrm{d}w(s)\right\|^2$$

and notice that

$$\lambda_n \to \lambda = \sup_{0\le t\le T} \|\xi(t)\|^2\,.$$

Let θ be a positive constant

$$j_k(\theta) = \begin{cases} 1, & \text{if } \mu(t_{n1}) < \theta, \ldots, \mu(t_{nk-1}) < \theta,\ \mu(t_{nk}) > \theta, \\ 0, & \text{in other cases.} \end{cases}$$

Since $\Sigma j_k(\theta) = 1$ for $\theta \le \lambda$ and $\Sigma j_k(\theta) = 0$ for $\theta > \lambda$, we have that for $\alpha > 1$ $\lambda^\alpha = \alpha\int_0^\infty \theta^{\alpha-1}\,\Sigma_{k=1}^n\, j_k(\theta)\,\mathrm{d}\theta$.

It results from the definition of $j_k(\theta)$ that

$$\theta j_k(\theta) \le \mu(t_{nk})\, j_k(\theta)\,.$$

As $j_k(\theta)$ is a measurable function of the random variables $\mu(t_{n1}), \ldots, \mu(t_{nk})$ it follows that

$$E\,\mu\,(t_{nk})\,j_k\,(\theta) = E\,j_k\,(\theta)\,E\,\{\mu\,(t_{nk})/\mu\,(t_{n1})\,,\dots,\mu\,(t_{nk})\}$$

and, hence,

$$E\;\theta^{\alpha-1}\sum_k j_k\,(\theta) \le E\;\Sigma\,\theta^{\alpha-2}\;\;j_k\,(\theta)\,\mu\,(t_{nn})\,.$$

By computing the integral of both sides of this relation over the interval $[0, \infty]$, we derive

$$E\,\frac{1}{\alpha}\;\lambda^{\alpha} \le E\;\;\frac{1}{\alpha-1}\;\lambda^{\alpha-1}\,\mu\,(t_{nn}),$$

which leads to

$$E\;\lambda^{\alpha} \le \frac{1}{\alpha-1}\;(E\,\lambda^{\alpha})^{\alpha-1/\alpha}\;\;[E\,\mu^{\alpha}\,(t_{nn})]^{1/\alpha}$$

due to Hölder inequality and, hence,

$$E\,\lambda^{\alpha} \le \left(\frac{1}{\alpha-1}\right)^{\alpha}\;E\,\mu^{\alpha}\,(t_{nn})\,.$$

Formula (2.38) follows from the last relation after passing to the limit for $n \to \infty$.

Notice finally that $P\,\{\lambda > \theta\} = \Sigma\,e\,j_k\,(\theta)$. Then the relation

$$\theta^{\alpha}\;\Sigma\,j_k\,(\theta) \le \sum_{k=1}^{n}\;[\,\mu\,(t_{nn})]^{\alpha}\;\,j_k\,(\theta)\,,\qquad \alpha = 1, 2,$$

implies that

$$P\,\{\lambda > \theta\} \le \frac{1}{\theta^2}\;E\;\left\|\;\int_0^t A\,(s)\;\,\mathrm{d}\,w\,(s)\;\right\|^2\,.$$

PROPOSITION 2.10. *Let* $\sigma_2^2\,(A\,(t)) \le K$, *where* K *is a positive constant. Then for* $\varepsilon > 0$, $\beta = \sup \|\,\varphi''\,(x)\,\|\;(\varphi\,(x) = \|\,x\,\|^2)$, *the estimate*

$$P\left\{\;\sup_{0\le\tau\le t}\;\left\|\;\int_0^t A\,(s)\;\mathrm{d}\;w\;\,(s)\;\right\|^2 > \varepsilon\right\} \le \frac{2\,K^2\,\beta^2}{\varepsilon^4}\;t^2 \tag{2.39}$$

holds if $t < \varepsilon/K\,\beta$.

Proof. Ito's formula applied to the smooth function $\varphi\,(x)$ leads to

$$\left\|\;\int_0^t A\,(s)\;\mathrm{d}\;w\;(s)\;\right\|^2 = \alpha_1 + \alpha_2, \tag{2.40}$$

$$\alpha_1 = \int_0^t \left(A^*\,(\tau)\;\varphi'\left(\;\int_0^\tau A\;\,(s)\;\,\mathrm{d}\;w\;\,(s)\right)\;\mathrm{d}\;w\;\,(\tau)\right),$$

$$\alpha_2 = \frac{1}{2}\int_0^t\;\sum_{k=1}^{\infty}\;\varphi''\left(\;\int_0^\tau A\;\,(s)\;\,\mathrm{d}\;w\;\,(s)\right)(A\,(\tau)\,e_k, A\,(\tau)\,e_k)\;\mathrm{d}\,\tau\,.$$

α_2 satisfies the estimate

$$\alpha_2 \le \frac{1}{2}\, t K \beta \le \frac{\varepsilon}{2} \qquad \text{for } t < \frac{\varepsilon}{K\beta}\,.$$

Hence, for $t < \varepsilon / K\beta$, it follows from (2.40) that

$$P\left\{ \sup_{0 \le \tau \le t} \left\| \int_0^\tau A(s)\; d w(s) \right\| > \varepsilon \right\}$$

$$\le P\left\{ \sup \alpha_1 > \frac{\varepsilon}{2} \right\} + P\left\{ \sup \alpha_2 > \frac{\varepsilon}{2} \right\}.$$

Keeping in mind (2.37), we obtain, due to $\sigma_2^2(A(t)) \le K$,

$$P\left\{ \sup \left\| \int_0^\tau A(s)\; d w(s) \right\| > \varepsilon \right\}$$

$$\le P\left\{ \sup \alpha_1^2 > \frac{\varepsilon^2}{2} \right\}$$

$$\le 4\,\varepsilon^{-4} \int_0^\tau K\beta^2 \int_0^S E\; \sigma_2^2(A(\tau))\, d\tau\; d s$$

$$\le 2 K^2 \beta^2 \varepsilon^{-4}\, t^2 .$$

3. Stochastic Equations

3.1. EXISTENCE AND UNIQUENESS THEOREM FOR SOLUTIONS OF STOCHASTIC EQUATIONS IN BANACH SPACES

Let $H_+ \subset H_0 \subset H_-$ be a rigged Hilbert space, $w(t)$ a Wiener process corresponding to H, and B a separable Banach space. Below, we assume that B belongs to the τ_2 class and thus there is defined a stochastic integral of a nonanticipating function

$$\int_0^t A(s)\; d w(s)$$

satisfying the estimate

$$E \left\| \int_0^t A(s, \omega)\, d w(s) \right\|^2 \le C \int_0^t E\; \sigma_2^2\; (A(s, \omega))\; d s.$$

Let

$$a(t, x) \in B, \quad A(t, x) \in L_{12}\;(H, B)$$

be measurable functions over $[0, t] \times B$.

The equation

$$d\xi = a(t, \xi(t))\; d t + A(t, \xi(t))\; d w \tag{3.1}$$

is called a stochastic differential equation.

We say that a random process $\xi(t)$ is a solution of (3.1) if it is adapted to the flow $\mathcal{F}_t = \{w(s), s \leq t\}$ (that is, $\mathcal{F}_t$-measurable) and has the stochastic differential (3.1) at the point $t \in [0, T]$.

Thus, the solution $\xi(t)$ of (3.1) ought to satisfy the relation

$$\xi(t) = \xi(s) + \int_s^t a(\tau, \xi(\tau))\, d\tau + \int_s^t A(\tau, \xi(\tau))\, dw(\tau) \tag{3.2}$$

$(0 \leq s \leq t \leq T)$ with probability 1.

Let us point out that for $a \equiv 0$, Equation (3.1) comes to be an ordinary differential equation; namely, a vector field, and its solution which is an integral path of this vector field, may be random only due to randomness of the initial value $\xi(s)$. If $A(t, x)$ is not equal to zero, Equation (3.1) loses vector field properties. Anyway, we postpone the discussion of its differential geometric nature to the next chapter.

We shall call the solution $\xi(t)$ of Equation (3.1) with nonrandom coefficients a diffusion process.

Choose a measurable space $(\mathcal{H}_-, Z_-)$ with the canonical Gaussian measure μ introduced in Chapter 2 as a probability space.

Consider the Banach space $S_p([0, T] \times \mathcal{H}_-, B)$ of measurable $\mathcal{F}_t$-adapted random functions $\xi(t, \omega)$ valued in B with the norm

$$|||\xi|||^p = \sup_{0 \leq t \leq T} E \|\xi(t)\|^p.$$

We shall say that a function $f(t, x)$, $(t \in [0, T], x \in B)$ valued in the Banach space Y, satisfies the condition $R(Y)$ and denote this fact as $f \in R(Y)$ if the inequalities

$$\|f(t, x)\|_Y \leq C_1 + C_2 \|x\|_B, \tag{3.3}$$

$$\|f(t, x) - f(t, y)\|_Y \leq C_2 \|x - y\|_B$$

hold for positive constants C_1, C_2.

PROPOSITION 3.1. *Let $a \in R(B)$, $A \in R(L_{12}(H, B))$. Then the formula*

$$\eta(t) = (\Phi \xi)(t) = \int_s^t a(\theta, \xi(\theta))\, d\theta + \int_s^t A(\theta, \xi(\theta))\, dw \tag{3.4}$$

determines a continuous mapping from S_p into itself.

Proof. It follows from (3.4) that if $\xi(t)$ is an $\mathcal{F}_t$-measurable process, then $\eta(t)$ is $\mathcal{F}_t$-measurable as well. Then, thanks to (3.3), it is easy to verify that the estimates

$$E \|\eta(t)\|^{2m} \leq C\left(1 + \int_s^t E \|\xi(\theta)\|^{2m}\, d\theta\right), \tag{3.5}$$

$$E \, \| \eta(t) - \eta_1(t) \|^2 \leq C \int_s^t E \, \| \xi(\theta) - \xi_1(\theta) \|^{2m/d\theta} \tag{3.6}$$

hold with the constant C depending only on m, $(T-s)$ and the constants in (3.3). These estimates imply the desired assertion.

THEOREM 3.1. *Let* $a \in R(B)$, $A \in R(L_{12}(H, B))$. *Then the Cauchy problem for* (3.1) *with the initial value*

$$\xi(s) = \xi_0 \tag{3.7}$$

has a unique (up to stochastic equivalence) solution $\xi(t)$ *in the space* S_{2m}. *This solution has a continuous path modification with a probability of one.*

Proof. Consider the equation

$$\xi(t) = \xi_0 + (\Phi \xi)(t) = (\Gamma \xi)(t), \tag{3.8}$$

which is equivalent to the Cauchy problem (3.1) (3.7). Due to Proposition 3.1, we know that Γ acts from S_{2m} into itself as a continuous map and, thus, the estimate

$$E \, ||| \Gamma(\xi) - \Gamma(\xi_1) |||^{2m} \leq C \int_s^t E \, \| \xi(\theta) - \xi_1(\theta) \|^{2m} \, d\theta$$

holds, due to (3.6).

The iteration procedure leads to the inequality

$$E \, ||| \Gamma(\xi) - \Gamma(\xi_1) |||^{2m}$$

$$\leq C \int_s^t E \, \| \Gamma(\xi)(\theta) - \Gamma(\xi_1)(\theta) \|^{2m} \, d\theta$$

$$\leq C^2 \int_s^t d\theta \int_s^\theta E \, \| \Gamma(\xi)(\tau) - \Gamma(\xi_1)(\tau) |||^{2m} \, d\tau$$

$$\leq \frac{C^2 (t-s)^2}{2!} \, ||| \xi - \xi_1 |||^{2m},$$

and by further induction, we have

$$||| \Gamma^r(\xi) - \Gamma^r(\xi_1) |||^{2m} \leq \frac{C^r (t-s)^r}{r!} \, ||| \xi - \xi_1 |||^{2m}.$$

Since for large enough r the estimate

$$\frac{C^r (t-s)^r}{r!} < 1$$

holds, the mapping Γ^r is a contraction map. This implies the existence of a unique fixed point of the mapping Γ or, what is the same, the existence of a unique (up to stochastic equivalence, which means that $\xi_1(t) \equiv \xi_2(t)$ if $||| \xi_2 - \xi_1 ||| = 0$) solution of the problem (3.1) (3.7).

The estimate

$$E \, \| \xi (t) - \xi (s) \|^{2m} \le C (t-s)^m \tag{3.9}$$

which is easily obtained from (3.2) due to (3.3) and (3.6) implies as a consequence of the Kolmogorov criteria, that the process $\xi (t)$ possesses a continuous modification.

Denote $\mathcal{H}_t$ as the subspace of S_2 consisting of $\mathcal{F}_\tau$-measurable functions $(\tau \le t)$ with the norm

$$\| \xi \|_t^2 = \sup_{\tau \le t} E \, \| \xi (\tau) \|^2 .$$

Notice that $\mathcal{H}_\tau \subset \mathcal{H}_t$ for $\tau \le t$.

The solution of (3.2) gives rise to a two-parameter family of mappings $T(t, s) : \mathcal{H}_s \to \mathcal{H}_t$ defined by the relation

$$\xi (t) = T (t, s) \, \xi (s). \tag{3.10}$$

In the sequel, we shall extensively use the following well-known assertion.

LEMMA (Gronwall). *Let* $\varphi (t), \alpha (t)$ *be measurable bounded functions, and* $h(t)$ *be a continuous nonnegative function such that*

$$\varphi (t) \le C + \alpha (t) + L \int_0^t h (\tau) \, \varphi (\tau) \, d\tau .$$

Then

$$\varphi (t) \le C \, e^{[\int_0^t h(s) \, ds]} + \alpha (t) + \int_0^t e^{[\int_\tau^t h(s) \, ds]} \alpha (\tau) \, h (\tau) \, d\tau . \tag{3.11}$$

PROPOSITION 3.2. *The mapping* (3.10) *satisfies the estimates*

$$E_s \, \| T (t, s) \, \xi (s) \|^2 \le \exp C (t-s) \left\{ \| \xi \|_s^2 + 3 \, C_1 (t-s) \right\} , \tag{3.12}$$

$$E_s \, \| T (t, s) \, \xi (s) - T (t, s) \, \xi_1 (s) \|^2 \le \exp C (t-s) \, \| \xi_1 (s) - \xi (s) \|^2 \tag{3.13}$$

and possesses the evolution property

$$T (t, s) \, \xi = T (t, \tau) \, T (\tau, s) \, \xi (s), \qquad T (t, t) \, \xi = \xi .$$

Proof. Let us compute the stochastic differential of the function $\varphi (\xi) = \| \xi (t) \|^2$ by applying Ito's formula. We immediately obtain that

$$\begin{aligned} &\| T (t, s) \, \xi (s) \|^2 \\ &\quad = \| \xi (s) \|^2 + \int_s^t 2 \langle \varphi' (\xi (\theta)), a (\theta, \xi (\theta)) \rangle \, d\theta + \\ &\qquad + \int_s^t \sigma_2^2 \left(A (\theta, \xi (\theta)) \right) d\theta + \\ &\qquad + \int_s^t 2 \langle \varphi' (\xi (\theta)), A (\theta, \xi (\theta)) \, dw (\theta) \rangle , \end{aligned}$$

which implies, due to (3.3), that the estimate

$$E_s \, \| \xi(t) \|^2 \le \| \xi(s) \|^2 + 3\, C_1 \, (t-s) + \int_s^t 3\, C_2\, E_s \, \| \xi(\tau) \|^2 \, d\tau$$

is valid and, thus, the desired result follows from the Gronwall inequality. The second estimate may be proved in a similar way.

Let us show that $\beta(t) = T(t, \tau)\, T(\tau, s)\, \xi(s)$ satisfies (3.2) as well. Then the equality

$$T(t, s)\, \xi(s) = T(t, \tau)\, T(\tau, s)\, \xi(s)$$

will result from uniqueness of the solution.

To estimate the difference $\delta(t) = \xi(t) - \tilde{\xi}(t)$ we use the following representation of $\tilde{\xi}(t)$.

$$\tilde{\xi}(t) = \xi(\tau) + \int_\tau^t a(\theta, \tilde{\xi}(\theta))\, d\theta + \int_\tau^t A(\theta, \tilde{\xi}(\theta))\, dw(\theta),$$

Now

$$\begin{aligned} &E \, \| \delta(t) \|^2 \\ &\le E \, \Big\| \int_\tau^t a(\theta, \xi(\theta))\, d\theta + \\ &\quad + \int_\tau^t A(\theta, \xi(\theta))\, dw(\theta) - \int_\tau^t a(\theta, \tilde{\xi}(\theta))\, d\theta - \\ &\quad - \int_\tau^t A(\theta, \tilde{\xi}(\theta))\, dw(\theta) \Big\|^2 \\ &\le C \int_\tau^t E \, \| \delta(\theta) \|^2 \, d\theta, \end{aligned}$$

and thus, $\delta(t) = 0$ with probability 1.

THEOREM 3.2. *Under the conditions of Theorem* 3.1, *the stochastic processes* $\xi(t)$ *which solve* (3.1) *is a Markov process with transition probility*

$$P(s, x, t\, \Gamma) = P\, \{ \xi_{s,x}(t) \in \Gamma \}, \quad \Gamma \in Z_B,$$

where $\xi_{s,x}(t) = T(t, s) \circ x$ *is the solution of the equation*

$$\xi(t) = x + \int_s^t a(\theta, \xi(\theta))\, d\theta + \int_s^t A(\theta, \xi(\theta))\, dw(\theta). \tag{3.14}$$

Proof. Let $\mathcal{F}_t$ be the minimal σ-algebra of events which, for $\tau \le t$, $\xi(s)$ and $w(\tau)$ are measurable. Denote $\mathcal{F}^t$ as the σ-algebra generated by $w(\tau) - w(t)$ for $\tau > t$. Obviously, events in $\mathcal{F}^t$ do not depend on events in $\mathcal{F}_t$. The variable $\xi_{s,x}(t)$ is defined by increments $w(\tau) - w(s)$ for $\tau \ge s$ and is $\mathcal{F}^s$-measurable. Since $\xi_{s,\xi(s)}(t)$ satisfies the equation

$$\xi(t) = \xi(s) + \int_s^t (\theta, \xi(\theta))\, d\theta + \int_s^t A(\theta, \xi(\theta))\, dw(\theta),$$

we obtain $\xi(t) = \xi_{s,\xi(s)}(t)$ due to the uniqueness of the solution of the equation. Hence, $\xi(t) = f(\xi(s), \omega)$, where $f(x, \omega)$ is a random function which does not depend on $\mathcal{F}_s$.

Let $f(x, \omega)$ be a bounded measurable random function which does not depend on $\mathcal{F}_s$ and γ be a $\mathcal{F}_s$-measurable random variable. We shall prove that the equality

$$E(f(\gamma, \omega)/\mathcal{F}_s) = \varphi(\gamma), \quad \varphi(x) = Ef(x, \omega) \tag{3.15}$$

holds. To verify it, assume first that $f(x, \omega)$ is of the form $\Sigma\, m_k(x)\, \lambda_k(x)$, where $m_k(x)$ is a nonrandom function. Then, for any $\mathcal{F}_s$-measurable random variable γ_1,

$$Ef(\gamma, \omega)\, \gamma_1 = E\, \Sigma\, m_k(\gamma)\, \lambda_k(\omega)\, \gamma_1 = E\, \varphi(\gamma)\, \gamma_1,$$

since $\varphi(\gamma) = \Sigma\, m_k(x)\, E\, \lambda_k(\omega)$ in this case. Taking into account that any bounded measurable function may be approximated by sums of the above form, we may prove (3.15) in a general case by means of a limiting procedure.

Equation (3.15), in turn, permits us to verify the equality

$$E\left\{j_\Gamma(\xi(t))/\mathcal{F}_s\right\} = E\left\{j_\Gamma(\xi_{s,x}(t))\right\}\Big|_{x=\xi(s)} = P(s, x, t, \Gamma),$$

which implies the assertion of the theorem.

We shall need below some a-priori estimates of stochastic equation solutions.

PROPOSITION 3.3. *Let*

$$\sigma_2^2(A(t,x)) \le C_1 + C_2\, \varphi(x),$$

$$\langle \varphi'(x), a(t,x)\rangle \le \| x \| \left\{\beta + C_2 \| x \|\right\}, \tag{3.16}$$

where $\varphi(x) = \| x \|^2$.

Then the solution $\xi(t)$ *of* (3.2) *satisfies the estimates*

$$E\, \| \xi(t) \|^{2p} \le \| \xi(s) \|^{2p} \exp C(t-s), \quad (p = 1, 2). \tag{3.17}$$

Proof. Ito's formula applies to the function $\varphi(\xi(t)) = \| \xi(t) \|^2$ yields

$$\| \xi(t) \|^2 = \| \xi(s) \|^2 + \int_s^t \langle a(\theta, \xi(\theta)), \varphi'(\xi(\theta))\rangle\, d\theta +$$

$$+ \int_s^t \langle A(\theta, \xi(\theta))\, dw(\theta), \varphi'(\xi(\theta))\rangle +$$

$$+ \frac{1}{2} \int_s^t \sum_{k=1}^{\infty} \varphi''(\xi(\theta)) \left(A(\theta, \xi(\theta))\, e_k,\ A(\theta, \xi(\theta))\, e_k\right) d\theta,$$

which leads to (3.17) due to (3.3) and the Gronwall Lemma. By a similar argument, one may prove the estimate for $p = 2$.

COROLLARY. *If the absolute value of* $\beta < 0$ *is large enough, then the estimate*

$$E \ \| \xi_{s,x}(t) \|^2 \leq K < \infty$$

is valid for any interval $[s, T]$.

Remark 3.1. One may weaken the conditions in Theorem 3.1, in the following way: the statement of Theorem 3.1. is valid if we change the global Lipschitz condition in (3.3) to a local one, that is, if we assume that for any $N > 0$, there exists a constant $C_N > 0$ such that

$$\| a(t,x) - a(t,y) \|_B^2 + \sigma_2^2 (A(t,x) - A(t,y)) \leq C_N \| x - y \|_B^2$$

for any $x, y \in B$ such that $\| x \|_B \leq N$ and $\| y \|_B \leq N$.

This weaker variant of assumptions appears to be admissible due to local dependence of the solutions of (3.1) on coefficients. We shall now prove this property of solutions. Let us mention that it will play an important part in the construction of manifold-valued stochastic processes.

THEOREM 3.3. *Let the coefficients of the stochastic equations*

$$d\,\xi_i(t) = a_i(t, \xi_i(t))\, dt + A_i(t, \xi_i(t))\, dw, \quad i = 1, 2, \tag{3.18}$$

satisfy Theorem 3.1 *conditions. Assume, in addition, that for some* $N < 0$ *and* $\| x \| \leq N$, $a_1(t,x) = a_2(t,x)$ *and* $A_1(t,x) = A_2(t,x)$. *Let* $\xi_1(t)$ *and* $\xi_2(t)$ *solve* (3.18) *respectively for* $i = 1, 2$, *and their initial values coincide* $\xi_1(s) = \xi_2(s) = \xi_0 \in B$. *Then*

$$P\{\tau_2 = \tau_1\} = 1 \quad and \quad P\left\{ \sup_{s \leq \theta \leq \tau_1} \| \xi_1(\theta) - \xi_1(\theta) \| = 0 \right\} 1,$$

where τ_i *is the largest* t *for which* $\sup_{0 \leq s \leq t} \| \xi_i(s) \| \leq N, i = 1, 2.$

Proof. Let $j_1(t)$ be defined by

$$j_1(t) \begin{cases} 1, & \text{if} \quad \sup_{0 \leq s \leq t} \| \xi_1(s) \| \leq N \\ 0, & \text{if} \quad \sup_{0 \leq s \leq t} \| \xi_1(s) \| > N \end{cases},$$

that is, $j_1(t)$ is the characteristic function of the interval $[0, \tau_1]$. Then

$$j_1(t)\,[\xi_1(t) - \xi_2(t)]$$

$$= j_1(t) \int_0^t [a_1(s, \xi_1(s)) - a_2(s, \xi_2(s))]\, ds +$$

$$+ j_1(t) \int_0^t [A_1(s, \xi_1(s)) - A_2(s, \xi_2(s))]\, dw(s)$$

$$= j_1(t) \left\{ \int_0^t [a_1(s, \xi_1(s)) - a_1(s, \xi_2(s))]\, ds \quad + \right.$$

$$+ \int_0^t [a_1(s, \xi_2(s)) - a_2(s, \xi_2(s))]\, \mathrm{d}s +$$

$$+ \int_0^t [A_1(s, \xi_1(s)) - A_1(s, \xi_2(s))]\, \mathrm{d}w(s) +$$

$$+ \int_0^t [A_1(s, \xi_2(s)) - A_2(s, \xi_2(s))]\, \mathrm{d}w \Big\}.$$

Notice that if $j_1(t) = 1$, then

$$a_1(s, \xi_1(s)) = a_2(s, \xi_1(s)) \quad \text{and} \quad A_1(s, \xi_1(s)) = A_2(s, \xi_1(s)).$$

It follows then that

$$j_1(t) = \| \xi_1(t) - \xi_2(t) \|^2$$

$$\le 2 j_1(t) \Big\| \int_0^t [a_2(s, \xi_1(s)) - a_2(s, \xi_2(s))]\, \mathrm{d}s \Big\|^2 +$$

$$+ 2 j_1(t) \Big\| \int_0^t [A_2(s, \xi_1(s)) - A_2(s, \xi_2(s))]\, \mathrm{d}w(s) \Big\|^2.$$

Since $j_1(t) = 1$ implies $j_1(s) = 1$ for $s \le t$, it then follows from stochastic integral properties and (3.3), that

$$E j_1(t) \| \xi_1(t) - \xi_2(t) \|^2$$

$$\le 2t \int_0^t E j_1(s) \| \xi_1(s) - \xi_2(s) \|^2\, \mathrm{d}s +$$

$$+ 2 \int_0^t E j_1(s) \| \xi_1(s) - \xi_2(s) \|^2\, \mathrm{d}s.$$

Gronwall's lemma then implies that

$$E j_1(t) \| \xi_1(t) - \xi_2(t) \|^2 = 0$$

and the path continuity property of $\xi_1(t)$ and $\xi_2(t)$ permits us to obtain the equality

$$P\Big\{ \sup_{0 \le t \le \tau_1} j_1(t) \| \xi_1(t) - \xi_2(t) \|^2 = 0 \Big\} = 1,$$

which implies that $\xi_1(t)$ coincides with $\xi_2(t)$ over the interval $[0, \tau_1]$ with probability 1. Hence $P\{\tau_2 \ge \tau_1\} = 1$. Changing the index 1 to 2 in the proof of the theorem, we verify that $P\{\tau_1 \ge \tau_2\} = 1$.

To conclude this section, we shall derive another estimate for the solution of a stochastic equation which will be crucial in the next chapter.

PROPOSITION 3.4. *Assume, in addition to Theorem* 3.1 *conditions, that*

$$\sigma_2^2(A(t)) \le K < \infty. \tag{3.19}$$

Then

$$P\Big\{ \sup_{s \le \tau \le t} \| \xi_{s,x}(\tau) - x \| > \delta \Big\} \le C\,(s - t)^2 \tag{3.20}$$

for $|s - t| < \alpha$, *where the constants* α *and* C *depend only on* δ, K *and the constants in* (3.3).

Proof. Let us estimate

$$\beta = \Big\{ \sup_{s \le \tau \le t} \| \xi_{s,x}(\tau) - x \| > \delta \Big\}$$

$$= P\Big\{ \sup_{s \le \tau \le t} \Big\| \int_s^\tau a(\theta, \xi(\theta))\, d\theta +$$

$$+ \int_s^\tau A(\theta, \xi(\theta))\, dw(\theta) \Big\| > \delta \Big\} = \alpha_1 + \alpha_2,$$

where

$$\alpha_1 = P\Big\{ \sup_{s \le \tau \le t} \Big\| \int_s^\tau a(\theta, \xi(\theta))\, d\theta \Big\| > \frac{\delta}{2} \Big\},$$

$$\alpha_2 = P\Big\{ \sup_{s \le \tau \le t} \Big\| \int_s^\tau A(\theta, \xi(\theta))\, dw\,\theta \Big\| > \frac{\delta}{2} \Big\}.$$

As follows from Proposition 2.7,

$$\alpha_2 \le C_1\,(s - t)^2.$$

The estimate $\alpha_1 \le C_2\,(s - t)^2$ is a consequence of the Hölder inequality and the estimate $E\,\| \xi(t) \|^2 \le C \le \infty$ which had been proved in proposition 3.1.

Let us mention one more remark.

Remark 3.2. Let the coefficients $a(s, x) \in B$, $A(s, x) \in L_{12}(H, B)$ be random $\mathcal{F}_s$-adapted functions. Theorem 3.1 is still valid in this case if $a \in R(B)$ and $A \in R(L_{12}(H, B))$ for almost all $\omega \in \Omega$ and the constants C_1 and C_2 in (3.3) do not depend on ω.

4. Multiplicative Functionals of Stochastic Processes

4.1. THE SIMPLEST SITUATION. MAIN DEFINITIONS

Consider the set of Borel bounded functions $x(t)$ defined over an interval $[s, T]$ and valued in a certain Banach space B. Let $a(t, x)$ be also a Borel bounded function defined on $[0, T] \times B$ and valued in R^1.

The two-parameter function family

$$y_x(t,s) = \exp\left\{\int_s^t a(\theta, x(\theta))\, d\theta\right\} \quad (0 \le s \le t \le T)$$

possesses an evident multiplicative property

$$y_x(t,s) = y_x(t,\tau) \cdot y_x(\tau,s) \quad (s \le \tau \le t). \tag{4.1}$$

We call this family a multiplicative functional over the considered function set.

The case in which $a(t,x) \in R^1$ is changed to $c(t,x) \in L(B_1)$, where B_1 is a Banach space, may be easily treated in a similar way.

Let $c : [s,T] \times B \to L(B_1)$ and $Y(t,s)$ be a solution of the problem

$$\frac{d\, Y(t,s)}{d\, t} = c(t, x(t))\, Y(t,s) \qquad Y(s,s) = I_{B_1}. \tag{4.2}$$

Assume that some conditions granting the existence and uniqueness of the solution of (4.2) are fulfilled (for instance, let us assume that $c(t,x)$ and $x(t)$ are continuous). It is obvious then that (4.1) holds. In that case, we shall speak about an operator multiplicative functional (o.m.f.).

If the considered function set $\{x(t)\}$ is a set of stochastic process paths, i.e., a function set which supports its distribution, we say that $Y(t,s;\xi(\cdot))$ is multiplicative functional of a stochastic process.

Below, we shall give a definition of this notion which permits us to pass from (4.2) to a linear stochastic equation. It will be shown in the sequel that multiplicative functionals of Markov processes are useful in different cases. In particular, they play an important part in the construction of measures on trajectory spaces as well as evolution families of linear operators acting on function spaces on B.

Let $\xi(t), (s \le t \le T)$ be a random process valued in a Banach space B. Denote by $\mathcal{F}_t$ the minimal σ-algebra with respect to which all random variables $\xi(s), (s \le t)$ are measurable.

Denote $\mathcal{R}$ as the Banach space of bounded random functions defined on the considered probability space $(\Omega, \mathcal{F}, P)$ and valued in the Banach space B_1

$$\mathcal{R} = \left\{\xi : \Omega \to B_1;\ |||\xi||| = \sup_{\omega \in \Omega} \|\xi(\omega)\|_{B_1}\right\}.$$

Let $\mathcal{R}_t$ be the subspace of $\mathcal{R}$ consisting of all the $\mathcal{F}_t$ -measurable elements of $\mathcal{R}$.

DEFINITION . A family of linear mappings

$$Y(t,s) : \mathcal{R}_s \to \mathcal{R}_t$$

satisfying the conditions

$$Y(t,\tau)\, Y(\tau,s) = Y(t,s), \qquad Y(s,s) = I \quad \text{(a.s.)}, \tag{4.3}$$

$$E\, \|Y(t,s)\eta\|^2_{B_1} / \mathcal{F}_s \le C\, \|\eta\|^2_{B_1}, \quad \eta \in \mathcal{R}_s \tag{4.4}$$

is called an operator multiplicative functional of the random process $\xi(t)$.

If, in addition, there exists a random element $\overline{Y}(t,s)$ belonging to $L(B_1)$ and such that

$$Y(t,s)\,\eta = \widetilde{Y}(t,s)\,\eta \qquad (\eta \in \mathcal{R}_s),$$

then the multiplicative functional is said to be uniform.

We shall see below that an o.m.f. is uniform only in some special cases.

EXAMPLE. Let $\xi(t)$ be a random process with a.e. continuous paths and $c : [s, T] \times B \to L(B_1)$ be a continuous bounded function. Then the operator which solves (4.2)

$$\widetilde{Y}(t,s) = \overleftarrow{\exp} \int_s^t c(\tau, \xi(\tau))\, d\tau$$

gives rise to a uniform operator multiplicative functional.

4.2. LINEAR STOCHASTIC EQUATIONS

Let B be a Banach space in the τ_2 class and $\xi(t)$ be a B-valued Markov process.

Consider now a more general situation, treating a linear stochastic differential equation in a Banach space B_1

$$d\eta = c(t, \xi(t))\, \eta(t)\, dt + C(t, \xi(t))\, (\eta(t), dw(t)) \tag{4.5}$$

instead of (4.2). Assume that

$$c : [s, T] \times B \to L(B_1),$$

$$C : [s, T] \times B \to L\big(B_1, L_{12}(H, B_1)\big) \tag{4.6}$$

are continuous functions in the respective norms and H is a Hilbert space with a Wiener process $w(t)$. Denote by $\mathcal{F}_t$ the flow of σ-algebras generated by $w(t)$.

Notice that (4.5) is a stochastic equation with random coefficients which satisfy the assumptions of Section 3, hence due to the results of Section 3, it has a unique solution $\eta(t)$ for $t > s$ if an initial $\mathcal{F}_s$-measurable value η is given.

Consider the mapping

$$S(t, s; \xi(\cdot)) : \eta(s) \to \eta(t). \tag{4.7}$$

This mapping is linear a.e. due to linearity of (4.5) and possesses the evolution property (4.3) as it results from the uniqueness of solution. Finally, the estimate (4.4) is a consequence of (3.12).

In this way, we may state the following assertion.

THEOREM 4.1. *Assume that* (4.6) *are valid; then the linear stochastic equation* (4.5) *defines the operator multiplicative functional* (4.7) *of the Markov process* $\xi(t) \in B$.

Suppose that the Markov process $\xi(t)$ is given as a solution of the stochastic equation

$$d\xi = a(t, \xi(t))\, dt + A(t, \xi(t))\, dw(t) \tag{4.8}$$

as well. If the coefficients $a : [s, T] \times B \to B$, $A : [s, T] \times B \to L_{12}\ (H, B)$ of this equation are Lipschitz mappings, then we may treat (4.5) and (4.8) as a system of stochastic equations and simultaneously find both $\xi(t)$ and $\eta(t)$. This approach appears to be fruitful in many cases.

Let us try to clarify which conditions grant uniformity to the operator multiplicative functional defined by (4.5). In other words, under what conditions (4.5) gives rise to a random linear operator acting on B_1. Put $\eta(s) = h$ and define a random function

$$h \mapsto Y(t, s; \xi(\cdot))\, h$$

which is a.e. linear

$$\alpha_1 Y(t, s; \xi(\cdot))\, h_1 + \alpha_2 Y(t, s; \xi(\cdot))\, h_2$$
$$= Y(t, s; \xi(\cdot))\, (\alpha_1 h_1 + \alpha_2 h_2)\,.$$

Notice that the set of full measure on which this relation is valid, generally speaking, depends on $\alpha_1, \alpha_2, h_1, h_2$. In the general case, we do not know any way to estimate the above mapping in the uniform norm topology.

Nevertheless, we succeed in doing this when considering (4.5) in a Hilbert space H_1, under the following assumptions.

Let $\{e_k\}$ be a orthonormal basis in H. Define the random variables

$$\eta_i(t, s) = y(t, s; \xi(\cdot))\, e_i \quad (t \ge s)\,.$$

Then

$$Y((t, s; \xi(\cdot))\, h = \Sigma\, h_i\, \eta_i(t, s)\,,$$

for $h \in \mathcal{D}$ where

$$\mathcal{D} = \left\{ h : h \sum_{i=1}^{N} h_i\, e_i,\ N < \infty \right\}$$

is a dense set of finite elements. For those elements $h, Y(t, \tau)$ is a linear mapping for each $\omega \in \Omega_1$, where $\Omega_1 \subset \Omega$ is a fixed set of total measure $P(\Omega_1) = 1$.

THEOREM 4.2. *Let*

$$\| c(t, x) \|_{L_{12}(H_1)} \le \gamma,\ \| C(t, x) \|_{L_{22}(H_1 \times H, H_1)} \le \gamma, \quad (0 < \gamma < \infty)\,.$$

Then

$$Y(t, s; \xi(\cdot)) = I + V(t, s; \xi(\cdot)),$$

where

$$V(t, s; \xi(\cdot)) \in L_{12}\ (H_1) \quad \text{a.s.}$$

Proof. Write the expression for η_i in the form

$$\eta_i(t) - e_i = \int_s^t c(\tau, \xi(\tau))\, [\eta_i(\tau) - e_i]\ d\tau +$$

$$+\int_s^t c(\tau,\xi(\tau))e_i \, d\tau +$$

$$+\int_s^t C(\tau,\xi(\tau))(\eta_i(\tau)-e_i, dw) +$$

$$+\int_s^t C(\tau,\xi(\tau))(e_i, dw) . \tag{4.9}$$

A number of simple transformations using stochastic integral estimates lead to the inequality

$$E\,\|\eta_i(t)-e_i\|^2_{H_1} \le$$

$$\le 4\left\{ \gamma^2 (t-s+1) \int_s^t E\,\|\eta_i(\tau)-e_i\|^2_{H_1} d\tau + \right.$$

$$\left. + (t-s)\int_s^t E\,\left\{ \|c(\tau,\xi(\tau))\,e_i\|^2_{H_1} + \sigma_2^2\left(C(\tau,\xi(\tau))\,e_i\right)\right\} d\tau \right\}.$$

Summing over i, we obtain for

$$z_n(t) = \sum_{i=1}^{\infty} E\,\|\eta_i(t)-e_i\|^2_{H_1}$$

the relation

$$z_n(t) \le 4\left[\gamma^2(T+1)\int_\tau^t z_n(s)\, ds + T(T+1)\gamma^2\right].$$

The Gronwall inequality permits us to prove the estimate $z_n(t) \le \text{const}$ with the left-hand side independent of n and, hence, to prove the convergence of the series

$$E\sum_{i=1}^{\infty} \|V(t,s;\xi(\cdot))\,e_i\|^2_{H_1} = \sum_{i=1}^{\infty} \|\eta_i(t)-e_i\|^2_{H_1} < \infty .$$

In particular,

$$P\left\{\sum_{i=1}^{\infty} \|V(t,\tau;\xi(\cdot))\,e_i\|^2_{H_1} < \infty\right\} = 1.$$

4.3. STOCHASTIC EQUATION SOLUTION DEPENDENCE ON PARAMETERS

Consider a family of stochastic equations

$$\xi_\lambda(t) = \varphi_\lambda + \int_s^t a_\lambda(\tau,\xi_\lambda(\tau))\, d\tau + \int_s^t A_\lambda(\tau,\xi_\lambda(\tau))\, dw , \tag{4.10}$$

where $\lambda \in \Lambda$, (Λ is a certain Banach space), $\varphi_\lambda \in B$ and the coefficients $\alpha_\lambda(t,x) \in B$, $A_\lambda(t,x) \in L_{12}(H,B)$ satisfy (3.3) with constants which do not depend on λ. As had been shown in the previous section, the solution $\xi_\lambda(t)$ of (4.10) may be constructed as a fixed point of the contraction Γ_λ which is given by (3.8). The mapping Γ_λ acts from $\mathcal{L}_2([0,T]\times\Omega, B)$ into itself and continuously depends on λ as also its fixed point does, due to the continuous dependence of (4.10) coefficients on the parameter λ.

In this section, we shall describe those conditions on coefficients of the equation

$$\xi_x(t) = x + \int_s^t a\,(\delta, \xi_x(\theta))\; d\theta + \int_s^t A\,(\theta, \xi_x(\theta))\; dw\,, \tag{4.11}$$

which grants to the solution $\xi_x(t)$ a continuous dependence on x in the mean square sense, i.e. as an element of the space $\mathcal{L}_2([0,T]\times\Omega B) = \mathcal{R}$.

Denote by $\mathcal{R}_t^s$ the Banach space of $\mathcal{F}_t^s$ measurable functions $\xi(\tau)$, $(s \le \tau \le t)$ valued in B. Define a norm in this space as follows

$$\langle\langle \xi \rangle\rangle = \{ \sup_{s\le\tau\le t} E\,\|\,\xi(\tau)\,\|^2 \}^{1/2}.$$

Put $\mathcal{R}_t = \mathcal{R}_t^0$. Let $f : [\tau, t]\ \omega\times B \to$ be such that for any $y \in B$, $f(\cdot,\cdot,y) \in \mathcal{R}_t^s$. Given $g(s,\omega,y) \in \mathcal{R}_s$ for each $y \in B$, we obtain that

$$h(t,\omega,y) = f(t,\omega, g(s,\omega\, y)) \tag{4.12}$$

belongs to $\mathcal{R}_t$ under the following assumption

$$h(s,\omega,y) = f(s,\omega, g(s,\omega,y)) \equiv g(s,\omega,y)\,.$$

Suppose now that $g(s,\omega,y)$ and $f(s,\omega,y)$, being elements of $\mathcal{R}_s$ smoothly depend on $y \in B$. We shall prove now that $h(t,\omega,y)$ given by (4.12) smoothly depends on y once again as an $\mathcal{R}_t$ element.

PROPOSITION 4.1. *Let*

$$f \in C_1(\mathcal{R}_s, \mathcal{R}_t^s)\,, \quad g \in C_1(B, \mathcal{R}_s)\,.$$

Then

$$h = f \circ g \in C_1(B, \mathcal{R}_t)$$

and

$$h'_y(t\,\omega,y)\,z = f'_g(t,\omega,g)\,g'_y(s,\omega,y)\,z,\ \ z \in B. \tag{4.13}$$

Proof. First we point out that the estimates

$$\langle\langle f'_g(\cdot,\cdot,g)\,u\rangle\rangle^2 \le C_1\langle\langle u\rangle\rangle^2\,, \qquad u \in \mathcal{R}_s\,, \tag{4.14}$$

$$\langle\langle g'_y(\cdot,\cdot,y)\,z\rangle\rangle^2 \le C_2\langle\langle z\rangle\rangle^2, \qquad z \in B, \tag{4.15}$$

hold under the above conditions and with the constants in (4.14), (4.15) depending on t, s only. To estimate the expression

$$\delta_t(\alpha) = E\,E_s\,\|\,\frac{1}{\alpha}\,[\,f(t,\omega, g(s,\omega, y+\alpha z)) -$$

$$- f(t,\omega,g(s,\omega,y)) - f'_g\,(t,\omega,g(s,\omega,y))\,\times$$

$$\times\, g_y' \,(s, \omega, y)\, z \,\|^2 ,$$

we use properties of f and g, which yield

$$f(g(y+\alpha z)) - f(g(y))$$

$$= \int_0^1 f'(g(y)) + \theta_1 (g(y+\alpha z) - g(y)) \times$$

$$\times\, [g(y+\alpha z) - g(y)]\; d\theta_1 .$$

and, therefore,

$$\delta_t(\alpha) \leq 2 E_s \int_0^1 \| f'(g(y) + \theta_1 [g(y+\alpha z) - g(y)]) -$$

$$- f'(g(y)) \|^2 E_0 \| g'(y) z \|^2 \; d\theta_1 +$$

$$+ 2 E_s \int_0^1 \| f'(g(y) + \theta_1 [g(y+\alpha z) - g(y)]) \|^2 \times$$

$$\times\, E_0 \left\| \frac{g(y+\alpha z) - g(y)}{\alpha} - g'(y) z \right\|^2 d\theta_1 .$$

Further conditional mean properties and estimates (4.14), (4.15) imply (4.13).

Consider now a pair of stochastic equations

$$\xi(t) = x + \int_s^t a(\theta, \xi(\theta))\; d\theta + \int_s^t A(\theta, \xi(\theta))\; dw,$$

$$\eta(t) = h + \int_s^t c(\theta, \xi(\theta))\, \eta(\theta)\; d\theta + \int_s^t C(\theta, \xi(\theta))\, (\eta(\theta), dw), \tag{4.16}$$

where the coefficients $a(s,x)$, $A(s,x)$, $c(s,x)$, $C(s,x)$ are functions valued, respectively, in B, $L_{12}(H,B)$, $L(B_1)$ $L(B_1, L_{12}(H, B_1))$. Assume that all those functions are continuous with respect to $s \in [0,T]$ and C_1-differentiable with respect to $x \in B$. Consider the equations which are derived from (4.13) and (4.16) by differentiation with respect to x along a direction $y \in B$

$$\alpha(t) = y + \int_s^t a_x'(\theta, \xi(\theta))\, \alpha(\theta)\; d\theta + \int_s^t A_x'(\theta, \xi(\theta))\, (\alpha(\theta),\; dw), \tag{4.17}$$

$$\beta(t) = \int_s^t c_x'(\theta, \xi(\theta))\, (\alpha(\theta), \eta(\theta))\, d\theta +$$

$$+ \int_s^t C'_x(\theta, \xi(\theta))\, (\alpha(\theta), \eta(\theta)\, dw) +$$

$$+ \int_s^t c(\theta, \xi(\theta))\, \beta(\theta)\, d\theta + \int_s^t C(\theta, \xi(\theta))\, (\beta(\theta)\, dw). \tag{4.18}$$

Notice that (4.16) – (4.18) may be treated as one linear equation in the new phase space $\tilde{B} = B_1 \oplus (B \otimes B_1)$ with respect to the function $\tilde{\eta}(t) = (\eta_1(t), \eta_2(t))$, where $\eta_1(t) = \beta(t)$ and $\eta_2(t) = \alpha(t) \otimes \eta(t)$.

Introduce a tensor sum operation for the operators

$$\Phi \oplus \Psi = \Phi \otimes I + I \otimes \Psi$$

such that

$$(\Phi \oplus \Psi)(x, y) = \Phi_x \otimes y + x \otimes \Psi y.$$

Using these notations, we write down the stochastic equations for $\eta_1(t)$ and $\eta_2(t)$ in the form

$$\eta_1(t) = \int_s^t c(\theta, \xi(\theta))\, \eta_1(\theta)\, d\theta + \int_s^t C(\theta, \xi(\theta))\, (\eta_1(\theta), dw) +$$

$$+ \int_s^t \hat{c}'_x(\theta, \xi(\theta))\, \eta_2(\theta)\, d\theta + \int_s^t \hat{C}'_x(\theta, \xi(\theta))\, (\eta_2(\theta), dw),$$

$$\eta_2(t) = (y \otimes h) + \int_s^t \big(a'_x(\theta, \xi(\theta)) \oplus c(\theta, \xi(\theta))\big)\, \eta_2(\theta)\, d\theta +$$

$$+ \int_s^t \big(A'_x(\theta, \xi(\theta)) \oplus C(\theta, \xi(\theta))\big)\, (\eta_2(\theta)\, dw),$$

where $\hat{c}$, $\hat{C}$ are linear operators from $B \otimes B_1$ into B_1, given by the relations

$$\hat{c}(x \otimes y) = c(x, y), \quad \hat{C}(x \otimes y) = C(x, y).$$

The resulting pair of equations can be written in the form

$$\gamma(t) = \gamma_0 + \int_s^t q(\theta, \xi(\theta))\, \gamma(\theta)\, d\theta + \int_s^t Q(\theta, \xi(\theta))\, (\gamma(\theta), dw), \tag{4.19}$$

where q and Q are linear and bilinear operators, respectively, given by the following expressions

$$q\begin{pmatrix} y \\ z \otimes y_2 \end{pmatrix} = \begin{pmatrix} c\, y_1 + c'_x(x, y_2) \\ a'_x\, z \otimes y_2 + z \otimes c\, y_2 \end{pmatrix} \tag{4.20}$$

$$= \begin{pmatrix} c & c'_x \\ 0 & a'_x \oplus c \end{pmatrix} \begin{pmatrix} y_1 \\ z \otimes y_2 \end{pmatrix},$$

$$Q\left(\begin{pmatrix} y_1 \\ z \otimes y_2 \end{pmatrix} \upsilon\right) = \begin{pmatrix} C(y_1, \upsilon) + C'_x(z, y_2, \upsilon) \\ A'_x(z, \nu) \otimes y_2 + z \otimes C(y_2, \upsilon) \end{pmatrix} \tag{4.21}$$

$$= \left(\begin{pmatrix} C & C'_x \\ 0 & A'_x \oplus C \end{pmatrix} \begin{pmatrix} y_1 \\ y_2 \otimes z \end{pmatrix} \upsilon\right).$$

As it follows from (4.20) and (4.21),by formal differentiation of the system (4.15) (4.16) with smooth coefficients, we obtain the system (4.15) (4.19) of the same kind as the former ans possessing similar properties. It is obvious, therefore, that the system (4.15) to (4.19) has coefficients with $(k-1)$ derivatives, while the coefficients of (4.15) to (4.16) k bounded derivatives. Hence, we may differentiate (4.15) (4.16) k times.

Let us give a justification of the above formal procedure.

THEOREM 4.3. *Let the coefficients* $a(s,x)$, $A(s,x)$, $c(s,x)$, $C(s,x)$ of (4.15) (4.16) *possess the following properties*

(1) $a(s,x), A(s,x), c(s,x), C(s,x)$ *are functions defined on* $[0,T]\times B$ *which belong, respectively, to* $C_k(B), C_k(B, L_{12}(H,B)), C_k(B, L(B_1))$, $C_k(B, L(B_1 L_{12}(H_1, B_1)))$.

(2) *Both the functions* a, A, c, C *and their derivatives belong, respectively, to* $R(B), R(L_{12}(H,B)), R(L(B_1)), R(L(B_1, L_{12}(H, B_1)))$.

Then there exists a unique up-to stochastic equivalence-solution $\zeta(t)=(\xi_x(t),\eta(t))$ *of the system* (4.15) (4.16) *which possesses* k *bounded Fréchet derivatives with respect to* x

$$\xi_x^{(k)}(t): B\times\ldots\times B\to \mathcal{L}_2([0,t]\times\mathcal{H}_-, B).$$

Proof. Under the above conditions, the existence and uniqueness of the (4.15) (4.16) solution is a consequence of Theorem 3.1. To prove the smoothness of the solution $\zeta(t)$ with respect to the initial value $\xi(s)=x$, we shall use a step-by-step procedure. First notice that the system considered satisfies the contraction map theorem assumption and, thus, its solution may be constructed by means of a successive approximation method.

Let $\zeta_n(t)=(\xi_n(t),\eta_n(t))$ be successive approximations of the solution $\zeta(t)=(\xi(t),\eta(t))$ which satisfy the following relations

$$\xi_n(t)=x+\int_s^t a(\tau,\xi_{n-1}(t))\,d\tau+\int_s^t A(\tau,\xi_{n-1}(t))\,dw(\tau), \tag{4.22}$$

$$\eta_n(t)=h+\int_s^t c(\tau,\xi_n(\tau))\,\eta_n(\tau)\,d\tau+$$

$$+\int_s^t C(\tau,\xi_n(\tau))\,(\eta_n(\tau),dw). \tag{4.23}$$

By differentiating (4.22) (4.23), we obtain the following relations

$$\xi_n'(t)=y+\int_s^t a_x'(\tau,\xi_{n-1}(\tau))\,\xi'_{n-1}(\tau)\,d\tau+$$

$$+\int_s^t A_x'(\tau,\xi_{n-1}(\tau))\,\xi'_{n-1}(\tau),\,dw(\tau),$$

$$\eta_n'(t)=\int_s^t c_x'(\tau,\xi_n(\tau))\,(\xi_n'(\tau),\,\eta_n(\tau))\,d\tau+$$

$$+\int_s^t C'_x(\tau, \xi_n(\tau))(\xi'_n(\tau), \eta_n(\tau), dw(\tau)) +$$

$$+\int_s^t c(\tau, \xi_n(\tau))\, \eta'_n(\tau)\, d\tau +$$

$$+\int_s^t C'(\tau, \xi_n(\tau))(\eta'_n(\tau), dw(\tau))$$

for $\xi'_n(t)$ and $\eta'_n(t)$. It is easy to verify, using stochastic integral estimates, that the processes $\xi'_n(t)$ and $\eta'_n(t)$ converge in a mean square sense to the solution $(\alpha(t), \beta(t))$ of the system (4.17) (4.18). In fact, let

$$\varphi_n(t) = E\, \| \xi'_n(t) - \alpha(t) \|^2 ,$$

$$\psi_n(t) = E\, \| \eta'_n(t) - \beta(t) \|^2 .$$

Using coefficients properties, the Hölder inequality and stochastic integral estimates, one may easily obtain that

$$\varphi_n(t) \le \text{const} \int_s^t \varphi_n(\tau)\, d\tau +$$

$$+ C \int_s^t E\, \| \xi_n(\tau) - \xi(\tau) \|^2\, \| \eta_n(\tau) \|^2\, d\tau$$

$$\le \text{const} \int_s^t E\, \| \xi_n(\tau) - \xi(\tau) \|^2\, \| \eta_n(\tau) \|^2\, d\tau\ \exp C(t-s) .$$

The last inequality is an immediate consequence of the Gronwall lemma.

In a similar way, one may prove the estimate

$$\psi_n(t) \le \text{const} \int_s^t E\, \| \xi_n(\tau) - \xi(\tau) \|^2\, \| \eta_n(\tau) \|^2\, d\tau\ \exp C(t-s) .$$

Finally, the results of Section 3 permit us to conclude that

$$\lim_{n\to\infty} \varphi_n(t) = 0 , \qquad \lim_{n\to\infty} \psi_n(t) = 0 .$$

Thus, we have proved that the process $\zeta(t) = (\xi(t), \eta(t))$ which solves (4.15) (4.16) has mean square derivatives with respect to x and its first derivative $\zeta'(t) = (\xi'(t), \eta'(t))$ solves the system (4.15) to (4.18).

5. Stochastic Flow

Consider a stochastic equation

$$d\xi = a(t, \xi(t))\, dt + A(t, \xi(t))\, dw \tag{5.1}$$

in a Banach space B. Assume that there are fulfilled conditions (see Theorem 3.1) which guarantee the existence of a unique solution $\xi(t, s; \xi_0)$ of (5.1) satisfying the relation

$$\xi(s, s; \xi_0) = \xi_0 .$$

This solution possesses the evolution property

$$\xi(t, s; \xi_0) = \xi(t, \tau; \xi(\tau, s, \xi_0)), \tag{5.2}$$

which holds with probability one. Notice that an exclusive elementary event set over which this property comes not to be satisfied depends in general on ξ_0. In particular, if $\xi_0 = x$ is a nonrandom element, the random variable $\xi(t, s; x)$ is defined on a set $\Omega_0 \subset \Omega$ depending on x. It is difficult to study the map

$$T(t, s) : x \mapsto \xi(t, s; x) \tag{5.3}$$

for given values $\omega \in \Omega$.

To overcome this obstacle, we consider a probability space $\Omega = B \times \mathcal{H}_-$ with measure $P = \nu \times \mu$, where μ is the canonic Gaussian measure in $\mathcal{H}_-$ which defines a Wiener process and ν is an arbitrary initial distribution. Let the initial random variable be defined for $\omega = x \times \kappa$

$$\xi_s(\omega) = x$$

and $P\{\xi_s \in A\} = \nu(A)$. In this case, and exclusive set in $B \times \mathcal{H}_-$ on which (5.2) ceases to be valid, depends on the initial measure ν only. In particular, there exists a set $\tilde{\Omega}_\nu \subset \mathcal{H}_-$, $\mu(\tilde{\Omega}_\nu) = 1$ (which depends on ν) such that for each $\kappa \in \tilde{\Omega}_\nu$ (and thus for almost all Wiener process paths), there exists a set $B_\kappa \subset B$, $\nu(B_\kappa) = 1$ on which the map (5.3) is well defined. In a similar way, the map

$$y \mapsto \xi(t, \tau; y), \qquad (t > \tau > s)$$

is defined on a set $B_\kappa \subset B$, $\nu_\tau(B_\kappa) = 1$, where $\nu_\tau = T(\tau, s)\nu = \nu \circ T^{-1}(\tau, s)$ is the distribution of a random variable $\xi(\tau, s; \xi_0)$. Thus, for $\kappa \in \tilde{\Omega}_\nu$, $x \in B$,

$$\xi(t, s; x) = \xi(t, \tau; \xi(t, s; x)). \tag{5.4}$$

Recall that if $A \equiv 0$, then (5.1) turns out to be an ordinary differential equation which determines an evolution family acting in B and, thus, no special probabilistic assumptions are needed. Moreover, this family is defined for both forward and backward movements in the time direction; hence, the inequality $s < \tau < t$ may not be valid. Let us try to reverse the stochastic map corresponding to the stochastic equation.

Consider a reversed stochastic equation

$$\xi_1(\tau) = \xi_1(\theta) + \int_\theta^\tau a_1(u, \xi_1(u))\, du +$$

$$+ \int_\theta^\tau A_1(u, \xi_1(u))\, dw(u), \qquad \tau < \theta. \tag{5.5}$$

A stochastic process $\xi_1(\tau)$ is said to be a solution of this equation if it is adapted to the reversed flow $\tilde{\mathcal{F}}_\tau = \mathcal{F}_{[\tau,\theta]}$ and satisfies (5.5) with probability 1.

It is evident that all the above results on stochastic equations are valid for (5.5) as well. The simplest way to verify this is to change the variables in (5.5). As a result, (5.5) is changed to

$$\xi(\tau) = \xi(\theta) + \int_\theta^\tau a_1(\theta - u, \xi(u))\, du + \int_\theta^\tau A_1(\theta - u, \xi(u))\, dw_1,$$

$$t = \theta - \tau, \quad w_1(t) = w(\theta) - w(t), \quad \xi(t) = \xi_1(\theta - t),$$

since $w_1(t)$ is a Wiener process as well.

As it was done for (5.1), for (5.5) one may define a map

$$y \mapsto \xi_1(\theta, t; y) \tag{5.6}$$

which has the above sense.

Namely, let $\tilde{\Omega} \subset \mathcal{H}_-$ be a set, depending on the distribution $\tilde{\nu}$ of the random variable $\xi_1(\theta)$ and such that $\mu(\tilde{\Omega}) = 1$. Then for each $\kappa \in \tilde{\Omega}$, the map (5.6) is determined for $y \in \tilde{B}_\kappa$, where $\tilde{\nu}(\tilde{B}_\kappa) = 1$.

Due to the above reason, we conclude the following.

PROPOSITION 5.1. *Let $s \le t \le t_1 \le \tau \le \theta$ and ν be a probability measure on B. There exists a set $\tilde{\Omega} \subset \mathcal{H}_-$ depending on ν (and on the chosen time moments as well) such that $\mu(\tilde{\Omega}) = 1$ and for each $\kappa \in \tilde{\Omega}$, the map*

$$T_\kappa(t, s) : x \mapsto \xi(t, s; x)$$

is defined for all $x \in B_\kappa$. Moreover, $\nu(B_\kappa) = 1$ and the map

$$T_{1\kappa}(t_1, t) : y \mapsto \xi_1(t_1, t; y)$$

is defined for all $y \in \tilde{B}_\kappa$, where $\nu(\tilde{B}_\kappa) = 1$. Finally the composition

$$T_{1\kappa}(t_1, t) \circ T_\kappa(t, \tau)$$

is correctly defined.

As an example we may consider the case $\tau = t_1$. In this case, we will try to choose coefficients a_1 and A_1 in such a way that the relation

$$T_{1\kappa}(t, \tau) \circ T_\kappa(\tau, t) = I \quad (\nu \text{ a.e.})$$

holds for μ almost all $\kappa \in \mathcal{H}_-$.

If $A \equiv 0$, then the stochastic equation turns out to be an ordinary differential equation,

$$T(t, \tau) \circ T(\tau, t) = I$$

and, hence, in order to guarantee that $T_1(t, \tau)$ is equal to $T^{-1}(\tau, t)$, we must set $a_1 = a$.

To investigate the situation in the general case, we shall use a special approximation technique, namely we shall approximate a stochastic equation by a series of ordinary ones.

Let $\varphi(t)$ be a smooth enough even function supported on $(-1, 1)$ and such that

$$\int_{-1}^{1} \varphi(t)\, \mathrm{d}t = 1.$$

Denote

$$\varphi_\varepsilon(t) = \frac{1}{\varepsilon}\, \varphi\left(\frac{t}{\varepsilon}\right).$$

Then

$$\varphi_\varepsilon(t) = 0 \quad \text{for} \quad |t| \geq \varepsilon \quad \text{and} \quad \int_{-\varepsilon}^{\varepsilon} \varphi_\varepsilon(t)\, \mathrm{d}t = 1,$$

and the inequality $\varepsilon\, \varphi_\varepsilon(t) \leq 1$ holds.

Consider the stochastic process

$$w_\varepsilon(t) = \int_{t-\varepsilon}^{t+\varepsilon} \varphi_\varepsilon(t-s)\, w(s)\, \mathrm{d}s.$$

This process is a Gaussian process in H_- and almost all of its paths are as smooth functions as $(\varphi(t)$ are. One may easily verify that

$$w'_\varepsilon(t) = \int_{t-\varepsilon}^{t+\varepsilon} \varphi_\varepsilon(t-s)\, \mathrm{d}w(s).$$

Let us change in (5.1) the Wiener process $w(t)$ to a process

$$w_{\varepsilon, G}(t) = G\, w_\varepsilon(t),$$

where $G \in L_{12}(H_0)$. In this way, (5.1) is reduced to an ordinary differential equation

$$\frac{\mathrm{d}\xi_\varepsilon}{\mathrm{d}t} = a(t, \xi_\varepsilon) + A(t, \xi_\varepsilon)\, G\, w'_\varepsilon(t) \tag{5.7}$$

with random coefficients in the space B. Notice that the above conditions enable us to consider this equation for almost all Wiener process paths.

Let us investigate how a solution of this equation behaves when $\varepsilon \to 0$, $G \to I$. Notice that the random processes $w_\varepsilon(t)$ and $\xi_\varepsilon(t)$ are adapted to $\mathcal{F}_{t+\varepsilon}$. That is why it is convenient to deal with the $\mathcal{F}_t$-adapted random process

$$\eta_\varepsilon(t) = \xi_\varepsilon(t-\varepsilon).$$

The equation

$$\xi_\varepsilon(t-\varepsilon) = \xi_0 + \int_0^{t-\varepsilon} a(s, \xi_\varepsilon(s))\, \mathrm{d}s + \int_0^{t-\varepsilon} A(s, \xi_\varepsilon(s))\, G\, w'_\varepsilon(s)\, \mathrm{d}s \tag{5.8}$$

is an immediate consequence of (5.7). Notice that (5.8) may be transformed to

$$\eta_\varepsilon(t) = \xi_0 + \int_0^t a(s, \eta_\varepsilon(s))\, \mathrm{d}s + \int_0^t A(s, \eta_\varepsilon(s))\, G\, \mathrm{d}w(s) +$$

$$+ \int_0^t b(s, \eta_\varepsilon(s))\, \mathrm{d}s + \theta(\varepsilon, G, t). \tag{5.9}$$

Here b is an additional drift ensuring the approximation of $\xi(t)$ by $\eta_\varepsilon(t)$ and θ is a variable which compensates the difference between the right-hand side of (5.8) and the right-hand side of the equation

$$\eta_\varepsilon(t) = \xi_0 + \int_0^t a(s, \eta_\varepsilon(s))\, ds + \int_0^t A(s, \eta_\varepsilon(s))\, G\, dw(s).$$

The drift b will be computed below. To investigate θ in detail, we write it as a sum

$$\theta = \theta_1 + \theta_2 + \theta_3 + \theta_4,$$

where

$$\theta_1 = \int_0^t [a(\tau, \xi_\varepsilon(\tau)) - a(\tau, \xi_\varepsilon(\tau - \varepsilon))]\, d\tau - $$
$$- \int_0^t (\tau, \xi_\varepsilon(\tau))\, d\tau,$$

$$\theta_2 = \int_0^t [A(\tau, \xi_\varepsilon(\tau)) - A(\tau, \xi_\varepsilon(\tau - 3\varepsilon))]\, G\, w'_\varepsilon(\tau)\, d\tau -$$
$$- \int_0^t (\tau, \xi_\varepsilon(\tau - \varepsilon))\, d\tau,$$

$$\theta_3 = \int_0^t [A(\tau, \xi_\varepsilon(\tau - 3\varepsilon))\, G\, w_\varepsilon(\tau) -$$
$$- \int_0^t [A(\tau, \xi_\varepsilon(\tau - 3\varepsilon))\, G\, w(\tau),$$

$$\theta_4 = \int_0^t [A(\tau, \xi_\varepsilon(\tau - 3\varepsilon)) - A(\tau, \xi_\varepsilon(\tau - \varepsilon))]\, G\, dw(\tau) -$$
$$- \int_{t-\varepsilon}^t A(\tau, \xi_\varepsilon(\tau))\, G\, w'_\varepsilon(\tau)\, d(\tau). \tag{5.10}$$

To calculate the mean square estimates of the above expressions, we shall need some simple inequalities. First, for a random operator $K \in L_{12}(H_0)$ we may obtain, due to (1.16), the following estimate of the fourth moment of the Gaussian vector

$$E\, \| K\, G\, w'_\varepsilon(t) \|^2$$
$$\le E\, \| K \|^2 \left\| \int_{t-\varepsilon}^{t+\varepsilon} \varphi_\varepsilon(t - s)\, G\, dw(s) \right\|^2$$
$$\le (E\, \| K \|^4)^{1/2} \left(E \left\| \int_t^{t+\varepsilon} \varphi_\varepsilon(t - s)\, G\, dw(s) \right\|^4 \right)^{1/2}$$
$$\le \frac{\sqrt{12}\, \sigma_2^2(G)}{\varepsilon}.$$

This inequality yields the estimate

$$E \left\| \int_{t_1}^{t_2} K(t)\, G\, w_\varepsilon'(t)\, dt \right\|^2$$

$$\leq \frac{\text{const}}{\varepsilon}\, \sigma_2^2(G) \int_{t_1}^{t_2} \left(E\, \sigma_2^4(K(t))\right)^{1/2} dt. \tag{5.11}$$

PROPOSITION 5.2. *Let the coefficients of* (5.1) *be bounded*

$$\| a(\tau, x) \| \leq C, \quad \sigma_2^2(A(\tau, x)) \leq C^2.$$

Then

$$E \| \xi_\varepsilon(t_2) - \xi_\varepsilon(t_1) \|^2 \leq \text{const} \frac{(t_2 - t_1)^2}{\varepsilon}\, \sigma_2^2(A). \tag{5.12}$$

Proof. Equation (5.8) and estimates (5.11) result in the following

$$\frac{1}{2} E \| \xi_\varepsilon(t_2) - \xi_\varepsilon(t_1) \|^2$$

$$\leq E \left\| \int_{t_1}^{t_2} (\tau, \xi_\varepsilon(\tau))\, d\tau \right\|^2 +$$

$$+ E \left\| \int_{t_1}^{t_2} A(\tau, \xi_\varepsilon(\tau))\, G\, w_\varepsilon'(\tau)\, d\tau \right\|^2$$

$$\leq C^2 (t_2 - t_1)^2 + \frac{\text{const}}{\varepsilon} (t_2 - t_1)\, \sigma_2^2(G)\, (t_2 - t_1)$$

$$\leq \text{const} \frac{(t_2 - t_1)}{\varepsilon}\, \sigma_2^2(G).$$

COROLLARY. *For an equation with bounded coefficients we have*

$$E \| \theta_1 \|^2 \leq \text{const}\ \varepsilon, \; E \| \theta_4 \|^2 \leq \text{const} + \sigma_2^2(G)\, \varepsilon,$$

where the constant depends only on the coefficient.

The first estimate is evident, while to prove the second one, it is worth noticing that the random function under the first integral sign in θ_4 is nonanticipating and, thus, the Ito integral estimate may be used. The second integral estimate results from (5.10).

PROPOSITION 5.3. *Assume the random function* $S(t)$ *to be* $\mathcal{F}_{t-\varepsilon}$ *adapted. Then*

$$E \left\| \int_0^t S(\tau)\, w_\varepsilon'(\tau)\, d\tau - \int_0^t S(\tau)\, dw(\tau) \right\|^2$$

$$\leq t \left\{ \max_{|\tau - s| < \varepsilon} \varepsilon \, \sigma_2^2 (S(\tau) - S(t)) + 4 \varepsilon \max_{\tau} \sigma_2^2 (S(\tau)) \right\}.$$

Proof. Let us transform the difference

$$\int_0^t S(\tau) w_\varepsilon'(\tau) \, d\tau - \int_0^t S(\tau) \, dw(\tau)$$

$$= \int_0^t S(\tau) \int_{t-\varepsilon}^{t+\varepsilon} \varphi_\varepsilon(\tau - s) \, dw(s) - \int_0^t S(\tau) \, dw(\tau)$$

$$= \int_\varepsilon^{t-\varepsilon} \int_{s-\varepsilon}^{s+\varepsilon} [S(\tau) - S(s)] \, \varphi_\varepsilon(\tau - s) \, d\tau \, dw(s) +$$

$$+ \int_0^\varepsilon \left[\int_0^{s+\varepsilon} S(\tau) \, \varphi_\varepsilon(\tau - s) \, d\tau - S(s) \right] dw(s) +$$

$$+ \int_{t-\varepsilon}^t \left[\int_{s-\varepsilon}^t S(\tau) \, \varphi_\varepsilon(\tau - s) \, d\tau - S(s) \right] dw(s).$$

Considering the first term in the sum, we obtain

$$E \left\| \int_\varepsilon^{t-\varepsilon} \int_{s-\varepsilon}^{s+\varepsilon} [S(\tau) - S(s)] \, \varphi_\varepsilon(\tau - s) \, d\tau \, dw(s) \right\|^2$$

$$\leq \int_\varepsilon^{t-\varepsilon} E \, \sigma_2^2 \left(\int_{s-\varepsilon}^{s+\varepsilon} [S(\tau) - S(s)] \, \varphi_\varepsilon(\tau - s) \, d\tau \right) ds$$

$$\leq \int_\varepsilon^{t-\varepsilon} \int_{s-\varepsilon}^{s+\varepsilon} E \, \sigma_2^2 (S(\tau) - S(s)) \, \varphi_\varepsilon(\tau - s) \, d\tau \, ds$$

$$\leq t \max_{|\tau - s| - \varepsilon} E \, \sigma_2^2 (S(\tau) - S(s)).$$

Similar arguments enable us to estimate the last two terms.

COROLLARY. *Consider Equation* (5.8) *with bounded coefficients. The estimates obtained in the last two propositions permit us to prove the inequality*

$$E \, \| \theta_3 \|^2 \leq \text{const} \left\{ \max_{\substack{\kappa \\ |\tau - s| < \varepsilon}} \sigma_2^2 (A(\tau, \kappa) - A(s, \kappa)) + \varepsilon \, \sigma_2^2 (G) \right\},$$

where the constant depends on equation coefficients only.

PROPOSITION 5.4. *Let* $A \in C_2(B, L_{12}(H, B))$. *Then for*

$$b(\tau, \kappa) = \frac{1}{2} \operatorname{Tr} A'(\tau, \kappa) \, (A(\tau, \kappa) \, G \, \cdot, G \, \cdot),$$

the following estimate

$$E \| \theta_2 \| \le \text{const } \varepsilon$$

holds with a constant which depends only on $\sigma_2^2(G)$ *and the coefficients of Equation* (5.1).

Proof. Using the representation

$$A(\tau, \xi_\varepsilon(\tau)) - A(\tau, \xi_\varepsilon(\tau - 3\varepsilon))$$

$$= \int_0^1 A'\big(\tau, \xi_\varepsilon(\tau - 3\varepsilon) + \theta[\xi_\varepsilon(\tau) - \xi_\varepsilon(\tau - 3\varepsilon)]\big) \times$$

$$\times (\xi_\varepsilon(\tau) - \xi_\varepsilon(\tau - 3\varepsilon))\, d\theta$$

and the equality

$$\xi_\varepsilon(\tau) - \xi_\varepsilon(\tau - 3\varepsilon)$$

$$= \int_{\tau-3\varepsilon}^{\tau} a(s, \xi_\varepsilon(s))\, ds + \int_{\tau-3\varepsilon}^{\tau} A(s, \xi_\varepsilon(s))\, G\, w_\varepsilon'(s)\, ds ,$$

we may transform the variable θ_2 to the following

$$\theta_2 = \int_0^t A'(\tau, \xi_\varepsilon(\tau - 3\varepsilon))\big((\xi_\varepsilon(\tau) - \xi_\varepsilon(\tau - 3\varepsilon), G\, w_\varepsilon'(\tau)\, d\tau +$$

$$+ \int_0^t \int_0^1 [A'(\tau, \xi_\varepsilon(\tau - 3\varepsilon) + \lambda(\xi_\varepsilon(\tau) - \xi_\varepsilon(\tau - 3\varepsilon))) -$$

$$- A'(\tau, \xi_\varepsilon(\tau - 3\varepsilon))]\big(\xi_\varepsilon(\tau) - \xi_\varepsilon(\tau - 3\varepsilon)), G\, w_\varepsilon'(\tau)\big)\, d\lambda\, d\tau -$$

$$- \int_0^t b(\tau, \xi_\varepsilon(\tau - \varepsilon))\, d\tau = \theta_5 + \theta_6 + \theta_7 + \theta_8 ,$$

where

$$\theta_5 = \int_0^t A'(\tau, \xi_\varepsilon(\tau - 3\varepsilon)) \times$$

$$\times A(\tau, \xi_\varepsilon(\tau - 3\varepsilon)) \int_{\tau-\varepsilon}^{\tau} G\, w_\varepsilon'(s)\, ds \times$$

$$\times G\, w_\varepsilon'(\tau)\, d\tau - \int_0^t b(\tau, \xi_\varepsilon(\tau - \varepsilon))\, d\tau,$$

$$\theta_6 = \int_0^t A'(\tau, \xi_\varepsilon(\tau - 3\varepsilon)) \times$$

$$\times \Bigg(\Bigg[\int_{\tau-3\varepsilon}^{\tau} A(s, \xi_\varepsilon(s)) G\, w_\varepsilon'(s)\, ds -$$

$$- \int_{\tau-3\varepsilon}^{\tau} A\,(\tau, \xi_\varepsilon (\tau - 3\,\varepsilon))G\, w'_\varepsilon\,(s)\;\mathrm{d}\,s\,\Big],\; G\; w'_\varepsilon\,(\tau)\Big)\,\mathrm{d}\,\tau,$$

$$\theta_7 = \int_0^t A'\,(\tau\,, \xi_\varepsilon\,(\tau - 3\,\varepsilon))\left(\int_{\tau-3\varepsilon}^{\tau} a\,(s, \xi_\varepsilon\,(s))\;\mathrm{d}\,s,\; G\, w'_\varepsilon\,(\tau)\right)\mathrm{d}\,\tau\,,$$

$$\theta_8 = \int_0^t \int_0^1 (1-\lambda)\; A''\,\big(\tau, \xi_\varepsilon\,(\tau - 3\,\varepsilon) +$$

$$+\,\lambda\,[\xi_\varepsilon\,(\tau) - \xi_\varepsilon\,(\tau - 3\,\varepsilon)]\big) \times$$

$$\times\,\big((\xi_\varepsilon\,(\tau) - \xi_\varepsilon\,(\tau - 3\,\varepsilon)),$$

$$(\xi_\varepsilon\,(\tau) - \xi_\varepsilon\,(\tau - 3\,\varepsilon))\,\mathrm{d}\,\lambda\, G\, w'_\varepsilon\;(\tau)\,\mathrm{d}\,\tau\,. \tag{5.13}$$

The last three terms θ_i, $i = 6, 7, 8$ satisfy the estimate

$$E\;\|\,\theta_i\,\|^2 \le \text{const}\,, \qquad i = 6, 7, 8$$

with a constant depending only on $\sigma_2\,(G)$ and the equation coefficients. This fact is proved in a way similar to the above for θ_2, θ_3 and θ_4, but with more complicated computations. This is due to the necessity of using higher moments estimates for the variables $\xi_\varepsilon\,(\tau) - \xi_\varepsilon\,(\tau - 3\,\varepsilon)$ and $G\,w'_\varepsilon\,(\tau)$. Nevertheless, no major obstacle arises on this way, and we omit the corresponding arguments.

Consider at last the first item in (5.10). Notice that $\xi_\varepsilon\,(\tau - 3\,\varepsilon)$ is $\mathcal{F}'_{\tau-2\varepsilon}$-measurable and $w'_\varepsilon\,(s)$, $(\tau - \varepsilon \le s \le \tau)$ does not depend on this σ-algebra. Due to this,

$$E \int_0^t A'\,(\tau, \xi_\varepsilon\,(\tau - 3\,\varepsilon)) \times$$

$$\times\left(A\,(\tau, \xi_\varepsilon\,(\tau - 3\,\varepsilon))\int_{\tau-\varepsilon}^{\tau} w'_\varepsilon\,(s)\;\mathrm{d}\,s,\; G\; w'_\varepsilon\,(\tau)\right)\mathrm{d}\,\tau$$

$$= E \int_0^t A'\,(\tau, \kappa)\left(A\,(\tau, \kappa)\int_{\tau-\varepsilon}^{\tau} G\; w'_\varepsilon\;\mathrm{d}\,s,\; G\; w'_\varepsilon\,(\tau)\right)\Bigg|_{x=\xi_\varepsilon(\tau-3\varepsilon)} \mathrm{d}\,\tau\,.$$

For the sake of brevity, denote

$$\Phi\,(f, g) = A'\,(\tau, \kappa)\;(A\,(t, \kappa)\,G\,f,\; G\,g)$$

and prove that for $\varepsilon \to 0$

$$E\;\Phi\left(\int_{\tau-\varepsilon}^{\tau} w'_\varepsilon\;\mathrm{d}\,s,\; w'_\varepsilon\,(\tau)\right) \to \frac{1}{2}\;\operatorname{Tr}\Phi\,(\cdot\,,\cdot). \tag{5.14}$$

Indeed, setting $w = \Sigma\, w_k\, e_k$ we obtain

$$E\,\Phi\left(\int_{\tau-\varepsilon}^{\tau} w_\varepsilon'(s)\,\mathrm{d}\,s,\, w_\varepsilon'(\tau)\right)$$

$$= E\,\Phi\left(\int_{\tau-\varepsilon}^{\tau}\int_{s-\varepsilon}^{s+\varepsilon} \varphi_\varepsilon(s-u)\,\mathrm{d}\,w(u)\,\mathrm{d}\,s,\, \int_{t-\varepsilon}^{t+\varepsilon}\varphi_\varepsilon(\tau-u)\,\mathrm{d}\,w(u)\right)$$

$$= \sum_{k,\,j}\Phi(e_k, e_j)\,E\int_{\tau-\varepsilon}^{\tau}\int_{s-\varepsilon}^{s+\varepsilon}\varphi_\varepsilon(s-u)\,\mathrm{d}\,w_k(u)\,\mathrm{d}\,s\,\times$$

$$\times\int_{\tau-\varepsilon}^{\tau+\varepsilon}\varphi_\varepsilon(\tau-u)\,\mathrm{d}\,w_j(u) = \mathrm{Tr}\;\Phi(\cdot,\cdot)\,\tilde{\varphi}_\varepsilon(\tau),$$

where

$$\tilde{\varphi}_\varepsilon(\tau) = \int_{\tau-\varepsilon}^{\tau}\int_{s-\varepsilon}^{s+\varepsilon}\varphi_\varepsilon(s-u)\,\varphi_\varepsilon(\tau-u)\,\mathrm{d}\,u\;\mathrm{d}\,s\,.$$

It easily results that $\lim_{\varepsilon\to 0}\tilde{\varphi}_\varepsilon(\tau) = 1/2$, which yields (5.14). Hence, if

$$b(\tau,\kappa) = \frac{1}{2}\;\mathrm{Tr}\;\Phi(\cdot,\cdot),$$

then

$$E\,\theta_5 \to 0, \qquad \varepsilon\to 0\,.$$

The estimate for the variable $E\,\|\,\theta_5\,\|^2$ may be proved in a similar way.

Now we may prove the following important result.

THEOREM 5.1. *Let*

$$a : [0, T]\times B\to B, \qquad A : [0, T]\times B\to L_{12}\;(H, B),$$

$$A' : [0, T]\,B\to L_{22}\;(B\times H_0, B),$$

be bounded Lipschitz functions and $G\in L_{12}\;(H_0)$.

Then the solution of the ordinary differential equation

$$\xi_\varepsilon'(t) = a\,(t, \xi_\varepsilon(t)) + A\,(t, \xi_\varepsilon(t))\,G\,w_\varepsilon'(t), \qquad \xi_\varepsilon(0) = \xi_0 \tag{5.15}$$

approximates in the mean square sense the solution of the stochastic equation

$$\mathrm{d}\,\xi = \left[a\,(t, \xi(t)) = \frac{1}{2}\;\mathrm{Tr}\,A'(t, \xi(t))\,(A\,(t, \xi(t))\,\cdot\,,\cdot)\right]\;\mathrm{d}\,t\,+$$

$$+\,A\,(t, \xi(t))\;\mathrm{d}\,w\,(t), \qquad \xi(0) = \xi_0\,, \tag{5.16}$$

that is

$$E\;\|\,\xi(t) - \xi_\varepsilon(t)\,\|^2 \to 0$$

if $\varepsilon\to 0$, *and* G *tends to* I *in the sense of strong operator topology.*

Proof. Thanks to the above arguments, the equation solved by

$\eta_\varepsilon(t) = \xi_\varepsilon(t-\varepsilon)$ may be transformed to

$$\eta_\varepsilon(t) = \xi_0 + \int_0^t \tilde{a}(\tau, \eta_\varepsilon(\tau))\, d\tau + \int_0^t A(\tau, \eta_\varepsilon(\tau))\, dw(\tau) + Q(\xi, G, t),$$

where

$$\tilde{a} = a + \frac{1}{2} \operatorname{Tr} A'(\cdot, \cdot), \quad E \| Q(\xi, G, t) \|^2 \le C\varepsilon,$$

and the constant C depends on a, A and G only. For the difference $\xi(t) - \eta_\varepsilon(t)$, due to the last equation and the integral form of (5.16), we have

$$\xi(t) - \eta_\varepsilon(t) = \int_0^t [\tilde{a}(\tau, \xi(\tau)) - \tilde{a}(\tau, \eta_\varepsilon(\tau))]\, d\tau + \int_0^t [A(\tau, \xi(\tau)) - A(\tau, \eta_\varepsilon(\tau))]\, G\, dw(\tau) + \int_0^t A(\tau, \xi(\tau))[I - G]\, dw(\tau) + Q(\xi, G, t).$$

The Lipschitz condition yields the integral inequality

$$E \| \xi(t) - \eta_\varepsilon(t) \|^2 \le 4L^2 C \int_0^t E \| \xi(t) - \eta_\varepsilon(t) \|^2\, d\tau\, (t+1) + 4E \| Q(\xi, G, t) \|^2 + 4 \int_0^t E\, \sigma_2^2 \big(A(\tau, \xi(\tau))(I - G)\big)\, d\tau.$$

Notice now that

$$E\, \sigma_2^2 \big(A(\tau, \xi(\tau))(I - G)\big) = \operatorname{Tr}\, (T - G)\, E A^* A\, (I - G) \tag{5.17}$$

and the right-hand side of (5.17) tends to zero if $G \to I$ in the strong operator topology, since $E A^* A$ is a nonnegative nuclear operator.

The Gronwall inequality yields $E \| \xi(t) - \eta_\varepsilon(t) \|^2 \to 0$ for $\varepsilon \to 0$, $G \to I$ and, at last, the proof of the relation

$$E \| \xi_\varepsilon(t) - \eta_\varepsilon(t) \|^2 \to 0, \quad \varepsilon \to 0$$

is the final point of our arguments.

COROLLARY. *There exists a set of elementary events* $\tilde{\Omega} \subset \Omega$, $P(\tilde{\Omega}) = 1$ *which depends on initial distribution and a sequence* $\varepsilon_k \to 0$, $G_k \to I$ *(in strong topology) such that*

$$\lim_{k \to \infty} \xi_{\varepsilon_k, G_k}(t) = \xi(t) \quad (\omega \in \tilde{\Omega}),$$

where $\xi(t)$ *solves* (5.16) *and* $\xi_{\varepsilon, G}(t)$ *solves* (5.15).

Remark. If we replace the requirement of coefficient boundedness with the estimate

$$\| a(t,x) \|^2 + \sigma_2^2 \, (A(t,x)) \le C_0 + C_1 \, \| x \|^2 , \tag{5.18}$$

we may use the following arguments.

Consider a sequence of bounded functions such that $a_n(t,x) \to a(t,x)$, $A_n(t,x) \to A(t,x)$, $A'_n(t,x) \to A'(t,x)$ and notice that the solutions of stochastic equations with coefficients a_n, A_n, A'_n converge in a mean square sense to the solution of the stochastic equation with coefficients a, A. Let us apply the results of Theorem 3.5. and its corollary. We then obtain a sequence $\xi_n(t)$ of solutions of ordinary differential equations similar to (5.15) for which

$$\xi(t) = \lim \xi_n(t) \qquad \text{(a.e.)} .$$

As a result of the last theorem and of Propositions 5.2 to 5.4, we deduce the main result of this section.

THEOREM 5.2. *Let the coefficients of the equation*

$$d\xi = a(t, \xi(t)) \, dt + A(t, \xi(t)) \, dw$$

meet the requirement of Theorem 3.1 *and*

$$T(t,s) : x \mapsto \xi(t, s; x), \qquad (0 \le s \le t \le T).$$

Let $\tilde{T}(s,t) : y \mapsto \xi_1(t, s; y)$ *where* $\xi_1(t, s; y)$ *is a solution of the reversed equation*

$$d\xi_1(s) = a_1(s, \xi_1(s)) \, ds + A(s, \xi_1(s)) \, dw_1(s) \tag{5.19}$$

with

$$a_1(t,x) = a(t,x) - \operatorname{Tr} A'(t,x) (A(t,x) \cdot , \cdot).$$

Then the stochastic mapping family $T(t,s)$ *is an evolution family*

$$T(t,\tau) \, T(\tau, s) = T(t,s) , \qquad T(t,s) \, \tilde{T}(s,t) = \mathrm{id} \qquad \text{(a.e.)} , \tag{5.20}$$

in the following sense. Given the initial distribution ν, *there exists a set* $\tilde{\Omega} \subset \mathcal{H}_-$ *such that* $P(\tilde{\Omega}) = 1$ *and for each* $\omega \in \tilde{\Omega}$, $T(t, \tau; \omega) \circ x$ *is well defined for* $T(s,\tau) \circ \nu$ *almost all* x. *Moreover, the relation* (5.20) *holds for these* x.

Proof. Due to what has been mentioned in the remark, it is enough to deal with an equation having bounded coefficients. Consider a smoothed equation

$$\xi'_\varepsilon(t) = \tilde{a}(t, \xi_\varepsilon(t)) + A(t, \xi_\varepsilon(t)) \, G \, w'_\varepsilon(t)$$

corresponding to (5.1), where $\tilde{a} = a - 1/2 \operatorname{Tr} A'(A \cdot , \cdot)$. The analogous equation for the reversed stochastic equation (5.15) looks like

$$\xi'_{1\varepsilon}(t) = a_1(t, \xi_{1\varepsilon}(t)) + A(t, \xi_{1\varepsilon}(t)) \, G \, w'_\varepsilon(t), \tag{5.21}$$

where

$$\tilde{a}_1 = a_1 + \frac{1}{2} \text{ Tr } A'(A\,\bullet\,,\,\bullet).$$

Those two equations coincide for $a_1 = a - \text{Tr } A'(A\,\bullet\,,\,\bullet)$ and, thus, give rise to an evolution operator family $T_{\varepsilon,G}\,(t,s)$ $(0 \le t \le s \le T)$ acting in B. Finally, we may show that the above assumptions imply convergence

$$T_{\varepsilon,G}\,(t,s) \circ x \mapsto T\,(t,s) \circ x, \qquad \nu_s \text{ a.e. } (s \le t),$$

$$T_{1\varepsilon,G}\,(s,t)\,y \mapsto \tilde{T}\,(s,t)\,y, \qquad \nu_t \text{ a.e.}$$

for each ω in a certain set of total measure 1.

CHAPTER 4

Stochastic Equations on Smooth Manifolds

In this chapter, we shall give an invariant construction of stochastic differential equations on smooth manifolds equipped with linear connections and investigate their solutions. We shall pay special attention to the case in which a manifold possesses a vector bundle total space structure and a stochastic equation is compatible with this structure. Finally, we shall construct formal differential extensions of stochastic equations and prove that the solutions of the equations on the considered manifold are smooth with respect to the initial values under some assumptions.

1. Stochastic Differentials

1.1. ITO'S BUNDLE

It has been mentioned in Chapter 2 that ordinary differential equations on a manifold are merely vector fields. Nevertheless, if we are going to extend the notion of stochastic differential equation to nonlinear phase space, we immediately find an obstacle. Indeed, a stochastic differential equation cannot be treated as a random vector field since, due to Ito's formula, the transformation rule for stochastic differential under nonlinear transformations of the phase space differs from the transformation law for vector field. That is why we need to construct a special bundle which we call Ito's bundle such that its sections may be identified with stochastic differentials.

Below, we shall need a certain extension of the notion of the vector bundle which will be called a quadratic bundle. It is a bundle whose fibres are still vector spaces while morphisms may be both linear and quadratic mappings.

Let X be a Banach manifold equipped with a connection modeled on B. Since a stochastic equation in a linear space is defined by a pair (a, A), where a is a vector function, and A is an operator function, it is natural to construct a bundle over X whose sections are formed by similar pairs. Consider a bundle

$$\Pi = \tau \oplus L_{12}\ (\theta, \tau_X)$$

which is the Whitney sum of the tangent bundle $\tau_X : TX \to X$ and the bundle $L_{12}\ (\theta, \tau_X)$, where $\theta : X \times H \to X$ is a trivial bundle over X with a Hilbert fibre H. Recall that the typical fibre of $L_{12}\ (\theta, \tau_X)$ is the space $L_{12}\ (H, B)$.

Let φ_1, φ_2 be a pair of morphisms mapping X into local manifolds U_1, U_2

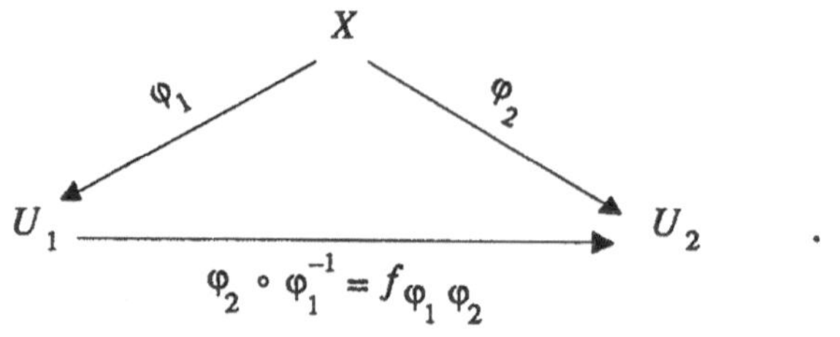

The corresponding trivializations of Π look like the following

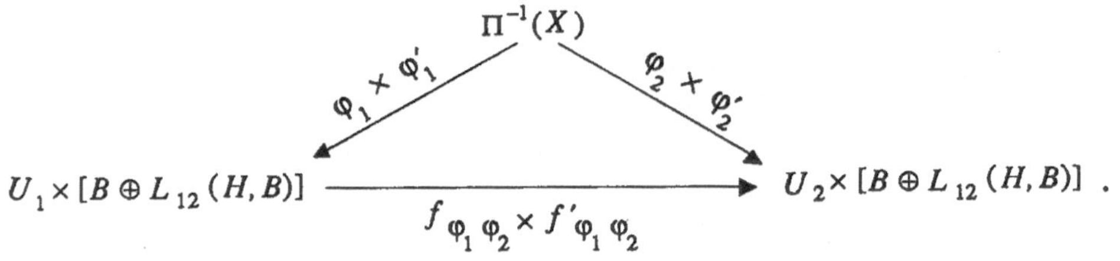

Here $f'_{\varphi_1\varphi_2}$ acts as

$$f'_{\varphi_1\varphi_2} : (a^{\varphi_1}, A^{\varphi_1}) \mapsto \left(f'_{\varphi_1\varphi_2} a^{\varphi_1}, f'_{\varphi_1\varphi_2} A^{\varphi_1}\right), \tag{1.1}$$

which corresponds to tangent bundle trivializations.

Construct a new bundle I over X with the same typical fibre as Π but with changed transition functions

$$I\,(f_{\varphi_1\varphi_2}) : (a^{\varphi_1},\ A^{\varphi_1}) \mapsto$$

$$\mapsto \left(f'_{\varphi_1\varphi_2} a^{\varphi_1} + \frac{1}{2}\ \mathrm{Tr}\ f''_{\varphi_1\varphi_2} (A^{\varphi_1},\ A^{\varphi_1}),\ f'_{\varphi_1\varphi_2} A^{\varphi_1}\right). \tag{1.2}$$

Let us show that the mapping $I : X \to T\,(X)$ which transforms manifold morphisms in accordance with (1.2) is a functor from the Man category into the (quadratic) bundle category.

Consider a local manifold $U \subset B$ with a distinguished point $x \in U$ and put in correspondence to it a linear space

$$I_X : U \to B \oplus L_{12}\ \ (H, B)\,. \tag{1.3}$$

Let the quadratic mapping

$$I_X\,(\varphi) : B_1 \oplus L_{12}\ \ (H, B_1) \to B_2 \oplus L_{12}\ \ (H, B_2)\,,$$

$$I_X\,(\varphi) : (a, A) \mapsto \left(\varphi'\,(x)\,a + \frac{1}{2}\ \mathrm{Tr}\ \varphi''\,(x)\,(A, A),\ \varphi'\,(x)\,A\right) \tag{1.4}$$

corresponds to the morphism $\varphi : U_1 \to U_2$ of local manifolds U_1 and U_2.

PROPOSITION 1.1. *Equations* (1.3), (1.4) *define a functor depending on* $x \in U$ *acting in the local manifold category if the relations*

$$I_x(g \circ f) = I_{f(x)}(g) \circ I_x(f),$$

$$I_x(\mathrm{id}_x) = \mathrm{id}_x$$

hold.

Proof. Let U_1, U_2, U_3 be local manifolds and $f : U_1 \to U_2$, $g : U_2 \to U_3$, $\varphi = g \circ f$. Then

$$\varphi'(x)\,h = g'(f(x))f'(x)\,h, \qquad h \in B_1$$

and

$$\varphi''(x)(h, h_1) = g''(f(x))(f'(x)h, f'(x)h) + g'(f(x))f''(x)\;(h, h_1).$$

Hence, for an orthonormal basis $\{e_k\}$ in H,

$$I_{f(x)}(g)\,I_x(f)(a, A)$$

$$= I_{f(x)}(g)\left(f'(x)\,a + \frac{1}{2}\sum_{k=1}^{\infty} f''(x)(Ae_k, Ae_k), f'(x)A\right)$$

$$= g'(f(x))f'(x)\,a + \frac{1}{2}\sum_{k=1}^{\infty} g'(f(x))f''(x)(Ae_k, Ae_k) +$$

$$+ \frac{1}{2}\sum_{k=1}^{\infty} g''(f(x))(f'(x)Ae_k, f'(x)Ae_k), g'(f(x)f'(x)A)$$

$$= \left(\varphi'(x)\,a + \frac{1}{2}\sum_{k=1}^{\infty}\varphi''(x)(Ae_k, Ae_k), \varphi'(x)A\right) = I_x(g \circ f).$$

Consider now a local manifold U and two trivializations of its tangent bundle

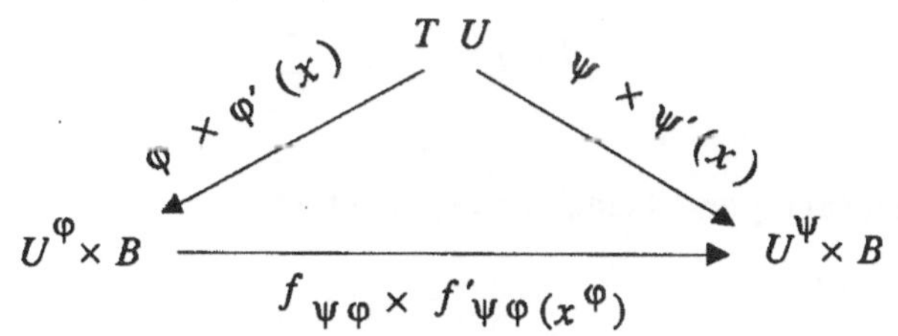

where $x^{\varphi} = \varphi(x)$, $f_{\psi\varphi} = \psi \circ \varphi^{-1}$. The bundle

$$I = I(\tau) : I(U) \to U$$

with trivializations

$$
\begin{array}{ccccc}
 & & I(U) & & \\
 & \swarrow{\scriptstyle \varphi \times I(\varphi)} & & {\scriptstyle \psi \times I(\psi)}\searrow & \\
U^{\varphi} \times I(U^{\varphi}) & & \xrightarrow{\quad f_{\psi\varphi} \circ I(f_{\psi\varphi}) \quad} & & U^{\psi} \times I(U^{\psi})
\end{array}
\tag{1.5}
$$

may be put in correspondence to it. It follows from Proposition 1.1, that for an arbitrary manifold X the above procedure defines a bundle over X with typical fibre $I(B)$. Denote $I(X)$ the total space of this bundle.

We have, hence, defined a functor I from the manifold category into the quadratic bundle category, which we call Ito's functor. The bundle $I: I(X) \to X$ will be called Ito's bundle over the manifold X.

Remark 1.1. Let $U = B$ be a Banach space and $\varphi(x) = Ax$ be a linear map. Then $\varphi'(x) = A$, $\varphi''(x) = 0$ so that the quadratic part of (1.4) vanishes, and this reduces to (1.1). This implies that the restriction of $I(X)$ to the category of linear spaces coincides with a 'direct sum' functor $I(B) = B \oplus L_{12}(H, B)$. Extending this functor to the tangent bundle τ_X in the way described in Chapter 2, we obtain the bundle $I\tau: I(TX) \to X$ which coincides with the Whitney sum

$$I\tau = \tau_X \oplus L_{12}(\theta, \tau_X).$$

Consider a pair of manifolds X_1, X_2 equipped with connections and let $f: X_1 \to X_2$ be a morphism compatible with those connections. Recall that this means that the diagram

$$
\begin{array}{ccc}
T_x X_1 & \xrightarrow{\quad f'(x) \quad} & T_{f(x)} X_2 \\
\Big\downarrow {\scriptstyle \exp_x^{X}} & & \Big\downarrow {\scriptstyle \exp_{f(x)}^{X_2}} \\
X_1 & \xrightarrow{\quad f(x) \quad} & X_2
\end{array}
\tag{1.6}
$$

is commutative. In a local representation, it corresponds to

$$
\begin{aligned}
&\Gamma_{f(x)}^{X_2}(f'(x)\, h_1, f'(x)\, h_2) \\
&\quad = f'(x)\, \Gamma_x^{X_1}(h_1, h_2) - f''(x)(h_1, h_2).
\end{aligned}
\tag{1.7}
$$

Let us treat the mappings in (1.6) as local manifold mappings and apply the functor I to them. In this way, we construct a new commutative diagram

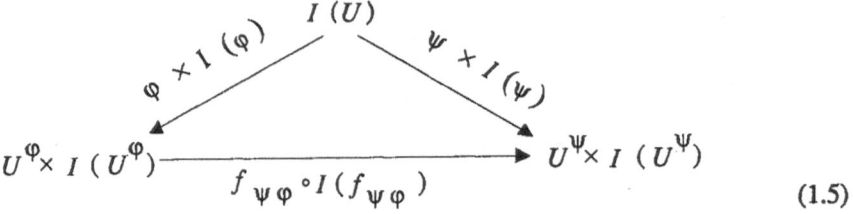

$$\begin{array}{ccc} I\,(T_x X_1) & \xrightarrow{\quad f'(x)\quad} & I\,(T_{f(x)} X_2) \\ {\scriptstyle I\,(\exp_x^{X_1})}\downarrow & & \downarrow{\scriptstyle I\,(\exp_{f(x)}^{X_2})} \\ I\,(X_1) & \xrightarrow{\quad f(x)\quad} & I\,(X_2) \end{array} \qquad . \tag{1.8}$$

As a result, we may state the following assertion.

THEOREM 1.1. *Let a manifold* X *be equipped with a connection and* $\exp^x : T X \to X$ *be the exponential map generated by this connection. Then*

$$I\,(\exp^x) : I\,(T X) \to I\,(X)$$

is a bundle morphism. If, in addition, (U, φ) *is a chart of* X *at a point* $x \in X$, *then the map*

$$I\left(\exp^{\varphi}_{\varphi(x)}\right) : B \oplus L_{12}\quad (H, B) \to B \oplus L_{12}\quad (H, B)$$

is given by

$$I\left(\exp^{\varphi}_{\varphi(x)}\right) : (a^{\varphi}_x\,,\ A^{\varphi}_x) \mapsto$$

$$\mapsto \left(a^{\varphi}_x - \frac{1}{2}\ \mathrm{Tr}\ \Gamma^{\varphi}_{\varphi(x)}\,(A^{\varphi}_x\,,\ A^{\varphi}_x),\ A^{\varphi}_x\right) . \tag{1.9}$$

Proof. It follows from (1.6) that $I\,(\exp) : I\,(T X) \to I\,(X)$ is a bundle morphism which associated Ito's bundle over X to the Whitney sum $\tau_X \oplus L_{12}\ \ (\theta, \tau_X)$.

Assume $\varphi : U \to U^{\Phi} \subset B$ and

$$F^{\varphi}_x = \exp^{\varphi}_{\varphi(x)} = \varphi \circ \exp_x \circ [\varphi'(x)]^{-1}, \quad x \in U.$$

The local mapping $F^{\varphi}_x : B \to U^{\varphi}$ is then such that $y = F_x\,(z)$ may be defined as the solution of the following problem

$$\frac{\mathrm{d}^2\, y\,(z,\,t)}{\mathrm{d}\,t^2} + \Gamma_{y(z,t)}\ \left(\frac{\mathrm{d}\,y\,(z,\,t)}{\mathrm{d}\,t},\ \frac{\mathrm{d}\,y\,(z,\,t)}{\mathrm{d}\,t}\right) = 0,$$

$$\frac{\mathrm{d}\,y\,(z,\,t)}{\mathrm{d}\,t} = z, \quad y\,(z,\,1) = y,\ \ y\,(z,\,0) = x\,.$$

It had been shown in Chapter 2 that these relations easily result in the following

$$F'_x\,(z)\,\big|_{z=0} = \mathrm{id}\,, \quad F''\,(z)\,\big|_{z=0} = -\,\Gamma\,.$$

Substituting the above expressions in (1.4), we obtain that

$$I\left(\exp^{\varphi}_{\varphi(x)}\right) : \left(a^{\varphi}_{x}, A^{\varphi}_{x}\right) \mapsto$$

$$\mapsto \left(a^{\varphi}_{x} - \frac{1}{2} \operatorname{Tr} \Gamma^{\varphi}_{\varphi(x)} \left(A^{\varphi}_{x}, A^{\varphi}_{x}\right), A^{\varphi}_{x}\right) .$$

In the next section we shall see that the functor I transforms stochastic differentials in a way compatible with Ito's formula under stochastic process transformations.

1.2. STOCHASTIC DIFFERENTIALS ON MANIFOLDS

Here we show the links between the previous section constructions and stochastic differential calculus. In what follows, assume B is a Banach space of the class τ_2.

Let $U \subset B$ be a local manifold, $x \in U$, $a \in B$, $A \in L_{12}(H, B)$. Given a triple (x, a, A), we associate to it a collection $\mathcal{R}_x(a, A)$ of B-valued diffusion stochastic processes which do not leave U (with probability I) and coincide (up to stochastic equivalence) with solutions of the stochastic differential equations

$$d\xi = a(s, \xi(s))\, ds + A(s, \xi(s))\, dw, \tag{1.10}$$

satisfying the condition $\xi(0) = x$. Here

$$a : [0, T] \times B \to B, \quad A : [0, T] \times B \to L_{12}(H, B)$$

are smooth functions with support inside U, such that

$$a(0, x) = a, \quad A(0, x) = A \tag{1.11}$$

and $w(t)$ is a given Wiener process in H.

We call $\mathcal{R}_x(a, A)$ the germ of the diffusion processes at the point x defined by the pair (a, A).

Let now $f : U \to U_1$ be a local manifold isomorphism. Then, due to the results of Chapter 3, a random process $\eta(t) = f(\xi(t))$ is a diffusion process which satisfies the equation

$$d\eta = a_1(s, \eta(s))\, ds + A_1(s, \eta(s))\, dw$$

with coefficients

$$a_1 = f'(x)\, a + \frac{1}{2} \operatorname{Tr} f''(x)(A, A), \quad A_1 = f'(x)\, A.$$

Thus, the mapping f brings the germ $\mathcal{R}_x(a, A)$ of the diffusion processes at the point x into a germ $\mathcal{R}^f_x(a, A) = \mathcal{R}_{f(x)}(I(f)(a, A))$ at the point $f(x)$.

Recall that for $f(x) = Kx$, a linear map, we have

$$\mathcal{R}^K_x(a, A) = \mathcal{R}_{Kx}(Ka, KA)$$

for a germ mapping. Thus, the commutative diagram

$$\begin{array}{ccc} U & \xrightarrow{\quad f \quad} & U_1 \\ {\scriptstyle I_x}\downarrow & & \downarrow{\scriptstyle I_{f(x)}} \\ I_x(U) & \xrightarrow{\quad I_x(f) \quad} & I_{f(x)}(U_1) \\ {\scriptstyle \mathcal{R}_x}\downarrow & & \downarrow{\scriptstyle \mathcal{R}_{f(x)}} \\ \mathcal{R}_x(I_x(U)) & \xrightarrow{\quad \tilde{f}_- \quad} & \mathcal{R}_{f(x)}(I_{f(x)}(U_1)) \end{array} \tag{1.12}$$

is valid. Here, $\tilde{f}$ is a mapping of the random process defined by

$$(\tilde{f}\,\xi)(t) = f(\xi(t)) .$$

Diagram (1.12) yields the following assertion.

PROPOSITION 1.2. *The fibre $I_x(U)$ of Ito's bundle $I(X)$ at a point $x \in X$, may be identified with the space $\mathcal{R}_x(I(B))$ of the diffusion process germs. Under this identification, a bundle morphism $\mathcal{R}(f) = f \times \tilde{f}$ corresponds to a manifold morphism $f: U \to U_1$.*

Assume now that $f = \exp$ in (1.12).

Let $\mathcal{R}_{(x,0)}(a, A)$ be a diffusion process germ at a point $y = 0$ in the tangent space $T_x X$. His image under an exponential map

$$\widetilde{\exp}_x\ \mathcal{R}_{(x,0)}(a,\ A) = \mathcal{R}_x\big(I(\exp_x)\big)(a,\ A)$$

gives a stochastic process germ at $x \in X$. It follows from (1.9) that

$$\varphi \circ \widetilde{\exp}_x\ \mathcal{R}_{(x,0)}(a,\ A) = \mathcal{R}_{x\varphi}\left(a^{\varphi} - \frac{1}{2}\ \mathrm{Tr}\ \Gamma^{\varphi}_{\varphi(x)}(A^{\varphi}_x,\ A^{\varphi}_x),\ A^{\varphi}_x\right) \tag{1.13}$$

for a coordinate mapping φ.

We call the germ $\widetilde{\exp}_x\ \mathcal{R}_{(x,0)}(a, A)$ a stochastic differential and, as a consequence, we call Ito's bundle a stochastic differential bundle.

Below, we use the notation

$$\mathcal{R}_{(x,0)}(a,\ A) = a_x\ \mathrm{d}t + A_x\ \mathrm{d}w . \tag{1.14}$$

DEFINITION 1.1. A section $\mathfrak{A}$ of the vector bundle $\Pi = \tau_x \oplus L_{12}(\theta, \tau_x)$ is called Ito's field over the manifold X. Its principal part is exactly a pair (a_x, A_x) where $a_x \in T_x X$, $A_x \in L_{12}(H, T_x X)$ are principal parts of a vector and tensor field, respectively.

An Ito field $\mathfrak{A}$ defines a stochastic differential field

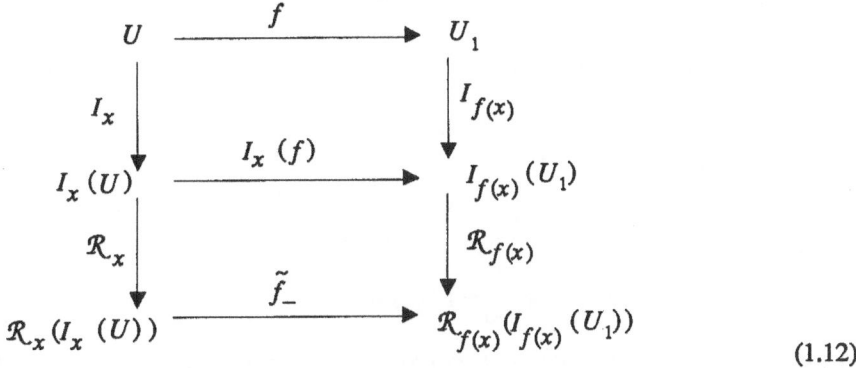

$$\mathcal{R}_x\left(I(\exp_x)(a, A)\right) = \widetilde{\exp}_x\ (a_x\ \mathrm{d}t + A_x\ \mathrm{d}w) \tag{1.15}$$

over X.

The above arguments lead to the following proposition.

PROPOSITION 1.3. *Ito's fields are transformed in a linear way under a manifold transformation* $f: X \to Y$, *i.e.*

$$(a_x, A_x) \mapsto (a^f_{f(x)}, A^f_{f(x)}),$$

where

$$a^f_{f(x)} = f'(x)\, a_x, \quad A^f_{f(x)} = f'(x)\, A_x,$$

and the stochastic differentials are transformed in a natural way

$$\tilde{f} : \widetilde{\exp}^X_x\ (a_x\ \mathrm{d}t + A_x\ \mathrm{d}w) \mapsto$$

$$\mapsto \widetilde{\exp}^Y_{f(x)}\ \left(a^f_{f(x)}\ \mathrm{d}t + A^f_{f(x)}\ \mathrm{d}w\right) \tag{1.16}$$

for f-connected mappings $\exp^X$ and $\exp^Y$.

Remark 1.2. Assume X_1, X_2 to be local manifolds with *f*-connected connections (see (1.6)). In this case, relation (1.16) may be rewritten as

$$\tilde{f} : [a_x - \frac{1}{2}\ \mathrm{Tr}\ \Gamma^X_x (A_x \cdot, A_x \cdot)]\ \mathrm{d}t + A_x\ \mathrm{d}w \mapsto$$

$$\mapsto [a^f_{f(x)} - \frac{1}{2}\ \mathrm{Tr}\ \Gamma^Y_{f(x)} (A^f_{f(x)} \cdot\, A^f_{f(x)} \cdot)]\ \mathrm{d}t +$$

$$+ A^f_{f(x)}\ \mathrm{d}w\,. \tag{1.17}$$

Notice that (1.17) is compatible with Ito's formula.

Remark 1.3. Consider a trivial bundle $\theta : X \times H \to X$ with a trivial connection $\Gamma^\theta = 0$. A tensor field $A_x \in L_{12}\ (H, T_x X)$ gives rise to a bundle mapping

$$X \times H \xrightarrow{\ \mathrm{id} \times A\ } TX$$

and, as a result (see Chapter 2), induces a connection on TX with local coefficients

$$\Gamma^X_x\ (A\, h_1, A\, h_2) = -A'_x\, (h_1,\ A_x\, h_2)\,. \tag{1.18}$$

Using this connection, we may rewrite (1.16) in the form

$$\tilde{f} : [a_x + \frac{1}{2}\ \mathrm{Tr}\ A'_x\ (A_x \cdot, \cdot)]\ \mathrm{d}t + A_x\ \mathrm{d}w \mapsto \tag{1.19}$$

$$\mapsto \left[a^f_{f(x)} - \frac{1}{2}\ \mathrm{Tr}\ [A^f_{f(x)}]'\ (A^f_{f(x)} \cdot, \cdot)\right] \mathrm{d}t + A^f_{f(x)}\ \mathrm{d}w\,.$$

Thus, the expression

$$\alpha^s(a, A; x) = \left[a_x + \frac{1}{2}\ \mathrm{Tr}\ A'_x\ (A_x \cdot, \cdot)\right]\ \mathrm{d}\,t + A_x\ \mathrm{d}\,w \qquad (1.20)$$

is invariant under manifold morphisms. This expression is called a Stratonovich differential. As the above speculations show, the invariance of this expression is of a differential geometric nature.

2. Stochastic Differential Equations on Manifolds

Let X be a smooth manifold.

DEFINITION 2.1. A random process $\xi(t)$ is said to possess a stochastic differential governed by an Ito field

$$\mathrm{d}\,\xi(s) = \widetilde{\exp}_{\xi(s)}\ \mathcal{R}\left(a_{\xi(s)},\ A_{\xi(s)}\right) = \exp\left[a_{\xi(s)}\ \mathrm{d}\,s + A_{\xi(s)}\ \mathrm{d}\,w\right]$$

if the following condition holds.

For $\nu_{\xi(s)}$ almost all $x \in X$, there exists a neighborhood V_x of the point x and a diffusion process belonging to the germ

$$\exp_x\ \mathcal{R}(a_x,\ A_x) = \mathcal{R}_x\left(I\,(\exp_x)\,(a_x,\ A_x)\right),$$

such that

$$P_{s,x}\left\{\xi(t) = \eta_x(t)/\xi(t) \in V_x,\ t \geq s\right\} = 1 \qquad (\nu_{\xi(s)}\ \text{a.e.}),$$

where

$$P_{s,x}(M) = P\,(M/\xi(s) = x), \qquad M \subset \Omega.$$

Next, we consider an Ito field

$$\mathfrak{A}(t) = (a_x(t), A_x(t))$$

depending on time t over an interval $[0, T]$.

An X-valued random process $\xi(t)$ is called an integral process of the Ito field $\mathfrak{A}(t)$ if for any $t \in [0, T]$ its stochastic differential exists and may be written in the form

$$\mathrm{d}\,\xi = \exp_{\xi(t)}\ (a_{\xi(t)}(t)\ \mathrm{d}\,t + A_{\xi(t)}(t)\ \mathrm{d}\,w). \qquad (2.1)$$

Relation (2.1) gives a stochastic differential equation over a manifold X corresponding to Ito's field $\mathfrak{A}(t) = (a_x(t),\ A_x(t))$, while $\xi(t)$ is a solution of this equation.

DEFINITION 2.2. An Ito field, as well as a differential stochastic equation over X corresponding to it, is said to be local if the support of this field is contained in the interior of a certain chart U of the manifold X. $\xi(t)$ is said to be a local solution of Equation (2.1) in a neighborhood U if it coincides inside U with a solution of a local stochastic differential equation which itself coincides with (2.1) inside U.

The following assertion is an immediate consequence of the definition.

PROPOSITION 2.1. *A random process $\xi(t)$ defined over $[0, T]$ is a solution of a stochastic differential equation over this interval iff at any $t \in [0, T]$ for $\nu_{\xi(t)}$ almost all*

$x \in X$, it coincides ($P_{t,x}$ a.e.) *with some local solution of this equation inside a neighborhood* U_x.

COROLLARY. *An almost surely continuous solution of* (2.1) *is defined in a unique way for* $t > t_0$ *by its value* $\xi(t_0)$.

Indeed for almost all x, two solutions $\xi_1(t)$ and $\xi_2(t)$ possessing the property $\xi_1(t_0) = \xi_2(t_0) = x$ coincide for $t > t_0$ with a solution of the local equation up to the moment of exit out of a certain neighborhood. Notice that due to the results of Chapter 3, the local equation solution is unique. Moreover, almost all paths of the solution, due to continuity, remain inside the neighborhood for a nonzero time.

On the other hand, if a solution is not unique, it means that there exists a time $s \in [0, T]$ such that $\xi_1(t)$ and $\xi_2(t)$ do not coincide for $t > s$, which is in contradiction with what was stated above.

It is worth mentioning that in the simplest case of $X = B$ a Banach space, the notions described above coincide with similar notions introduced in Chapter 3. Notice that the exponential mapping in that case may be considered to be trivial, as well as the tangent bundle. Given a coefficient field $(a_x(t), A_x(t))$ we obtain a stochastic equation

$$d\xi = a_{\xi(t)}(t)\, dt + A_{\xi(t)}(t)\, dw \tag{2.2}$$

and the conditions described in Chapter 3 guarantee that there exists a unique solution of the Cauchy problem for (2.2).

As has been shown in Chapter 3, a solution possesses a localization property; namely, the solutions of any two equations coincide if the coefficients of the equations are equal over a certain region. Thus, the solution of (2.2) meets the requirements of Definition 2.2.

Consider a smooth one-to-one mapping F from B into itself. As results from (2.10), the mapping F induces a nonzero connection in B with local coefficients

$$\Gamma_{F(x)}(F'(x)\,y,\ F'(x)\,z) = -F''(x)(y, z), \qquad x, y, z \in B. \tag{2.3}$$

Formulas (1.7) and (2.3) imply, that under this mapping, a stochastic differential transforms like

$$\mathcal{R}_{F(x)}\left(F'(x)\,a_x + \frac{1}{2}\ \mathrm{Tr}\ F''(x)(A_x\,\cdot,\ A_x\,\cdot),\ F'(x)\,A_x\right).$$

Notice that this is just the expression we obtain through Ito's formula, which once again proves that the above formalism is correct.

Consider a measurable space $(X, \mathcal{Z})$ and a flow of σ-algebras $\mathcal{F}_t$, $t \in [0, T]$ in the probability space $(\Omega, \mathcal{F}, P)$.

Define the random mapping $S(t, \tau)$ of X into itself for all $t, \tau \in [0, T]$ by

$$x \mapsto S(t, \tau; x) = S(t, \tau) \circ x$$

and assume that the following holds.

(1) The mapping

$$x \times \omega \mapsto S(t, \tau; x, \omega)$$

is $\mathcal{Z}\times\mathcal{F}$-measurable for all t, τ;

(2) The random variable $S(t, \tau; x)$ is $\mathcal{F}_t$-measurable and does not depend on $\mathcal{F}_t$ for all $t, \tau, x, (t > \tau)$. Moreover, given the $\mathcal{F}_\tau$-measurable random variable ξ, the $\mathcal{F}_t$-measurable random variable $S(t, \tau; \xi)$ is well defined.

(3) The relations

$$S(t, s) \circ S(s, \tau) \circ x = S(t, \tau) \circ x,$$

$$S(t, t) \circ x = x \tag{2.4}$$

hold for $t \geq s \geq \tau$ and $x \in X$.

In this case, we say that the above random mappings form a random evolution family (adapted to the flow $\mathcal{F}_t$).

Given $\tau \in [0, T]$ and an $\mathcal{F}_\tau$-measurable random variable $\xi(\tau)$, consider the stochastic process

$$\xi_{\tau,\xi(\tau)}(t) = S(t, \tau; \xi(\tau)). \tag{2.5}$$

It follows from (2.4) that

$$\xi_{\tau,\xi(\tau)}(t) = \xi_{\tau,\eta}(t), \quad \text{(a.e.)}, \tag{2.6}$$

$$\eta = \xi(\tau).$$

Notice that (2.6) implies that $\xi_{\tau,\xi(\tau)}(t)$ is a Markov process. Indeed, define a transition probability by

$$P(s, x, t, U) = P\{\xi_{s,x}(t) \in U\}.$$

Then, given a bounded $\mathcal{Z}$-measurable function f, we obtain

$$E\{f(\xi_{\tau,\xi(\tau)}(t))/\mathcal{F}_s\}$$

$$= E\{f(\xi_{s,y}(t))\}\big|_{y=\xi_{\tau,\xi(\tau)}(s)}$$

$$= E\{f(\xi_{\tau,\xi(\tau)}(t))/\xi_{\tau,\xi(\tau)}(s)\}$$

and, in particular,

$$P\{\xi_{\tau,\xi(\tau)}(t) \in U/\mathcal{F}_s\} = P(s, \xi_{\tau,\xi(\tau)}(s), t, U).$$

A random evolution family $S(t, \tau; x)$ is said to be generated by a stochastic differential equation if the stochastic process (2.5) satisfies this equation. We say this family is local if it is generated by a local equation.

We now aim at constructing a random evolution family generated by a stochastic differential equation on a manifold.

Consider first the local case.

Clearly, under conditions described in Chapter 3, the stochastic equation (2.2) in the Banach space B generates a random evolution family

$$S(t,\tau) : x \mapsto \xi_{\tau,x}(t),$$

where $\xi_{\tau,x}(t)$ is a solution of the equation

$$\xi_{\tau,x}(t) = x + \int_\tau^t a(\theta, \xi_{\tau,x}(\theta))\, d\theta + \int_\tau^t A(\theta, \xi_{\tau,x}(\theta))\, dw. \tag{2.7}$$

Assume next that (a, A) is a local Ito field over X with support inside a chart (U, φ). Under this assumption, a solution of the equation

$$d\xi = \exp_{\xi(t)}(a_{\xi(t)}(t)\, dt + A_{\xi(t)}(t)\, dw) \tag{2.8}$$

may be constructed in the following way.

The image of the solution $\xi(t)$ under the mapping

$$\varphi : U \to U,\ \varphi(\xi(t)) = \xi^\varphi(t)$$

may be defined as a solution of the equation

$$d\xi^\varphi(t) = a^\varphi(t, \xi^\varphi(t))\, dt + A^\varphi(t, \xi^\varphi(t))\, dw -$$

$$- \frac{1}{2} \operatorname{Tr} \Gamma^\varphi_{\xi^\varphi(t)} \left(A^\varphi(t, \xi^\varphi(t))\,\cdot,\ A^\varphi(t, \xi^\varphi(t))\,\cdot\right) dt, \tag{2.9}$$

with

$$a^\varphi(t, x^\varphi) = \varphi'(x)\, a_x(t), \quad A^\varphi(t, x^\varphi) = \varphi'(x)\, A_x(t).$$

As has been shown in the previous section, this equation is invariant under changes of chart and defines a solution of (2.8) in a correct way. As a result, the corresponding local random evolution family $S(t,\tau)$ also comes to be correctly determined by

$$S^\varphi(t,\tau) \circ \xi^\varphi(\tau) = \varphi(S(t,\tau) \circ \xi(\tau)),$$

where S^φ is generated by (2.9).

To construct a solution of a nonlocal stochastic equation, we need one more assumption concerning the manifold X. Namely, we assume that X is equipped with an atlas of a special kind, a uniform atlas.

By uniform atlas, we mean collection of charts possessing the following properties.

(a) At each point $x \in X$ one may choose a triple of the neighborhoods $U_x^2 \subset U_x^1 \subset U_x$ such $U_x^2 \subset U_y^1$ for $y \in U_x^2$.

(b) The image $\varphi_x(U_x^2) \subset B$ contains a ball of fixed size

$$\varphi_x(U_x^2) \supset S_r^\varphi = \{y \in U_x^\varphi,\ \|y\|_\varphi \le r\}.$$

(c) For each pair of intersecting charts (U_1, φ_1), (U_2, φ_2) in the considered atlas, patching mappings $F_{\varphi_2\varphi_1} = \varphi_2 \circ \varphi_1^{-1}$ possess the properties

$$\sup_x \|F'_{\varphi_2\varphi_1}\|_{\varphi_2} \le C, \qquad \sup_x \|F'_{\varphi_2\varphi_1}\|_{\varphi_1} \le C,$$

where C is a constant which does not depend on φ_1 and φ_2.

It seems that the existence of a uniform atlas is an additional assumption about the manifold which is still rather natural. A uniform atlas exists, for example, on a manifold equipped with a Riemannian structure. The desired charts may be constructed by putting

$$U_x = \exp_x S_r ,$$

where r is the radius of the ball S_r in $T_x X$ and

$$U_x^2 = \exp_x S_r , \quad U_x^1 = \exp_x S_{2r} ,$$

with exp being the exponential map generated by a connection compatible with the Riemannian structure. Nevertheless, in order to make our definitions correct, we must assume that there exists a lower limit of the above radii for which exp is correctly defined. The last assumption is valid for a compact manifold or a group with a uniform metric.

Given a manifold X equipped with a uniform atlas, we may correctly define a bounded vector field.

A vector field ξ is said to be bounded if there exists a constant C such that the principal part $\bar{\xi}$ of the vector field ξ satisfies the estimate

$$\sup \, \| \bar{\xi}_x^{\varphi} \| \leq C$$

for any chart in the uniform atlas.

Since norms in fibres of tensor bundles are defined by a tangent bundle fibre norm, similar definitions are sensible both for tensor fields and for connection coefficients.

In particular, we shall speak of C_k-smooth vector fields, keeping in mind that they are fields having bounded derivatives up to kth order.

Let $(a_x(t), A_x(t))$ be an Ito field and $\lambda_y(x)$, $(y \in X)$ be family of functions on X belonging to the class $C_1(X, B, R^1)$ which satisfies the conditions

$$\lambda_y(x) = 0 \quad \text{if } x \bar{\in} U_y , \quad \lambda_y(x) = 1 \quad \text{if } x \in U_y^1 .$$

Notice that if the model B of the manifold X has a smooth norm, those functions exist. Let us put in correspondence to the field $(a_x(t), A_x(t))$ a set of local fields $\left(a_x^y(t), A_x^y(t)\right)$, where

$$a_x^y(t) = \lambda_y(x)\, a_x(t), \quad A_x^y(t) = \lambda_y(x)\, A_x(t).$$

These fields are said to be localized (with respect to the initial ones). It follows from the results of the previous chapter that there exist local evolution families $S_y(t, \tau)$ generated by localized equations

$$d\,\xi^y(t) = \exp_{\xi^y(t)} \left(a^y_{\xi^y(t)}(t)\, d\,t + A^y_{\xi^y(t)}(t)\, d\,w\right) \tag{2.10}$$

such that

$$S_y(t, \tau) \circ x \in U_y , \quad x \in U_y .$$

It results from (3.3.4) that those random evolution families satisfy the estimate

$$P\left\{\, S_x(t, \tau) \circ x \bar{\in} U_x^2 \,\right\}$$

$$\leq P\left\{ \sup \| \varphi(S_x(t,\tau) \circ x) \|_\varphi > 1 \right\} \leq C(t-\tau)^2 \tag{2.11}$$

for small enough $(t-\tau)$, with a constant C which does not depend on the chart φ in a uniform atlas.

Let $q : \tau\, t_0 \leq t_1 \leq \ldots \leq t_n = t$ be a partition of the interval $[\tau, t]$.

Now construct X-valued random variables in accordance with the relations

$$S^q(t,\tau) \circ x = S_{\xi_{n-1}(t_{n-1})}(t, t_{n-1}) \circ \xi_{n-1}(t_{n-1}), \tag{2.12}$$

$$\xi_k(s) = S_{\xi_{k-1}(t_{k-1})}(t, t_{k-1}) \circ \xi_{k-1}(t_{k-1}),$$

$$\xi(\tau) = x .$$

THEOREM 2.1. *Assume that X is a manifold equipped both with a uniform atlas and a connection. Let both the Ito field (a, A) and the local connection coefficients Γ_x^X be bounded.*

Then there exists a unique (up to stochastic equivalence) random evolution family $S(t,\tau)$ generated by a solution $\xi(t)$ of (2.8) satisfying the condition $\xi(\tau) = x$ and such that

$$\xi(\tau) = S(t,\tau) \circ x = \lim_q S^q(t,\tau) \circ x . \tag{2.13}$$

Here $S^q(t,\tau)$ is given by (2.12) and the limit in a direction formed by an arbitrary countable set of partitions is meant in an a.s. sense.

Moreover, the limiting procedure includes a stabilization, which means that there exists a monotone increasing sequence of events $M_{q_n}, P(M_{q_n}) \to 1$ such that for $\omega \in M_{q_n}, \xi(t) = S^{q_n}(t,\tau) \circ x$.

Proof. Denote $\xi^q(t)$ as the random process given by

$$\xi^q(s) = \xi_k(s), \quad t_{k-1} \leq s < t_k ,$$

where $\xi_k(s)$ is solution of (2.7) for $x = \xi_{k-1}(t_{k-1})$. Let $S_y(s, t_{k-1})$ be an evolution family generated by (2.10) and $z(s) = S_y(s, t_{k-1}) \circ y \in U_y^2$ for all $s \in [t_{k-1}, t_k]$. Then for $t \in [t_{k-1}, t_k]$, we obtain

$$S_y(t, t_{k-1}) \circ y = S_{z(s)}(t, s) \circ z(s) .$$

Indeed, due to the properties of the localized equation solution,

$$S_y(t, t_{k-1}) \circ y = S_y(t, s) \circ S_y(s, t_{k-1}) \circ y .$$

Recall that the stochastic processes $S_y(t,s) \circ z(s)$ and $S_{z(s)}(t,s) \circ z(s)$ are defined by stochastic equations whose coefficients coincide in a region $U_{z(s)}^1$. Moreover, these processes possess a common initial point $z(s)$. Since $U_y^z \subset U_{z(s)}^1$, $S_y(t,s) \circ z(s)$ is inside $U_{z(s)}^1$ and, hence, $S_{z(s)}(t,s) \circ z(s)$ is inside it as well since due to

Theorem 3.3.3.

$$S_y(t, s) \circ z(s) = S_{z(s)}(t, s) \circ z(s) \quad \text{(a.s.)} \,. \tag{2.14}$$

Consider a set of elementary events

$$M^q = \bigcap_k M_k \,,$$

where

$$M_k = \left\{ \omega : S_{\xi_{k-1}(t_{k-1})}(s, t_{k-1}) \circ \xi_{k-1}(t_{k-1}) \in U^2_{\xi_{k-1}(t_{k-1})},\ s \in [t_{k-1}, t_k] \right\}.$$

Formula (2.14) implies that for $\omega \in M^q$ and a partition q', $q' \supset q$ the equality

$$S^{q'}(t, \tau) \circ x = S^q(t, \tau) \circ x$$

holds. This means that a stabilization takes place in (2.12).

Denote $\tilde{M}^q = \{\omega : S^{q'}(t, \tau) \circ x = S^q(t, \tau) \circ x\}$. Those sets $\tilde{M}^q$ form a directed family $\tilde{M}^{q'} \supset \tilde{M}^q$, $q' \supset q$ and , obviously, if $\omega \in M^q$, then

$$S(t, \tau) \circ x = \lim_{q' \supset q} S^{q'}(t, \tau) \circ x = S^q(t, \tau) \circ x \,,$$

since $\tilde{M}^q \supset M^q$. Thus, to prove (2.12) for a set of measure 1, it suffices to show that $P\{\cup M^q\} = 1$, i.e. $\lim_q P\{M^q\} = 1$. Recall that we consider a countable set of partitions q possessing the property $\lim_q \max_k |\Delta_k t| = 0$. Using (2.11), we obtain the estimate

$$\begin{aligned} P(\Omega \backslash M^q) &= \sum_{k=1}^{n} (\Omega \backslash M_k) = \sum_{k=1}^{n} P_{t_{k-1},\, \xi(t_{k-1})} \times \\ &\qquad \times \left\{ \sup \| \varphi (S_{\xi(t_{k-1})}(t, t_{k-1}) \circ \xi(t_{k-1})) \|_{\varphi} \geq 1 \right\} \\ &\leq C \sum_{k=1}^{n} (t - t_{k-1})^2 \leq C (t - \tau) \max_k |\Delta_k t| \,, \end{aligned}$$

which implies that $\lim_q P(M^q) = 1$.

Let $\theta \in q$ be a point in the partition q, then evidently

$$S^q(t, \tau) \circ x = S^q(t, \theta) \circ S^q(\theta, \tau) \circ x \,.$$

Stabilization in (2.12) makes it trivial to pass to the limit and, therefore, to prove that $S(t, \tau)$ possesses the evolution property

$$S(t, \tau) \circ x = S(t, \theta) \circ S(\theta, \tau) \circ x \,.$$

The uniqueness of the solution constructed in this way results from a corollary to Proposition 2.1.

Finally, notice that $S(t, \tau) \circ x$ is a measurable function of the argument x, since it is a superposition $S(t, \tau) \circ x = S^q(t, \tau) \circ x$ of local measurable functions.

Stabilization in (2.12) yields the following relations for the process $\xi_{\tau,x}(t) = S(t,\tau) \circ x$

$$\xi_{\tau,x}(t) = \xi_{s,y}(t), \quad y = \xi_{s,y}(s), \quad \tau < s < t.$$

One may obtain the Markovian property of the process $\xi_{\tau,x}(t)$ from those relations. Indeed, stochastic equation solution properties imply that $y = \xi_{\tau,x}(s)$ is an $\mathcal{F}_s$-measurable function and $\xi_{s,y}(t)$ does not depend on $\mathcal{F}_s$ for fixed y. Consider a measurable bounded function f defined on the manifold $C([\tau,t],X)$ of continuous X-valued functions defined on (τ,t) and denote $f(x_1,\dots,x_n)$, $(x_1,\dots,x_n \in X)$, its cylindrical projection. It follows from the properties of solution of (2.8) described above that

$$E\{f(\xi_{\tau,x}(t_1),\dots,\xi_{\tau,x}(t_n)/\mathcal{F}_s\}$$

$$= \{f(\xi_{\tau,x}(t_1),\dots,\xi_{\tau,x}(t_n)/\mathcal{F}_s\}/_{y=\xi_{\tau,x}(s)},$$

$$\tau < s \le t_1 \le \dots \le t_n \le t,$$

which implies that $\xi_{\tau,x}(t_n)$ is a Markov process.

Remark. The evolution family constructed in Theorem 2.1 satisfies (2.4) over a set of probability measure 1. Generally speaking, this set depends on x.

Nevertheless, arguments similar to those which have been used in the last section of Chapter 3, permits us to show the following. There exists a set $\Omega_0 \subset \Omega$, $P(\Omega_0) = 1$, which depends on an initial distribution $\nu = \nu_\tau$, and for each $\omega \in \Omega_0$ there exists a set $X_0^\omega \subset X$, $\nu_\tau(X_0^\omega) = \nu_\tau(X)$, such that (2.4) is satisfied for that and all $x \in X_0^\omega$.

In this case, we do not need any more to order the values t, s, τ, since an evolution family $S(t,\tau)$ with reverse time ordering may be constructed by means of the reversed stochastic equation

$$d\,\xi_1 = \exp_{\xi_1(t)}\left\{a_{1\xi_1(t)}(t)\, d\,t + A_{1\xi_1(t)}(t)\, d\,w_1\right\}$$

defined on the manifold.

In particular, for $x \in X_0^\omega$, we obtain

$$S(\tau,t) \circ S(t,\tau) \circ x = x. \tag{2.15}$$

Let $f: X \to B_1$ be a smooth, possibly degenerate, mapping from a manifold X into a Banach space B_1 of τ_2 class and $\xi(t)$ be a solution of (2.1).

Ito's formula and the representation (2.12) enable us to derive a relation which determines the random process $\eta(t) = f(\xi(t))$ in the space B_1.

PROPOSITION 2.2. *Let $f \in C_2(X, B, B_1)$ and let $\xi(t)$ solve (2.1) and satisfy the condition $\xi(\tau) = x$. Then the process $\eta(t) = f(\xi(t))$ is given by*

$$\eta(t) = f(x) + \int_\tau^t \nabla_{a_{\xi(s)}(s)} f(\xi(s))\, d\,s +$$

$$+\frac{1}{2}\int_{\tau}^{t}\operatorname{Tr}\left\{\nabla_{A_{\xi(s)}(s)}\nabla_{A_{\xi(s)}(s)}-\nabla_{\nabla_{A_{\xi(s)}(s)}A_{\xi(s)}(s)}\right\}f(\xi(s))\,\mathrm{d}s+$$

$$+\int_{\tau}^{t}\nabla_{A_{\xi(s)}(s)}f(\xi(s))\,\mathrm{d}w\,. \tag{2.16}$$

Proof. Relation 2.12) yields the following representation of $f(\xi(t))$:

$$f(\xi(t))$$

$$=f\left(\sum_{k=1}^{n}S_{\xi(t_{k-1})}(t_k,t_{k-1})\circ x\right)=f(S^q(t,\tau)\circ x)$$

$$=f\left(S_y(t,t_{n-1})\circ y\right)\Big|_{y=\Pi_{k=1}^{n-1}S_{\xi(t_{k-1})}(t_k,t_{k-1})\circ x},$$

and due to Ito's formula, we obtain

$$f(\xi(t))=f\left(\prod_{k=1}^{n}S_{\xi(t_{k-1})}(t_k,t_{k-1})\circ x\right)-$$

$$-\frac{1}{2}\int_{t_{n-1}}^{t}f'(\xi^q(s))\operatorname{Tr}\Gamma^X_{\xi^q(s)}\left(A_{\xi^q(s)}(s);A_{\xi^q(s)}(s)\cdot\right)\mathrm{d}s+$$

$$+\int_{t_{n-1}}^{t}f'(\xi^q(s))\,a_{\xi^q(s)}(s)\,\mathrm{d}s+\int_{t_{n-1}}^{t}f'(\xi^q(s))\,A_{\xi^q(s)}(s)\,\mathrm{d}w+$$

$$+\int_{t_{n-1}}^{t}\operatorname{Tr}f''(\xi^q(s))\left(A_{\xi^q(s)}(s)\cdot A_{\xi^q(s)}(s)\cdot\right)\mathrm{d}s\,.$$

Repeating the above arguments, we derive the relation

$$f(\xi(t))=f(x)+\sum_{k=1}^{n}\left\{\int_{t_{k-1}}^{t_k}f'(\xi^q(s))\,a_{\xi^q(s)}\,\mathrm{d}s\right.-$$

$$-\frac{1}{2}\int_{t_{k-1}}^{t_k}f'(\xi^q(s))\operatorname{Tr}\Gamma^X_{\xi^q(s)}\left(A_{\xi^q(s)}(s)\cdot,A_{\xi^q(s)}(s)\cdot\right)\mathrm{d}s+$$

$$+\int_{t_{k-1}}^{t_k}\operatorname{Tr}f''(\xi^q(s))\left(A_{\xi^q(s)}(s)\cdot,A_{\xi^q(s)}(s)\cdot\right)\mathrm{d}s+$$

$$\left.+\int_{t_{k-1}}^{t_k}f'(\xi^q(s))\,A_{\xi^q(s)}(s)\,\mathrm{d}w\right\}$$

and (2.16) may be derived from it thanks to stabilization.

3. Stochastic Equations in Vector Bundles

3.1. STOCHASTIC EQUATIONS ON A VECTOR BUNDLE TOTAL SPACE

The results of the above sections continue to be valid if the manifold possesses a vector bundle total space structure. It is natural in this case to expect that stochastic equation solutions possess some new additional properties if the equation coefficients are compatible with the bundle structure, i.e., are linear over each fibre.

Let $\pi : \mathcal{E} \to X$ be a vector bundle. Denote Γ_x^{π}, Γ_x^{τ} the local connection coefficients of the bundle π and of the tangent bundle τ_X, respectively. As has been shown in Chapter 2, given $\Gamma^{\pi}, \Gamma^{\tau}$, we may define a connection of the tangent bundle $\tau_{\mathcal{E}} : T\mathcal{E} \to \mathcal{E}$ with local connection coefficients $\Gamma^{\mathcal{E}}$ linear over bundle fibres. In a local trivialization, $\Gamma^{\mathcal{E}}$ is given by

$$\Gamma^{\mathcal{E}}_{\binom{x}{a}} \left(\binom{y}{b}, \binom{z}{c} \right) = \begin{pmatrix} \Gamma_x^{\tau}(y, z) \\ B_x^{\mathcal{E}}(a, y, b, z, c) \end{pmatrix}, \tag{3.1}$$

where

$$\begin{aligned} B_x^{\mathcal{E}} & (a, y, b, z, c) \\ & = \Gamma_x^{\pi}(y, b) - \Gamma_x^{\pi}(z, c) + \\ & \quad + (\Gamma_x^{\pi})'(y, z, a) + \Gamma_x^{\pi}(y, \Gamma_x^{\pi}(z, a)) - \\ & \quad - \Gamma_x^{\pi}(\Gamma_x^{\pi}(y, z), a) . \end{aligned} \tag{3.2}$$

Denote by $\exp^{\mathcal{E}}$ the exponential map generated by this connection.

A stochastic equation on the manifold $\mathcal{E}$ with respect to the process $\zeta(t)$ is written in the form

$$d\,\zeta(t) = \exp^{\mathcal{E}}_{\zeta(t)} \left(c_{\zeta(t)}(t)\, d\,t + C_{\zeta(t)}(t)\, d\,w \right), \tag{3.3}$$

where (c, C) is a vector Ito field on $\mathcal{E}$

$$c : \mathcal{E} \to T\,\mathcal{E}, \quad C : \mathcal{E} \to L(H_X, T\,\mathcal{E}) .$$

Under some conditions, resulting from the preceding section, there exists a random evolution family $V(t, \tau)$ of manifold $\mathcal{E}$ mappings generated by a solution of (3.3). To investigate the structure of this family, let us first consider a local manifold $\mathcal{E} = B \times F$.

Suppose that the principal part of the fields c, C has the following form

$$c_{\binom{x}{y}}(t) = a_x(t) \times b_x(t)\, y,$$

$$C_{\binom{x}{y}}(t) = A_x(t) \times B_x(t)\, y,$$

where

$$a_x(t) \in B, \quad b_x(t)\, y \in F, \quad A_x(t) \in L_{12}(H, B),$$

$B_X(t) \in L(F, L_{12}(H, F))$.

Notice that

$$c_{\binom{x}{y}} = a_X \times b_X \, y$$

transforms according to the rule

$$c_X(t) \mapsto (\varphi'(x)\, a_X(t),\ \Phi(x)\, b_X(t)\, y + \Phi'(x)\left(a_X(t), y\right))$$

under the action of a local manifold morphism $\Psi = \varphi \times \Phi : (x, y) \mapsto (\varphi(x), \Phi(x)\, y)$. This rule shows that the linear dependence of b on y is invariant under transformation of the bundle π. Applying the connection map K^{π} to c, we obtain

$$\ell_{\binom{x}{y}} = b_X\, y + \Gamma^{\pi}_{x}(a_X, y),$$

which determines a section of π depending linearly on y for each $x \in X$. That is why we may write

$$\ell_{\binom{x}{y}} = b_X\, y,$$

where b_X is the principal part of a section of the bundle $L(\pi) : U \times L(F) \to U$ and, hence, the component c of Ito's field (c, C) depending linearly on y, may be written as

$$c_{\binom{x}{y}} = b_X\, y - \Gamma^{\pi}_{x}(a, y).$$

Similar arguments lead to the construction of the component C. Finally, changing local bundles to global ones, we obtain the following result.

PROPOSITION 3.1. *An Ito field (c, C) on a vector bundle $\pi : \mathcal{E} \to X$ total space, compatible with the linear structure on the fibre, is defined by the fields a_X, b_X, A_X, B_X over X, which represents sections of the bundles $\tau_X, L(\pi), L_{12}(\theta, \tau_X), L_{12}(\theta, L(\pi))$ respectively. For a local bundle $\mathcal{E} = B \times F$*

$$c_{\binom{x}{y}} = \left(a_X,\ b_X\, y - \Gamma^{\pi}_{x}(a_X, y)\right),$$

$$C_{\binom{x}{y}} = \left(A_X,\ B_X\, y - \Gamma^{\pi}_{x}(A_X, y)\right). \tag{3.4}$$

The above formulas enable us to derive a local stochastic equation with respect to $\zeta(t) = (\xi(t), \eta(t))$ which is invariant under bundle morphisms and to write it in the form

$$d\,\xi(t) = a_{\xi(t)}(t)\, dt - \frac{1}{2}\ \mathrm{Tr}\ \Gamma^{\tau}_{\xi(t)}\left(A_{\xi(t)}(t)\,\cdot,\ A_{\xi(t)}(t)\,\cdot\right) dt + {} + A_{\xi(t)}(t)\, dw, \tag{3.5}$$

$$d\eta(t) = [b_{\xi(t)}(t)\eta(t) - \Gamma^{\pi}_{\xi(t)}(a_{\xi(t)}(t), \eta(t))]\, dt -$$

$$- \operatorname{Tr} \Gamma^{\pi}_{\xi(t)} \left(A_{\xi(t)}(t)\cdot, B_{\xi(t)}(t)(\eta(t), \cdot) -\right.$$

$$\left.- \Gamma^{\pi}_{\xi(t)} (A_{\xi(t)}(t)\cdot, \eta(t))\right) dt -$$

$$- \frac{1}{2} \operatorname{Tr} \left\{ (\Gamma^{\pi}_{\xi(t)})' (A_{\xi(t)}(t)\cdot, A_{\xi(t)}(t)\cdot, \eta(t)) + \right.$$

$$+ \Gamma^{\pi}_{\xi(t)} \left(A_{\xi(t)}(t)\cdot, \Gamma^{\pi}_{\xi(t)} (A_{\xi(t)}(t)\cdot, \eta(t))\right) -$$

$$\left. - \Gamma^{\pi}_{\xi(t)} \left(\Gamma^{\tau}_{\xi(t)} (A_{\xi(t)}(t)\cdot, A_{\xi(t)}(t)\cdot)\right)\right\} +$$

$$+ B_{\xi(t)}(t)(\eta(t), dw) - \Gamma^{\tau}_{\xi(t)} (A_{\xi(t)}(t)\, dw, \eta(t))$$

$$= N_{\xi(t)}(t)\eta(t)\, dt + M_{\xi(t)}(t)(\eta(t), dw). \tag{3.6}$$

The first equation is a stochastic equation on the manifold X which describes a random process $\xi(t)$. By substituting $\xi(t)$ in the second equation, we realize that is a linear (with respect to $\eta(t)$) stochastic equation on the fibre F corresponding to a chosen trivialization. Clearly, it defines a multiplicative operator functional $V(t, \tau; \xi(\cdot))$ of the Markov process $\xi(t)$.

Hence, a solution of a stochastic equation on a local vector bundle gives rise to a random evolution family of total space mapping $Y(t, \tau) : \zeta(\tau) \to \zeta(t)$ which possesses a.s. the following properties.

(1) A projection of $Y(t, \tau)$ onto the base coincides with the correspondent family $S(t, \tau)$ acting on the base and generated by the projection

$$\pi \circ Y(t, \tau) = S(t, \tau) \circ \pi. \tag{3.7}$$

(2) The mapping Y is linear on each fibre, i.e., for $\pi z_1 = \pi z_2$, $\alpha_1, \alpha_2 \in \mathcal{E}$, $z_1, z_2 \in R^1$ one has

$$Y(t, \tau)(\alpha_1 z_1 + \alpha_2 z_2) = \alpha_1 Y(t, \tau) z_1 + \alpha_2 Y(t, \tau) z_2. \tag{3.8}$$

DEFINITION 3.1. A random evolution family of mappings of the total space $\mathcal{E}$ of a vector bundle π which satisfies (3.7), (3.8) is called a π-multiplicative functional of the Markov process $\xi(t) = \pi\zeta(t)$ on the base X.

It follows from the definition that a solution of a local stochastic equation on a vector bundle with coefficients linear over its fibres generates a multiplicative functional of its own projection onto the base.

Notice that while dealing with a trivialization $\mathcal{E} = X \times F$ of the bundle π, we have the splitting

$$Y(t,\tau) = S(t,\tau) \otimes V(t,\tau;\xi(\cdot)) ,$$

where $V(t,\tau;\xi(\cdot)) \in L(F)$ is an operator multiplicative functional of the process $\xi(t)$ in the sense of Chapter 3.

Let us pass on to treating a nonlocal stochastic equation in the bundle π. Assume that the base X of the bundle has a uniform atlas and the equation coefficients are bounded with respect to this atlas as well as the connection coefficients Γ^{τ} and Γ^{π}. It must be mentioned that, formally, we cannot apply Theorem 2.1 in the present situation, since stochastic equation coefficients are linear over fibres and, hence, unbounded.

Nevertheless, it is easy to see that in the proof of Theorem 2.1, we needed the coefficients to be bounded only once. Namely, coefficient boundedness had been used in proving the estimate for the measure of the set M^q, which grants the process to be in the domain of a single chart. Since bundle charts are of a cylindrical nature, that is $\pi^{-1}(U) = U \times F$, it is the base projection of the process that governs its being located inside the chart. Hence, Theorem 2.1 is valid in the considered situation as well.

THEOREM 3.1. *Assume that the conditions of Theorem* 2.1 *hold and, moreover, the fields* b_X, B_X, Γ^{π}_X *are bounded in the uniform atlas of the base* X *of the bundle* π. *Then the stochastic equation* (3.3) *generates a random evolution family* $Y(t,\tau)$ *in the total space* $\mathcal{E}$ *of the bundle* π, *such that*

$$\zeta(t) = Y(t,\tau)\,\zeta$$

gives the unique solution of (3.1) *corresponding to an* $\mathcal{F}_\tau$*-measurable initial value* ζ. *The family* $Y(t,\tau)$ *may be represented as a local random evolution family composition*

$$Y(t,\tau) = \lim_q \; Y^q(t,\tau)\cdot\zeta, \tag{3.9}$$

where

$$Y^q(t,\tau)\,\zeta = \left(S^q(t,\tau)\circ x,\; V^q(t,\tau;\xi(\cdot))\,y\right),$$

$$V^q(t,\tau;\xi(\cdot))\,y = \prod_{k=1}^{n} V\left(t_k, t_{k-1};\, S_{\xi(t_{k-1})}(\cdot,\cdot)\,\xi_{\xi(t_{k-1})}\right), \tag{3.10}$$

and a stabilization takes place in (3.9) *in the same sense as in Theorem* 2.1.

The family $Y(t,\tau)$ *generates a* π*-multiplicative functional of the process* $\xi(t) = \pi\,\zeta(t)$ *on the base* X.

Proof. The last assertion is the only statement which is not yet proved. Indeed, due to the above remarks, all other theorem assertions may be drawn out of Theorem 2.1 by reformulating its statement under the corresponding conditions.

Notice that (3.7) and (3.8), included in the multiplicative functional definition, are surely valid for the local family composition V^q and are not violated by passing to the limit in (3.9).

Remark 3.1. Consider a normed bundle π, that is a bundle π, with a fibre $\pi^{-1}(x)$ which admits a norm $\|\cdot\|_x$, such that trivializing mappings connected with a uniform atlas of the base, should be isometric mappings. In particular, the tangent bundle of a Riemannian manifold with a uniform atlas generated by exponential images of sphere neighborhoods (see Section 2) possesses this property. Given a normed bundle π, consider the Banach space $\mathcal{H}_t$ of $\mathcal{F}_t$-measurable random variables $\eta(t) \in \pi^{-1}(\xi(t))$ with norm

$$|||\eta|||_t^2 = E\ \|\eta(t)\|^2_{\xi(t)}.$$

In this case the π-multiplicative functional $Y(t, \tau)$ gives a linear bounded operator

$$Y(t, \tau) : \mathcal{H}_\tau \to \mathcal{H}_t, \qquad |||Y\eta|||_t \le C\ |||\eta|||_\tau, \qquad \eta \in \mathcal{H}_\tau, \tag{3.11}$$

where $C = \exp\{\alpha(t-\tau)\}$ for a certain $\alpha > 0$.

Such an estimate has been proved in Chapter 3 for multiplicative operator functionals of processes in a linear space. Since trivializing mappings are isometric ones, this property may be extended to local processes on bundles. Finally, due to Theorem 3.1, it comes to be valid for nonlocal bundles as well due to the stabilization which takes place in (3.9).

An important situation is the following one. Let a multiplicative operator functional be generated by a random bounded operator acting from one fibre into another

$$Y(t\,\tau, \xi(\cdot)) : \pi^{-1}(\xi(\tau)) \to \pi^{-1}(\xi(t)).$$

Such a mapping may be constructed by the solution of Equation (3.6) under the assumption that its coefficients are Hilbert–Schmidt operators and that a typical fibre of the bundle is a Hilbert space. In particular, connection coefficients must be Hilbert–Schmidt operators as well. To grant this property to be global, it suffices to assume that patching mappings $\Phi_{\varphi\psi}(x)$ satisfy the condition

$$\Phi'_{\varphi\psi}(x) \in L_{12}(H). \tag{3.12}$$

Hence, those mappings can be written in the form

$$\Phi_{\varphi\psi}(x) = C_{\varphi\psi} + Z_{\varphi\psi}(x), \tag{3.13}$$

where $C_{\varphi\psi}$ is a constant invertible operator and

$$Z_{\varphi\psi}(x) \in L_{12}(H), \qquad x \in X.$$

Remark 3.2. The arguments mentioned in the remark Theorem 2.1 apply as well to an evolution mapping family $Y(t, \tau)$. In particular, one may deal with invertible mappings of bundle fibres generated by multiplicative functionals.

Consider a bundle satisfying (3.13). Let us fix an initial distribution ν of the projection $\xi(\tau) = \pi\zeta(\tau)$ and take the identity operator $I_{\pi^{-1}(\xi(\tau))}$ as the initial value of the multiplicative functional $Y(t, \tau)$. In this way, we obtain a set $\Omega_0 \subset \Omega$ depending on ν and satisfying $P(\Omega_0) = 1$. Next, given $\omega \in \Omega_0$, we obtain a set of initial values

X_0^ω, $\nu(X_0^\omega) = 1$ such that

$$Y_x(t, \tau; \omega)\, Y^{-1}_{\xi(t)}(\tau, t, \omega) = I_{\pi^{-1}(x)}, \tag{3.14}$$

where Y_x is an evolution operator corresponding to the initial value $\xi(\tau) = x$.

Example. Let $\mathcal{E}$ be the total space of the bundle $\pi : \mathcal{E} \to X$. Consider a random process $\zeta(t)$ generated by a solution of the following stochastic equation

$$d\zeta(t) = \exp^{\mathcal{E}}_{\zeta(t)}(\alpha_{\zeta(t)}(t)\, dt + \beta_{\zeta(t)}(t)\, dw), \tag{3.15}$$

where the principal part of the Ito's field (α, β) is given by

$$\alpha_{\binom{x}{y}} = (a_x, -\Gamma^\pi_x(a_x, y)),$$

$$\beta_{\binom{x}{y}} = (A_x, -\Gamma^\pi_x(A_x, y)).$$

In a local trivialization, (3.15) looks like

$$d\xi = a_{\xi(t)}(t)\, dt - \frac{1}{2}\operatorname{Tr}\Gamma^\tau_{\xi(t)}(A_{\xi(t)}(t)\cdot, A_{\xi(t)}(t)\cdot)\, dt + \\ + A_{\xi(t)}(t)\, dw, \tag{3.16}$$

$$d\eta = -\Gamma^\tau_{\xi(t)}(a_{\xi(t)}(t), \eta(t))\, dt + \\ + \operatorname{Tr}\Gamma^\tau_{\xi(t)}(A_{\xi(t)}(t)\cdot, \Gamma^\tau_{\xi(t)}(A_{\xi(t)}(t)\cdot, \eta(t)))\, dt - \\ - \frac{1}{2} B_{\mathcal{E}}(t)\eta(t)\, dt - \Gamma^\tau_{\xi(t)}(A_{\xi(t)}(t)\, dw, \eta(t)), \tag{3.17}$$

where $B_{\mathcal{E}}(t)$ is given by (3.2) for $b = c = 0$.

Equation (3.15), as has been proved above, generates a random evolution family of mappings

$$\tilde{Y}(t, \tau) : \pi^{-1}(\xi(\tau)) \to \pi^{-1}(\xi(t)).$$

We call the mapping $\tilde{Y}(t, \tau)$ a stochastic parallel displacement of a bundle section along a diffusion process path.

To explain this name, recall that, as has been shown in Chapter 2, a parallel displacement along a smooth curve generates an evolution family as well. Besides, a curve satisfying (3.17) may be constructed by the following procedure. Approximate a diffusion process $\xi(t)$ path by a curve which consists of parts of geodesic curves. In other words, given a partition $q : \tau = t_0 \le t_1 \le \ldots t_n = t$ of the interval $[\tau, t]$, let us connect points $\xi(t_{k-1})$ and $\xi(t_k)$ by geodesic curves $\gamma(s)$, $s \in [t_{k-1}, t_k]$. Next, construct a section of the bundle π consisting of sections $\beta_k(s)$ obtained by parallel translations of $\beta_{k-1}(s)$ along geodesic $\gamma(s)$.

Letting $n \to \infty$ in the partition and keeping in mind the diffusion process properties, one may prove that the limit section $\beta(t) = P - \lim \beta_q(t)$ satisfied (3.17), while the process $\zeta(t) = (\xi(t), \beta(t))$ satisfies (3.15) in the case $\beta_q(s) = \beta_k(s)$ for $s \in [t_{k-1}, t_k]$.

3.2. SMOOTH PROPERTIES OF STOCHASTIC EQUATION SOLUTIONS

Consider an Ito field (in general depending on time) $\mathfrak{A}(t) = (a(t), A(t))$ on a manifold X equipped with a connection and a stochastic equation

$$\mathrm{d}\,\xi = \exp_{\xi(t)}(a_{\xi(t)}(t)\,\mathrm{d}t + A_{\xi(t)}(t)\,\mathrm{d}w)\,. \tag{3.18}$$

Formally applying the functor T to $\mathfrak{A}$ gives rise to an Ito field $T\mathfrak{A}(t) = (T a(t), T A(t))$ on the total space TX of the tangent bundle. Notice that $T\mathfrak{A}$ is compatible with the bundle structure (that is, depends in a linear way on the fibre argument).

At the same time (see Chapter 2), TX is equipped with an induced connection having local coefficients Γ^{TX} of the form (2.2.60). This connection permits us to obtain the equality

$$T \exp_x^X \circ S = \exp^{TX}. \tag{3.19}$$

Given Γ^{TX} and the field $T\mathfrak{A}$, consider the stochastic equation

$$\mathrm{d}\,\zeta = \exp^{TX}_{\zeta(t)}((T a)_{\zeta(t)}(t)\,\mathrm{d}t + (T A)_{\zeta(t)}(t)\,\mathrm{d}w) \tag{3.20}$$

on TX. We call this equation a formal differential extension of (3.18).

In a local trivialization, (3.18) and (3.20) look like

$$\mathrm{d}\,\xi(t) = \left[a_{\xi(t)}(t) - \frac{1}{2}\,\mathrm{Tr}\,\Gamma^{\tau}_{\xi(t)}\left(A_{\xi(t)}(t)\,\cdot\,,\ A_{\xi(t)}(t)\,\cdot\right)\right]\mathrm{d}t +$$

$$+ A_{\xi(t)}(t)\,\mathrm{d}w,$$

$$\mathrm{d}\,\eta(t) = a'_{\xi(t)}(t)\,\eta(t)\,\mathrm{d}t -$$

$$- \frac{1}{2}\,\mathrm{Tr}\left\{\left(\Gamma^{\tau}_{\xi(t)}\right)'\left(A_{\xi(t)}(t)\,\cdot\,, A_{\xi(t)}(t)\,\cdot\,,\ \eta(t)\right) +\right.$$

$$+ \Gamma^{\tau}_{\xi(t)}\left(A'_{\xi(t)}(t)\,\eta(t)\,\cdot\,,\ A_{\xi(t)}(t)\,\cdot\right) +$$

$$\left. + \Gamma^{\tau}_{\xi(t)}\left(A_{\xi(t)}(t)\,\cdot\,,\ A'_{\xi(t)}(t)\,\eta(t)\,\cdot\right)\right\}\mathrm{d}t +$$

$$+ A'_{\xi(t)}(t)\,(\eta(t),\,\mathrm{d}w). \tag{3.21}$$

It is easy to see that the second equation may be derived from the first one by formal differentiation with respect to $x = \xi(\tau)$. Hence, given $\xi(\tau) = x$, $\eta(\tau) = h$, it describes a derivative $\eta(\tau) = \xi'_x(t)\, h$ along a direction h.

Recall that this differentiation is defined in a special sense. Namely, it must be treated in mean square sense

$$\lim_{\alpha \to 0} E \left\| \xi'_x(t)\, h - \frac{\xi_{x+\alpha h}(t) - \xi_x(t)}{\alpha} \right\|^2 = 0 .$$

To investigate nonlocal equations in a similar way, we need an extension of this notion to manifold valued stochastic processes.

DEFINITION 3.2. Let $\xi : \Lambda \to X$ be a random mapping of Banach manifolds. We have said that ξ is differentiable in the mean square sense if, given a bounded smooth mapping $f : X \to B$, the composition $f \circ \xi : \Lambda \to B$ valued in a Banach space B, is a mean square differentiable function.

THEOREM 3.2. *Let the manifold X and the field $\mathfrak{A}$ satisfy the assumption of Theorem* 2.1. *Then the solution $\xi(t) = S(t, \tau) \circ x$ of* (3.18) *is mean square differentiable and its derivative solves* (3.20).

Proof. It follows from the representation

$$\xi(t) = S(t, \tau) \circ x$$

that for a smooth mapping $f : X \to B$ we have

$$f(\xi(t)) = \lim_q f \circ S^q(t, \tau) \circ x.$$

It suffices to prove that this relation admits differentiation.

First, it follows from the locality of each family that

$$Y_x(t, \tau) = (S_x(t, \tau))' .$$

Notice that given the linear space mappings, a composition of independent smooth (in mean square sense) mappings admits a mean square differentiation of each factor and gives the product of those derivatives.

Since $f(\xi(t)) = f\left(\Pi_{k=1}^n S_{\xi(t_{k-1})}(t_k, t_{k-1}) \circ x\right)$ then taking $f_k = f \circ \varphi_k^{-1} \varphi_k \circ S_y$ we come to deal with the above situation because, in a neighborhood U_y of each point $y \in X$, the process $S_y(t_k, t_{k-1})\, y$ does not leave the chart on which φ_k is defined.

By induction, we obtain

$$(f \circ S^q(t, \tau))'(x)\, h = f'(S^q(t, \tau) \circ x)\, Y^q(t, \tau)\, h$$

and to complete the proof, we notice that passing to the limit in this relation is an easy task, thanks to the stabilization property proved in Theorem 3.1.

Remark 3.3. If b_x is a vector field, then the equation

$$d\,\xi = \exp_{\xi(t)}(b_{\xi(t)}(t)\; d\,t)$$

turns out to be an ordinary differential equation

$$\frac{d\xi}{dt} = b_{\xi(t)}(t). \tag{3.22}$$

Consider, as has been done in Chapter 3, a smooth random process

$$w_\alpha(t) = \int_{t-\alpha}^{t+\alpha} \varphi_\alpha(t-s)\, w(s)\, ds$$

and set

$$b_x(t) = a_x(t) + A_x(t)\, w'_\alpha(t).$$

Equation (3.22) with $b_x(t)$ having this form tends to the equation

$$d\xi = \left[a_{\xi(t)}(t) + \frac{1}{2} \operatorname{Tr} A'_{\xi(t)}(t)\, A_{\xi(t)}(t)\right] dt + A_{\xi(t)}(t)\, dw \tag{3.23}$$

for $\alpha \to 0$.

Notice that (3.23) is invariant under local manifold morphisms. Next, consider a tensor field c_x such to compensate the difference between the given connection on X and the connection (1.18) generated by the mapping A. Equation (3.23) may be rewritten in the form

$$d\xi(t) = \left(a_{\xi(t)}(t) + \frac{1}{2} c_{\xi(t)}(t)\right) dt + A_{\xi(t)}(t)\, dw -$$

$$- \frac{1}{2} \operatorname{Tr} \Gamma^\tau_{\xi(t)} \left(A_{\xi(t)}(t)\,\cdot\,,\ A_{\xi(t)}(t)\,\cdot\,\right) dt.$$

In this way, we put Equation (3.23) in correspondence to the Ito field $(a + 1/2\, c, A)$. Hence, changing the Ito field (a, A) to the vector field $(a - 1/2\, c + A\, w'_\alpha,\ 0)$ is an analogue of the $\mathcal{E}$-regularization of a stochastic equation described in Chapter 3. In particular, if $\Gamma(A, A) = -A' A$, then $c \equiv 0$ and (3.23) coincides with (2.9).

In this way, thanks to the results of Chapter 3, we may construct an $\mathcal{E}$-regularization of the stochastic flow $S(t, \tau)$ on X corresponding to an ordinary differential equation. This ordinary differential equation admits successive differential extensions up to an order which is equal to its coefficients smoothness order.

Evolution families generated by those extensions approximate stochastic families corresponding to differential extensions of the initial stochastic equation.

Let $\pi : \mathcal{E} \to X$ be a vector bundle equipped with a connection having local coefficients Γ^π. Let Γ^τ be local connection coefficients on the base X of the bundle π.

Consider the stochastic equation

$$d\zeta(t) = \exp^{\mathcal{E}}_{\zeta(t)} \left(c_{\zeta(t)}(t)\, dt + C_{\zeta(t)}(t)\, dw\right) \tag{3.24}$$

which had been investigated in Section 1.

Let

$$c_{\binom{x}{y}} = \left(a_x,\ a_x y - \Gamma^\pi_x(a_x, y)\right),$$

$$C_{\binom{x}{y}} = \left(A_x\,,\; B_x\,y - \Gamma^{\pi}_x\,(A_x, y)\right)$$

and $\exp^{\mathcal{E}}$ be an exponential mapping generated by the connection on $\mathcal{E}$ with local coefficients (2.2.4). Set $y = z$ and $b = c$ in (2.2.4) and denote

$$B^{\mathcal{E}}_x(a, y, b, z, c) = B^{\mathcal{E}}_x(a, y, b)$$

$$= 2\,\Gamma^{\pi}_x\,(y, b) - G_x\,(y, a)\,,$$

where

$$G_x\,(y, a) \;= (\Gamma^{\pi}_x)'\,(y, y, a) + \Gamma^{\pi}_x\,(y\,,\,\Gamma^{\pi}_x(y, a)) - \tag{3.25}$$

$$-\Gamma^{\pi}_x\,(\Gamma^{\pi}_x\,(y, y),\; a)\,.$$

As it has been discussed above, Equation (3.24) admits a formal differential extension

$$\mathrm{d}\,\gamma = \exp^{T\mathcal{E}}_{\gamma(t)}\left((T_c)_{\gamma(t)}\,(t)\;\mathrm{d}\,t + (T\,C\,)_{\gamma(t)}\,(t)\;\mathrm{d}\,w\right). \tag{3.26}$$

In a local trivialization, (3.26) looks like a system consisting of (3.5), (3.6), (3.21) and of the equation

$$\mathrm{d}\,\beta\,(t) \;= \Big[b'_{\xi(t)} = (t)\,(\alpha\,(t),\; \eta\,(t)) - (\Gamma^{\pi}_{\xi(t)})'\,(\alpha\,(t),$$

$$a_{\xi(t)}\,(t),\; \eta\,(t)) - \Gamma^{\pi}_{\xi(t)}\,(a'_{\xi(t)}\,(t)\,\alpha\,(t),\; \eta\,(t))\Big]\;\mathrm{d}\,t +$$

$$+ B'_{\xi(t)}\,(t)\,(\alpha\,(t),\; \eta\,(t),\; \mathrm{d}\,w) -$$

$$-(\Gamma^{\pi}_{\xi(t)})'\,(\alpha\,(t),\; A_{\xi(t)}\,(t)\;\mathrm{d}\,w,\;\; \eta\,(t)) -$$

$$-\;\Gamma^{\pi}_{\xi(t)}\,(A'_{\xi(t)}\,(t)\,(\alpha\,(t), \mathrm{d}\,w),\; \eta\,(t)) -$$

$$-\,\mathrm{Tr}\;\Big\{(\Gamma^{\pi}_{\xi(t)})'\;\Big(\alpha\,(t),\; A_{\xi(t)}\,(t)\,\cdot,\;\Big[B_{\xi(t)}\,(t)\,(\eta\,(t),\,\cdot)\; -$$

$$-\;\Gamma^{\pi}_{\xi(t)}\,(A_{\xi(t)}\,(t)\,\cdot\,, \eta\,(t))\Big]\Big) + \Gamma^{\pi}_{\xi(t)}\;\Big(A'_{\xi(t)}\,(t)\,(\alpha\,(t),\cdot)\,,$$

$$+\;\Big[B_{\xi(t)}\,(t)\,(\eta\,(t),\,\cdot) - \Gamma^{\pi}_{\xi(t)}\,(A_{\xi(t)}\,(t)\,\cdot\,, \eta\,(t))\Big]\Big) +$$

$$+\,\Gamma^{\pi}_{\xi(t)}\;\Big(A_{\xi(t)}\,(t)\,\cdot\,,\;\Big[B'_{\xi(t)}\,(t)\,(\alpha\,(t),\; \eta\,(t),\cdot) \;-$$

$$- \left(\Gamma^{\pi}_{\xi(t)}\right)' \ (\alpha(t), A_{\xi(t)}(t) \cdot, \eta(t)) -$$

$$- \Gamma^{\pi}_{\xi(t)} \left(A'_{\xi(t)}(t)\, \alpha(t) \cdot, \eta(t)\right)\big]\big)\Big\} \, \mathrm{d}t +$$

$$+ \mathrm{Tr} \left[(G_{\xi(t)})' \ (\alpha(t), A_{\xi(t)}(t) \cdot, \eta(t)) + \right.$$

$$\left. + G_{\xi(t)} \left(A'_{\xi(t)}(t) (\alpha(t), \cdot), \eta(t) \right) \right] \mathrm{d}t +$$

$$+ N_{\xi(t)}(t)\, \beta(t)\, \mathrm{d}t + M_{\xi(t)}(t)\, (\beta(t), \mathrm{d}w). \tag{3.27}$$

Assume that the connection coefficients Γ^{τ}, Γ^{π}, as well as a_X, A_X, b_X, B_X, belong to $C_1(X)$. Then, due to Theorem 3.2, Equation (3.26) has a unique solution which yields the derivative (in the mean square sense) of the process $\zeta(t)$.

In the next chapter, we will also need another interpretation of the system of equations (3.5), (3.6), (3.21), (3.27).

Let $\kappa : T\mathcal{E} \to X$ be a bundle over X. In general, it is not a vector bundle over X for a natural choice of $T\mathcal{E}$ trivialization (as a tangent trivialization to $U \times F$). Nevertheless, if we choose the following trivialization

$$\psi : \kappa^{-1}(U \to U \times Q, \quad Q = F \oplus B \otimes B,$$

then κ becomes a vector bundle and the process $\lambda(t) = (\xi(t, \theta(t))$ defines a section of the bundle. Here

$$\theta(t) = (\varphi(t), \delta(t)), \ \varphi(t) = \beta(t),$$

$\delta(t) = \alpha(t) \otimes \eta(t)$ while $\xi(t)$, $\eta(t)$, $\alpha(t)$, $\beta(t)$ solve, respectively, (3.5), (3.6), (3.21) and (3.27). Using the notations of Section 3.4, we can write down stochastic equations for $\varphi(t)$ and $\delta(t)$ in the form

$$\mathrm{d}\varphi(t) = b_{\xi(t)}(t)\, \varphi(t)\, \mathrm{d}t - \Gamma^{\pi}_{\xi(t)} \left(a_{\xi(t)}(t), \varphi(t)\right) \mathrm{d}t +$$

$$+ B_{\xi(t)}(t)\, (\varphi(t), \mathrm{d}w) - \Gamma^{\pi}_{\xi(t)} \left(A_{\xi(t)}(t)\, \mathrm{d}w, \varphi(t)\right) \mathrm{d}t +$$

$$+ \hat{F}_{\xi(t)}(t)\, \delta(t)\, \mathrm{d}t + \hat{S}_{\xi(t)}(t)\, (\delta(t), \mathrm{d}w),$$

$$\mathrm{d}\delta(t) = \left(n_{\xi(t)}(t) \oplus m_{\xi(t)}(t)\right) \delta(t)\, \mathrm{d}t +$$

$$+ \left(\tilde{N}_{\xi(t)}(t) \oplus \tilde{M}_{\xi(t)}(t)\right) (\delta(t), \mathrm{d}w),$$

where n_X, m_X, $\tilde{N}_X$, $\tilde{M}_X$ are defined in the obvious way from (3.5), (3.6), (3.21) and (3.27). The resulting pair of equations may be written down in the form

$$d\gamma = q_{\xi(t)}(t)\,\gamma(t)\,dt + Q_{\xi(t)}(t)\,(\gamma(t),\,dw),$$

where q and Q are linear and bilinear operators given respectively by

$$q\begin{pmatrix} y_1 \\ z \otimes y_2 \end{pmatrix} = \begin{pmatrix} by_1 - \Gamma^\pi(a, y_1) + \hat{F}(z, y_2) \\ n\, z \otimes y_2 + z \otimes m y_2 \end{pmatrix}$$

$$= \begin{pmatrix} b - \Gamma^\pi(a, \cdot\,) & \hat{F} \\ 0 & n \oplus m \end{pmatrix} \begin{pmatrix} y_1 \\ z \otimes y_2 \end{pmatrix},$$

$$Q\left(\begin{pmatrix} y_1 \\ z \otimes y_2 \end{pmatrix}, h\right)$$

$$= \begin{pmatrix} B(y, h) - \Gamma^\pi(A\,h, y) + \hat{S}(z, y_2, h) \\ \tilde{N}(z, h) \otimes y_2 + z \otimes \tilde{M}(y_2, h) \end{pmatrix}$$

$$= \begin{pmatrix} B\,h - \Gamma^\pi(A\,h, \cdot\,) & \hat{S}\,h \\ 0 & \tilde{N}\,h \oplus \tilde{M}\,h \end{pmatrix} \begin{pmatrix} y_1 \\ z \otimes y_2 \end{pmatrix}.$$

CHAPTER 5

Kolmogorov Equations

In this chapter, we deal with linear evolution families of operators acting on spaces of functions and measures defined on a manifold, which are generated by solutions of stochastic differential equations, studied in the previous chapter. We derive both backward and forward Kolmogorov equations and study the Cauchy problem for them.

Along with linear ones, we consider nonlinear evolution equations in the spaces of functions generated by solutions of stochastic equations with coefficients depending on their own solution distributions. We derive the quasilinear equations satisfied by those evolution families. As a consequence, we obtain the conditions of classical solvability of the Cauchy problem for quasilinear equations.

1. Backward Kolmogorov Equations

1.1. GENERAL ARGUMENTS

Consider a random evolution family $S(t, \tau)$ of mappings acting from a manifold X into itself

$$S(t, \tau) : x \mapsto S(t, \tau; x) \qquad (t > \tau)$$

adapted to a flow $\mathcal{F}_t$ of σ-algebras (that is, $\mathcal{F}_t$ -measurable). Recall that $S(t, \tau)$ does not depend on $\mathcal{F}_\tau$.

The evolution property of this family may be described by the relation

$$S(t, \tau) = S(t, \theta) \circ S(\theta, \tau) \quad \text{(a.s.)}. \tag{1.1}$$

As has been shown in Chapter 4, a solution of a stochastic differential equation

$$d\xi = \exp_{\xi(t)}\left(a_{\xi(t)}(t)\ dt + A_{\xi(t)}(t)\ dw\right) \tag{1.2}$$

generates such a family.

Later on in this chapter, we will assume that both the manifold X and the coefficients of Equation (1.2) meet all the requirements of Chapter 4.

Recall that the family $S(t, \tau)$ associated with (1.2) may be constructed from the solutions of localized equations in the following way.

For each point $x \in X$, define a pair of neighborhoods $V_x^1 \subset V_x$ and a random evolution family $S_x(t, \tau)$ which does not lead outside V_x and coincide with $S(t, \tau)$ on V_x^1

$$y \in V_x \mapsto S_x(t, \tau; y) \in V_x \qquad \text{(a.s.)},$$

$$S_x(t, \tau; y) \in V_x^1 \Rightarrow S_x(t, \tau; y) = S(t, \tau; y).$$

Let $q : \tau = t_1 \le t_2 \le \ldots \le t_n = t$ be a partition of the interval $[\tau, t]$ and

$$S_q(t, \tau) = \tilde{S}(t, t_{n-1}) \circ \tilde{S}(t_{n-1}, t_{n-z}) \circ \ldots \circ S(t_1, \tau), \tag{1.3}$$

where

$$\tilde{S}(t, \tau) : x \mapsto S_x(t, \tau; x). \tag{1.4}$$

Then, for any partition subsequence q_n, such that $\operatorname{diam} q_n \to 0$

$$S(t, \tau) \circ x = \lim_{q_n} S_{q_n}(t, \tau) \circ x \qquad \text{(a.s.)} \tag{1.5}$$

and

$$\lim_{q_n} P\{S(t, \tau; x) = S_{q_n}(t, \tau; x)\} = 1.$$

Now let $W(X)$ be the Banach space of real-valued bounded measurable functions, defined on X, with norm $\|f\| = \sup_{x \in X} |f(x)|$. Consider a linear operator on $W(X)$

$$U(\tau, t) : f \mapsto U(\tau, t) f$$

defined by the relation

$$U(\tau, t) f(x) = E f(S(t, \tau; x)). \tag{1.6}$$

This operator is clearly bounded

$$\| U(\tau, t) \| \le 1$$

and possesses the evolution property

$$U(\tau, t) = U(\tau, \theta)\, U(\theta, t), \qquad \tau \le \theta \le t.$$

In fact,

$$\begin{aligned} & E f(S(t, \tau; x)) \\ &\quad = E f\big(S(t, \theta; S(\theta, \tau; x))\big) \\ &\quad = E\left\{E_\theta f(S(t, \theta; y))\big|_{y = S(\theta, \tau; x)}\right\} \\ &\quad = E\, g(S(\theta, \tau; x)) = U(\tau, \theta)\, g(x), \end{aligned} \tag{1.7}$$

with

$$g(y) = E_\theta f(S(t, \theta; y)) = U(\theta, t) f(y).$$

Let us now consider some consequence of the localization relation (1.5).

Let $U_x(\tau, \theta)$ be the evolution family of the form (1.6), corresponding to the local random family $S_x(t, \tau)$ and

$$\tilde{U}(\tau, t) f(x) = U_x(\tau, t) f(x) = E f(S_x(t, \tau; x)). \tag{1.8}$$

THEOREM 1.1. *If the random evolution family* $S(t,\tau)$ *satisfies the localization relation* (1.5), *then the following multiplicative representation is valid*

$$U(\tau,t)=\lim_{q}\ \tilde{U}(\tau,t_1)\ldots\tilde{U}(t_{n-1},t)\,.$$

Proof. Put

$$U_q(\tau,t)f(x)=Ef(S_q(t,\tau;x))\,.$$

By the same argument as in the proof of (1.7), we may check that

$$U_q(\tau,t)=\ \tilde{U}(\tau,t_1)\ldots\tilde{U}(t_{n-1},t)\,.$$

It remains to note that

$$\begin{aligned}&U(\tau,t)f(x)\\&\quad=Ef(S(t,\tau;x))\\&\quad=\lim_{q}\ f(S_q(t,\tau;x))=\lim_{q}\ U_q(\tau,t)f(x)\,.\end{aligned}\tag{1.9}$$

Remark. An important peculiarity of (1.9) is that for each $x\in X$ one may calculate $\tilde{U}(\tau,t)f(x)$ locally, without leaving a neighborhood V_x located in the domain of a single chart of the manifold. Thus, we indicate a way to construct a nonlocal evolution family $U(\tau,t)$ by superposing local (but not evolution) families $\tilde{U}(\tau,t)$. It is worth recalling that although $U_x(\tau,t)$ is an evolution family, $\tilde{U}(\tau,t)$ is not. We say that relation (1.9) is the evolution family localization principle.

Consider the generator (an infinitesimal operator) $\mathfrak{A}(t)$,

$$\mathfrak{A}(t)f(x)=\lim_{\Delta t\to 0}\ \frac{U(t-\Delta t,t)-I}{\Delta t}f(x)\tag{1.10}$$

of an evolution family $U(\tau,t)$ and let $\mathfrak{A}_y(t)$ be the generator of the family $U_y(\tau,t)$, $y\in X$.

PROPOSITION 1.1. *Let* $\mathfrak{A}_y(t)f$ *be defined for all* $y\in X$ *and the inequality*

$$P\left\{S(t,\tau;y)\,\overline{\in}\,V_y^1\right\}\le c\,(t-\tau)^2\tag{1.11}$$

be valid.

Then for each $x\in X$

$$\mathfrak{A}(t)f(x)=\mathfrak{A}_x(t)f(x)\,.\tag{1.12}$$

Proof. Consider the set of elementary events

$$U_1=\{\omega:S^q(t,t-\Delta t;x)=S_x(t,t-\Delta t;x)\}\,.$$

Since $S(t,\tau;x)$ and, as a consequence, $S^q(t,\tau;x)$ coincide with $S_x(t,\tau;x)$ up to the moment of exit from V_x, it follows from (1.11) that

$$P(U_1) \geq 1 - c\,(\Delta t)^2 .$$

Therefore, we have

$$\begin{aligned}&| U_q (t - \Delta t, t) f(x) - \tilde{U}_q (t - \Delta t, t) f(x) | \\ &\quad \leq E\, | f(S_q (t, t - \Delta t; x)) - f(S_x (t, t - \Delta t; x)) | \\ &\quad \leq 2\, \| f \|\, c\,(\Delta t)^2 .\end{aligned}$$

On the other hand, the definition of the generator $\mathfrak{A}(t)$ implies the relation

$$| U_x (t - \Delta t, t) f(x) - f(x) - \mathfrak{A}_x (t) f(x)\, \Delta t | = 0\,(\Delta t)$$

for any given function f. Comparing the above relations, we obtain

$$| U_q (t - \Delta t, t) f(x) - f(x) - \mathfrak{A}_x (t) f(x)\, \Delta t | = 0\,(\Delta t),$$

which leads to

$$| U (t - \Delta t, t) f(x) - f(x) - \mathfrak{A}_x (t) f(x)\, \Delta t | = 0\,(\Delta t),$$

due to the fact that the estimate does not depend on q.

Consider now an important special case: the manifold $\mathcal{E}$ we are going to deal with, is equipped with a structure of total space of a normed vector bundle $\pi : \mathcal{E} \to X$ over the manifold X, and assume that all the constructions described above are compatible with this structure.

In this case, we start with a random evolution family $Y(t, \tau; z)$ acting from $\mathcal{E}$ to $\mathcal{E}$ with a projection $\pi(Y(t, \tau; z)) = S(t, \tau; z)$ to the base X. Suppose that $Y(t, \tau; z)$ possesses π-multiplicative functional properties, i.e., it has a linear restriction to each fibre $\mathcal{E}_x = F$

$$Y(t, \tau; z) = (S(t, \tau; x),\ V(t, \tau)\, y),$$

$$z = (x, y)$$

and

$$E_\tau\ \| V(t, \tau)\, y \|^2_{S(t,\tau;x)} \leq C\ \| y \|^2_x . \tag{1.13}$$

Introduce, moreover, the space $W(\mathcal{E})$ of real-valued functions defined on $\mathcal{E}$ which are linear over a fibre and satisfy the estimate

$$\| f \| = \sup_{x \in X\ \|y\|_x \leq 1} | f(z) | < \infty . \tag{1.14}$$

An evolution family of the type (1.6)

$$Q(\tau, t) f(z) = E f(Y(t, \tau; z)) \tag{1.15}$$

is acting on this space. Multiplicative functional properties imply that $Q(\tau, t)$ is a linear evolution family of bounded operators on $W(\mathcal{E})$.

Let us point out that the elements of the space $W(\mathcal{E})$ may be given in the form

$$f(z) = \langle y_x, \varphi_x \rangle_x, \quad x = \pi(z)$$

with $y : X \to \mathcal{E}$ a section of the bundle $\pi : \mathcal{E} \to X$, and $\varphi : X \to \mathcal{E}^*$ a section of the cobundle $\pi^* : \mathcal{E}^* \to X$. In this way, linear operations over these elements may be canonically identified with operations over the corresponding sections

$$(Bf)(z) = \langle y_{\pi(z)}, (\widetilde{B}\varphi)_{\pi(z)} \rangle_{\pi(z)} . \tag{1.16}$$

As a result, (1.15) leads to a description of the evolution family acting on the space $W(\pi^*)$ of bounded sections of the bundle π^*. Consider the dual random evolution family $V^*(\tau, t)$,

$$\langle V(t,\tau)\, y, \ \varphi(S(t,\tau;x)) \rangle_{S(t,\tau;x)} = \langle y \ (V^*(\tau,t)\, \varphi) \rangle_x ,$$

$$V^*(\tau, t) : \mathcal{H}_t^* \to \mathcal{H}_\tau^* ,$$

where $\mathcal{H}_\tau^*$ is the Banach space of $\mathcal{F}_t$-measurable random functions $\eta(\tau) = (\pi^*)^{-1}(x)$ with norm $\|\eta\|_\tau^2 = E\ \|\eta(\tau)\|_x^2$.

The equality

$$\begin{aligned} f(Y(t,\tau;z)) &= \langle V(t,\tau)\, y, \ \varphi(S(t,\tau;x)) \rangle_{S(t,\tau;x)} \\ &= \langle y \ (V^*(\tau,t)\, \varphi) \rangle_x , \end{aligned}$$

together with (1.16), yields the equality

$$Q(\tau, t) f(z) = \langle y, \ \mathcal{N}(\tau, t)\, \varphi \rangle_x , \tag{1.17}$$

with $\mathcal{N}(\tau, t)$ an evolution family on the section space $W(\pi^*)$, $\mathcal{N}(\tau, t) = \widetilde{Q}(\tau, t)$.

The generators of these families satisfy the relation

$$\mathfrak{A}_Q f(z) = \langle y \ \mathfrak{A}_{\mathcal{N}} \varphi \rangle_x \tag{1.18}$$

if both right and left-hand terms are defined.

Remark. In the case of a trivial bundle $\pi : X \times F \to X$, the above relations reduce to relations between the principal parts of the objects considered. In fact, in this case

$$\begin{aligned} f(Y(t,\tau;z)) &= \langle V(t,\tau)\, y, \ \varphi(S(t,\tau;x)) \rangle \\ &= \langle y \ (V^*(\tau,t)\, \varphi) \rangle \end{aligned} \tag{1.19}$$

with $y \in F$, $\varphi(x) \in F^*$ and (1.17) reads

$$\langle y \ \mathcal{N}(\tau, t)\, \varphi) \rangle = \langle y\, E\, (V^*(\tau, t)\, \varphi)\, (x) \rangle_x . \tag{1.20}$$

Here $V^*(\tau, t)$ is a multiplicative operator functional (on the space F^*) of the random process $\xi(t) = S(t, \tau; x)$.

These relations must be taken into account if we are going to use the localization principle (1.9) for constructing an evolution family on the space of sections of π^*.

1.2. CALCULATION OF THE INFINITESIMAL OPERATOR OF THE EVOLUTION FAMILY GENERATED BY A RANDOM PROCESS IN A LINEAR SPACE

First, consider a random family $S(t, \tau; x)$ acting from a Banach space B into itself, which is generated by a solution of a stochastic differential equation

$$d\xi = a(t, \xi(t)\, dt + A(t, \xi(t)\, dw. \tag{1.21}$$

Here and below in this chapter, we assume that B belongs to the τ_2 class and all the requirements of Theorem 3.3.1, which grant the existence of the family, are met.

Let $f \in C_2(B, B, R^1)$. Then Ito's formula implies that

$$f(S(t+\Delta t, t; x)) - f(x)$$

$$= \int_t^{t+\Delta t} \Big[f'(\xi(\tau))\, a(\tau, \xi(\tau)) +$$

$$+ \frac{1}{2} \operatorname{Tr} f''(\xi(\tau)) \big(A(\tau, \xi(\tau)) \cdot, A(\tau, \xi(\tau)) \cdot\big)\Big]\, d\tau +$$

$$+ \int_t^{t+\Delta t} f'(\xi(\tau))\, A(\tau, \xi(\tau))\, dw(\tau),$$

which yields $f \in \mathcal{D}_{\mathfrak{A}(t)}$ and

$$\mathfrak{A}(t) f(x) = \lim_{\Delta t \to 0} \frac{E f(S(t+\Delta t, t; x)) - f(x)}{\Delta t}$$

$$= \langle a(t, x), f'(x)\rangle + \frac{1}{2} \operatorname{Tr} f''(x) (A(t, x) \cdot, A(t, x) \cdot).$$

Using symbolic notations introduced in Chapter 1, we obtain the expression for the restriction of the generator to $C_2(B, R^1)$

$$\mathfrak{A}(t) = \langle a(t, x), D\rangle + \frac{1}{2} \langle A(t, x) A^*(t, x) D, D\rangle. \tag{1.22}$$

Let us pass on to consider a trivial bundle $\mathcal{E} = B \times F$ and a system of stochastic equations

$$d\xi = a(t, \xi(t))\, dt + A(t, \xi(t))\, dw(t),$$

$$d\eta = c(t, \xi(t))\, \eta(t)\, dt + C(t, \xi(t)) (\eta(t), dw(t)). \tag{1.23}$$

This system, along with $S(t, \tau)$, defines an operator multiplicative functional $V(t, \tau)$ on the space F.

We may treat the system (1.23) as an equation

$$d\zeta = a_1(t, \zeta(t))\, dt + A_1(t, \zeta(t))\, dw(t) \tag{1.24}$$

with respect to a process $\zeta(t) = (\xi(t), \eta(t))$ in the space $\mathcal{E} = B \times F$ with coefficients

$$a_1(t, \zeta) = (a(t, \xi), c(t, \xi)\, \eta),$$

$$A_1(t,\zeta) = (A(t,\xi), C(t,\xi)\eta).$$

The corresponding random family

$$Y(t,\tau) = (S(t,\tau), V(t,\tau))$$

gives rise to an evolution operator family $Q(\tau, t)$ on the space of functions $W(\mathcal{E})$ defined on $\mathcal{E}$. A generator of this later family has the form

$$\mathfrak{A}_Q(t) = \langle a_1(t,z), D_z\rangle_{\mathcal{E}} + \frac{1}{2}\langle A_1(t,z)A_1^*(t,z)D_z, D_z\rangle_{\mathcal{E}},$$

with $D_z = D_x \times D_y$ and, thus,

$$\langle a_1(t,z), D_z\rangle_{\mathcal{E}} = \langle a(t,x), D_x\rangle_B + \langle c(t,x)y, D_y\rangle_F,$$

$$\langle A_1(t,z)A_1^*(t,z)D_z, D_z\rangle_{\mathcal{E}}$$

$$= \left\langle \begin{pmatrix} A(t,x)A^*(t,x) & A(t,x)(C(t,x)y)^* \\ C(t,x)y\,A^*(t,x) & C(t,x)y\,(C(t,x)y)^* \end{pmatrix} \begin{pmatrix} D_x \\ D_y \end{pmatrix}, \begin{pmatrix} D_x \\ D_y \end{pmatrix} \right\rangle_{\mathcal{E}}$$

$$= \langle A(t,x)A^*(t,x)D_x, D_x\rangle_B +$$

$$+ \langle A(t,x)(C(t,x)y)^* D_y, D_x\rangle_B +$$

$$+ \langle C(t,x)y\,(C(t,x)y)^* D_y, D_y\rangle_F +$$

$$+ \langle C(t,x)A^*(t,x)D_x, D_y\rangle_F.$$

Let us compute the result obtained by applying those differential operators to a function $f(z) = \langle y, \varphi(x)\rangle_F$,

$$\langle a_1(t,x), D_z\rangle f(z)$$

$$= \langle y, \langle a(t,x), D_x\rangle_B\, \varphi(x)\rangle_F + \langle c(t,x)y, \varphi(x)\rangle_F$$

$$= \langle y, [\langle a(t,x), D_x\rangle + c^*(t,x)]\,\varphi(x)\rangle_F.$$

$$\langle A_1(t,z)A_1^*(t,z)D_z, D_z\rangle f(z)$$

$$= \langle y, \langle A(t,x)A^*(t,x)D_x, D_x\rangle\, \varphi(x)\rangle_F +$$

$$+ 2\langle C(t,x)(y, A^*(t,x)D_x), \varphi(x)\rangle_F.$$

For a bilinear mapping $C: F\times H \to F$, we introduce an adjoint mapping $C^*: H \times F^* \to F^*$ by the relation

$$\langle y, C^*(h,\varphi)\rangle_F = \langle C(y,h), \varphi\rangle_F.$$

This allows us to write

$$\langle C\,(y, A^*\,D_x), \varphi\,(x)\rangle = \langle y, C^*\,(A^*\,D_x\,, \varphi)\rangle\,.$$

Thus, we have obtained the following result.

PROPOSITION 1.2. *Let* a, A, c, C *be functions on* $[0,T]\times B$ *valued, respectively, in* $B, L_{12}\,(H,B), L\,(F),\ L\,(F, L_{12}\,(H,F))$ *and satisfying all the conditions of Theorems* 3.3.1 *and* 3.4.1.

Let $S\,(t,\tau) : \xi\,(\tau)\mapsto\xi\,(t)$ *be a random evolution family and* $V\,(t,\tau):\eta\,(\tau)\mapsto\eta\,(t)$ *be an operator multiplicative functional of* $\xi\,(t)$ *generated by system* (1.23).

Then the relation

$$\mathcal{N}(\tau, t)\,\varphi\,(x) = E\,V^*\,(\tau, t)\,\varphi\,(S\,(t,\tau; x)) \tag{1.25}$$

defines an evolution family of bounded operators on the space $V\,(B, F^*)$. *A restriction of the generator of this family to* $C_2\,(B, F^*)$ *has the form*

$$\begin{aligned}\mathfrak{A}_{\mathcal{N}}\,(t)\,\varphi\,(x) &= \frac{1}{2}\,\langle A\,(t,x)\,A^*\,(t,x)\,D_x\,,\ D_x\rangle\,\varphi\,(x) + \\ &+ \langle a\,(t,x), D_x\,\rangle\ \varphi\,(x) + C^*\,(t,x)\,(A^*\,(t,x)\,D_x, \varphi\,(x)) + \\ &+ C^*\,(t,x)\,\varphi\,(x).\end{aligned} \tag{1.26}$$

Proof. Let us consider the important case $F = R^1, C\equiv 0$. Given a scalar function c, consider the multiplicative functional

$$V\,(t,\tau) = \exp\left\{\int_\tau^t c\,(s, \xi\,(s))\ d\,s\right\} \tag{1.27}$$

corresponding to it, which gives rise to the evolution family

$$\mathcal{N}(t,\tau)\,\varphi\,(x) = E\ \exp\left\{\int_\tau^t c\,(s,\xi\,(s))\ d\,s\right\}\varphi\,(\xi\,(t)) \tag{1.28}$$

on the space of real-valued functions with generator

$$\mathfrak{A}_{\mathcal{N}} = \frac{1}{2}\,\langle A\,A^*\,D,\ D\rangle + \langle a, D\rangle + c\,. \tag{1.29}$$

The relation (1.28) is usually called the Feynmann–Kac formula.

If C does not vanish as well, we obtain, in a similar way, the relations

$$\begin{aligned}\mathcal{N}(t,\tau)\,\varphi\,(x) = E\ \exp\Big\{ &\int_\tau^t c\,(s,\xi\,(s))\ d\,s\ - \\ &-\frac{1}{2}\int_\tau^t C^2\,(s,\xi\,(s))\ d\,s + \\ &+\int_\tau^t C\,(s,\xi\,(s))\ d\,w\quad \varphi\,(\xi\,(t))\Big\}\end{aligned} \tag{1.30}$$

and

$$\mathfrak{A}_{\mathcal{N}} = \frac{1}{2} \langle A A^* D, D\rangle + \langle a, D\rangle + C A D + c. \tag{1.31}$$

1.3. INFINITESIMAL OPERATORS OF EVOLUTION FAMILIES GENERATED BY A MANIFOLD VALUED STOCHASTIC PROCESS

Consider a local stochastic differential equation

$$d\xi = \exp_{\xi(t)} (a_{\xi(t)}(t)\, dt + A_{\xi(t)}(t)\, dw) \tag{1.32}$$

on the manifold X. Let us first identify (without changing notations) a corresponding chart of a manifold with its image in the Banach space B. After this identification, Equation (1.32) is changed to the equation

$$d\xi = a_{\xi(t)}(t)\, dt - \frac{1}{2} \operatorname{Tr} \Gamma^{X}_{\xi(t)} \left(A_{\xi(t)}(t) \cdot, A_{\xi(t)}(t) \cdot\right) dt + $$

$$+ A_{\xi(t)}(t)\, dw. \tag{1.33}$$

It follows from (1.22) that the restriction of the generator $\mathfrak{A}(t)$ of the evolution family (1.6) to the space $C^2(X, B, R^1)$ has the form

$$\mathfrak{A}(t) = \frac{1}{2} \langle A A^*(t) D, D\rangle + \langle a, D\rangle - $$

$$- \frac{1}{2} \left\langle \operatorname{Tr} \Gamma^X (A(t) \cdot, A(t) \cdot), D\right\rangle. \tag{1.34}$$

As has been shown in Chapter 2, this expression may be changed to the following

$$\mathfrak{A}_{\mathcal{N}} = \frac{1}{2} \operatorname{Tr} \left[\nabla_A \nabla_A - \nabla_{\nabla_A A}\right] + \nabla_a,$$

where ∇ is the symbol of the corresponding covariant derivative. We shall sometimes use the same symbolic form as in Chapter 2, denoting

$$\langle A A^* \nabla, \nabla\rangle = \operatorname{Tr} \left[\nabla_A \nabla_A - \nabla_{\nabla_A A}\right],$$

$$\nabla_a = \langle a, \nabla\rangle. \tag{1.35}$$

Now consider a local bundle $\pi : X \times F \to X$ and a stochastic equation

$$d\xi(t) = \exp^{\mathcal{E}}_{\zeta(t)} \left(a_{1\zeta(t)}(t)\, dt + A_{1\zeta(t)}(t)\, dw\right)$$

compatible with its structure which has been treated in Chapter 4.

In a local trivialization, this equation is a system consisting of Equation (1.33) and an equation for the corresponding multiplicative functional

$$d\eta = c_{\xi(t)}(t)\, \eta(t)\, dt - \Gamma^{\pi}_{\xi(t)} (a_{\xi(t)}(t), \eta(t))\, dt -$$

$$- \frac{1}{2} \operatorname{Tr} \Big\{ (\Gamma^{\pi}_{\xi(t)}) \left(A_{\xi(t)}(t) \,\cdot\,, A_{\xi(t)}(t) \,\cdot\,, \eta(t)\right) +$$

$$+ \Gamma^{\pi}_{\xi(t)} \left(A_{\xi(t)}(t) \,\cdot\,, \Gamma^{\pi}_{\xi(t)} \left(A_{\xi(t)}(t) \,\cdot\, \eta(t)\right)\right) -$$

$$- \Gamma^{\pi}_{\xi(t)} \left(\Gamma^{\pi}_{\xi(t)} \left(A_{\xi(t)}(t) \,\cdot\,, A_{\xi(t)}(t) \,\cdot\,\right) \eta(t)\right) +$$

$$+ 2\, \Gamma^{\pi}_{\xi(t)} \left(A_{\xi(t)}(t) \,\cdot\,, C_{\xi(t)}(t) \left(\,\cdot\,, \eta(t)\right)\right) \Big\} \, \mathrm{d}t +$$

$$+ C_{\xi(t)}(t) \left(\eta(t), \mathrm{d}w\right) - \Gamma^{\pi}_{\xi(t)} \left(A_{\xi(t)}(t)\, \mathrm{d}w, \eta(t)\right). \tag{1.36}$$

Comparing this pair of equations with (1.23) and, using (1.26), we obtain the expression for the restriction of the evolution family (1.17) to $C_2(X, B, F^*)$

$$\langle y, \mathfrak{A}_{\mathcal{N}}(t)\, \varphi(x) \rangle$$

$$= \Big\langle y, \frac{1}{2} \operatorname{Tr} A^*_x(t)\, \varphi''_x A_x(t) -$$

$$- \frac{1}{2} \varphi'(x) \operatorname{Tr} \Gamma^X_x \left(A_x(t) \,\cdot\,, A_x(t) \,\cdot\,\right) + \varphi'_x a_x(t) \Big\rangle +$$

$$+ \Big\langle y, \operatorname{Tr} \big[C^*_x(t) \left(\,\cdot\,, \varphi'_x A_x(t) \,\cdot\,\right) +$$

$$+ \Gamma^{\pi *}_x \left(A_x(t) \,\cdot\,, \varphi'_x A_x(t) \,\cdot\,\right) \big] \Big\rangle +$$

$$+ \left\langle y, c_x(t)\, \varphi_x + \Gamma^{\pi *}_x \left(a_x(t), \varphi_x\right) \right\rangle +$$

$$+ \frac{1}{2} \Big\langle y, \operatorname{Tr} \big[\Gamma^{\pi *}_x \left(A_x(t) \,\cdot\,, \Gamma^{\pi *}_x \left(A_x(t) \,\cdot\,, \varphi_x\right)\right) -$$

$$- \Gamma^{\pi *}_x \left(\Gamma^X_x \left(A_x(t) \,\cdot\,, A_x(t) \,\cdot\,\right), \varphi_x \right) +$$

$$+ \left(\Gamma^{\pi *}_x\right)' \left(A_x(t) \,\cdot\,, A_x(t) \,\cdot\,, \varphi_x\right) \big] \Big\rangle +$$

$$+ \Big\langle y, \operatorname{Tr} \Gamma^{\pi *}_x \left(A_x(t) \,\cdot\,, C^*_x(t) \left(\,\cdot\,, \varphi'_x\right)\right) \Big\rangle +$$

$$+ \Big\langle y, \operatorname{Tr} \Gamma^{\pi *}_x \left(A_x(t) \,\cdot\,, \Gamma^{\pi *}_x \left(A_x(t) \,\cdot\,, \varphi(x)\right)\right) \Big\rangle . \tag{1.37}$$

It is easy to check that the last expression may be rewritten an invariant way if we change the symbol D to the symbol $\nabla^{\pi *}$ of the covariant derivative of the cobundle $\pi^* : \mathcal{E}^* \to X$ and set

$$\langle A A^* \nabla^{\pi^*}, \nabla^{\pi^*}\rangle = \nabla^{\pi^*}_A \nabla^{\pi^*}_A - \nabla^{\pi^*}_{\nabla_A A} .$$

In fact, by substituting

$$\nabla^{\pi^*}_a \varphi = \varphi'_x a + \Gamma^{\pi^*}_x (a, \varphi)$$

and

$$\nabla^{\pi^*}_b \nabla^{\pi^*}_a \varphi = \varphi'' (b, a) + (\Gamma^{\pi^*})' (b, a, \varphi) + \Gamma^{\pi^*} (a' b, \varphi) +$$

$$+ \Gamma^{\pi^*} (a, \varphi' b) + \Gamma^{\pi^*} (b, \Gamma^{\pi^*} (a, \varphi)) + \Gamma^{\pi^*} (b, \varphi' a)$$

in (1.37), we obtain the equality

$$\langle y, \mathfrak{A}_{\mathcal{N}}(t) \varphi (x)\rangle_x$$

$$= \Big\langle y, \frac{1}{2} (A_x (t) A^*_x (t) \nabla^{\pi^*} \nabla^{\pi^*}) \varphi\Big\rangle_x +$$

$$+ \Big\langle y, \operatorname{Tr} C^*_x (t) (\cdot , \langle A_x (t) \cdot , \nabla^{\pi^*}\rangle \varphi)\Big\rangle +$$

$$+ \Big\langle y, \langle a_x (t), \nabla^{\pi^*}\rangle \varphi\Big\rangle + \Big\langle y, c^*_x (t) \varphi \Big\rangle . \tag{1.38}$$

Proposition 1.1 gives the foundation for passing to the global case. Moreover, the above calculations permit us to state the following result.

THEOREM 1.2. *Let $\pi : \mathcal{E} \to X$ be a normed vector bundle and $Y(t, \tau)$ be a random evolution family generated by a stochastic equation*

$$d \zeta (t) = \exp^{\mathcal{E}}_{\zeta(t)} \big(c_{\zeta(t)} (t) \; d t + C_{\zeta(t)} (t) \; d t\big) , \tag{1.39}$$

where $c_z (t)$, $C_z (t)$ are sections of the bundles $\tau_{\mathcal{E}}$, $L_{12} (\theta, \tau_{\mathcal{E}})$ satisfying the assumptions of Theorem 4.3.1.

The evolution family

$$\mathcal{N}(\tau, t) \varphi (\pi (z)) = E\, V^* (\tau, t) \varphi \big(\pi (Y (t, \tau; z))\big) \tag{1.40}$$

on the space $W(\pi^*)$ of the bounded sections of the cobundle π^*, has a generator with the restriction

$$\mathfrak{A}_{\mathcal{N}}(t) \varphi (x) = \frac{1}{2} \operatorname{Tr} \nabla^{\pi^*}_{A_x(t)} \nabla^{\pi^*}_{A_x(t)} \varphi (x) -$$

$$- \frac{1}{2} \operatorname{Tr} \nabla^{\pi^*}_{\nabla_{A_x(t)} A_x(t)} \varphi (x) + \nabla^{\pi^*}_{a_x(t)} \varphi (x) +$$

$$+ \operatorname{Tr} C^*_x (t) \nabla^{\pi^*}_{A_x(t)} \varphi (x) + C^*_x (t) \varphi (x) \tag{1.41}$$

to $\sigma_2 (\pi^*)$.

1.4. CAUCHY PROBLEM FOR A PARABOLIC EQUATION

Evolution family generators considered above are second order elliptic operators acting on both a space of functions, defined on a manifold X, and sections of vector bundles over X.

First, we shall deal with the scalar case. Let $U(\tau, t)$ be the evolution family defined by (1.6), and $\mathcal{K}$ be a certain linear subset of $W(x)$ which is invariant under $U(\tau, t)$ and is contained in the domain of the generator $\mathfrak{A}_u(t)$ for any t.

For $f \in \mathcal{K}$ and $t < \tau$, the function

$$u(t, x) = U(\tau, t) f(x) \tag{1.42}$$

belongs to $\mathcal{K}$ as well, and there exists

$$\frac{\partial u}{\partial t} = -\lim_{\Delta t \to 0} \frac{U(t - \Delta t, \tau) - U(t, \tau)}{\Delta t} f(x)$$

$$= -\lim_{\Delta t \to 0} \frac{U(t - \Delta t, t) - I}{\Delta t} U(t, \tau) f(x) = -\mathfrak{A}_u(t) u(x).$$

Thus, the parabolic equation

$$\frac{\partial u}{\partial T} + \mathfrak{A}(t) u = 0 \tag{1.43}$$

appears in a natural way in the above context and the function (1.42) solves it over the interval $[t, \tau]$. Moreover, the equality

$$u(\tau, x) = f(x) \tag{1.44}$$

clearly holds at the point τ which is the right-hand end of the interval $[t, \tau]$.

Equation (1.43) is called the backward Kolmogorov equation for the Markov process defined by the corresponding stochastic equation. Thus, we have the possibility of constructing a solution of the Cauchy problem (1.44), (1.43) by formula (1.6), i.e., as a result of calculating the integral of the Cauchy data function with respect to a measure generated by the Markov process.

The same arguments lead to the construction of Cauchy problem solutions for equations in vector bundle sections. They are connected with corresponding evolution families, acting on the space of (linear over a fibre) functions defined on the total space of the bundle.

We apply all the above arguments to prove that the corresponding Cauchy problems are correct. First, we must choose a class of functions which are invariant under the considered evolution family.

PROPOSITION 1.3. *Let ξ_α be a random X-valued function which depends on a real parameter α and is differentiable up to the kth order with respect to this parameter in a neighborhood of the point α_0 in the mean square sense. If $f \in C_k(X, B, R^1)$, then the function*

$$u(\alpha) = E f(\xi_\alpha) \tag{1.45}$$

is differentiable as well up to order k in the neighborhood of the point α_0 *and*

$$u'_\alpha(\alpha_0) = E\langle f'(\xi_{\alpha_0}),\ \xi'_{\alpha_0}\rangle_{\xi_{\alpha_0}}\ . \tag{1.46}$$

Proof. Let $k=1$. Since ξ_α has a mean square derivative of the first order, then it follows immediately from the definition that $f(\xi_\alpha)$ possesses the same property. Next, $u(\alpha)$ has the first-order derivative, since we may differentiate with respect to a parameter under the integration sign. To prove (1.46), it is enough to take a local function for f. Then one may reduce the investigation to the case of a linear space.

Consider the difference

$$\frac{1}{\alpha-\alpha_0}\,[u(\alpha)-u(\alpha_0)] - E\,\langle f'(\xi_{\alpha_0})\ \xi'_{\alpha_0}\rangle$$

$$= E\left\{\int_0^1 \langle\, [f'(\xi_{\alpha_0}+\tau\,(\xi_\alpha-\xi_{\alpha_0})) -\right.$$

$$- f'(\xi_{\alpha_0})]\,\frac{\xi_0-\xi_{\alpha_0}}{\alpha-\alpha_0}\rangle\, d\tau\ +$$

$$\left. + \left\langle f'(\xi_{\alpha_0}),\ \frac{\xi_0-\xi_{\alpha_0}}{\alpha-\alpha_0} - \xi'_{\alpha_0}\right\rangle\right\}.$$

Under the above conditions, it is easy, letting $\alpha\to\alpha_0$ in the right-hand side of the last relation, to prove that (1.46) holds and all the other statements of the proposition are valid.

COROLLARY. *Under conditions which grant mean square differentiability up to the k-th order of the random evolution family* $S(\tau, t; x)$ *with respect to the initial value* x, *the class* $C_k(x, B, R^1)$ *is invariant under the action of the family* $U(t,\tau)$

$$f \mapsto U(t,\tau)f(x) = Ef(S(\tau, t; x)), \quad (t<\tau)\,.$$

Remark 1.1. Let $\varphi\in\sigma_k(\pi^*)$, which means that it is a C_k-section of the bundle $\pi^* : \mathcal{E}^*\to X$.

The above result may be applied to the function $f(z) = \langle y, \varphi(x)\rangle_x$ on the manifold $\mathcal{E}$ and the evolution family $Y(\tau, t, z)$, $z=(x,y)$, from $\mathcal{E}$ into itself, possessing π-multiplicative functional properties.

Thus, the mapping

$$\varphi \mapsto \mathcal{N}(t,\tau)\,\varphi(x) = \upsilon(t,x)$$

leaves the class $\sigma_k(\pi^*)$ invariant.

In accordance with (1.46), we may calculate the derivative of the function $u(t,x)$ and write it in the form

$$\left\langle \begin{pmatrix} y'_x\ h \\ h \otimes y \end{pmatrix}, \begin{pmatrix} u \\ u' \end{pmatrix} \right\rangle = E \left\langle \begin{pmatrix} \eta'(t) \\ \xi'(t) \otimes \eta(t) \end{pmatrix}, \begin{pmatrix} \varphi(\xi(t)) \\ \varphi'(\xi(t)) \end{pmatrix} \right\rangle , \tag{1.47}$$

$$x \in X, \quad h \in B, \quad y \in F, \quad \varphi(x) \in F^*, \quad \varphi'(x) \in (B \otimes F)^*.$$

As a consequence, we obtain

$$\langle y, u'(x)\, h \rangle = E\ \langle \eta'_x(t),\ \varphi(\xi_x(t)) \rangle +$$

$$+ E\ \langle \eta_x(t),\ \varphi'(\xi_x(t))\, \xi'_x(t) \rangle - \langle y'_x\ h, u(x) \rangle . \tag{1.48}$$

Notice that here we have used the vector bundle $\alpha : \mathcal{E} \otimes TX \oplus\ \mathcal{E} \to X$ introduced in Chapter 4, with local trivialization of the form

$$\Phi : \alpha^{-1}(U) \to U \times \{f \otimes B \oplus F^*\}$$

and its dual $\alpha^* : \mathcal{E}^* \otimes T^*X \oplus \mathcal{E}^* \to X$ with local trivialization

$$\tilde{\Phi} : (\alpha^*)^{-1}(U) \to U \times \{(F \otimes B)^* \oplus F^*\}$$

as well as pairing between them

$$\left\langle \begin{pmatrix} \kappa \\ \beta \otimes \gamma \end{pmatrix}, \begin{pmatrix} f \\ \varphi \end{pmatrix} \right\rangle = \langle \kappa, f \rangle + \langle \beta \otimes \gamma, \varphi \rangle ,$$

where

$$f \in F^*, \quad \kappa \in F, \quad \beta \otimes \gamma \in F \otimes B, \quad \varphi \in (F \otimes B)^* .$$

Remark. In the case of a trivial vector bundle $\mathcal{E} = X \times F$, we may consider a constant section y_x; in this case (1.48) becomes

$$\langle y, u'(x)\, h \rangle = E\,\langle \eta'(t), \varphi(\xi(t)) \rangle + \tag{1.49}$$

$$+\ E\,\langle \eta(t), \varphi'\ (\xi(t))\, \xi'(t) \rangle .$$

Consider the Cauchy problem for a scalar parabolic equation on the manifold X equipped with a uniform structure

$$\frac{\partial u}{\partial T} + \frac{1}{2}\, \mathrm{Tr}\left[\nabla_{A_x(t)\,\cdot}\, \nabla_{A_x(t)\,\cdot} - \nabla_{\nabla_{A_x(t)}\cdot A_x(t)\,\cdot}\right] u +$$

$$+ \nabla_{a_x(t)}\ u = 0, \quad u(\tau, x) = f(x) , \tag{1.50}$$

as well as the Cauchy problem for a more general equation in vector bundle sections

$$\frac{\partial \upsilon}{\partial t} + \frac{1}{2}\, \mathrm{Tr}\,[\nabla^{\pi^*}_{A_x(t)}\, \nabla^{\pi^*}_{A_x(t)} - \nabla^{\pi^*}_{\nabla_{A_x(t)} A_x(t)}]\, \upsilon +$$

$$+ \nabla_{a_x(t)}\, \upsilon + \mathrm{Tr}\ C^*_x(t)\, \nabla_{A_x(t)}\, \upsilon +$$

$$+ c_x^* (t)\, \upsilon = 0, \qquad \upsilon (\tau, x) = \varphi (x) . \tag{1.51}$$

Notice that although (1.50) is a special case of (1.51), as follows from the above arguments, the two problems are equivalent. In fact, the problem (1.51) may be treated as the problem (1.50) on the manifold $\mathcal{E}$ with specially chosen coefficients (linear over fibres) for the function

$$\Phi (z) = \langle y, \varphi (x)\rangle, \qquad z = (x, y) . \tag{1.52}$$

Let us state the main results of this paragraph.

THEOREM 1.3. *Let*

$$a_X (t) \in T_X X, \qquad A_X (t) \in L_{12} \; (H \, , T_X X \,)$$

be C_k*-smooth bounded fields on* X *along with local connection coefficients and the real function* $f(x)$ *. Then the Cauchy problem* (1.43), (1.44) *has a unique solution in the class* $C_k (X)$ *which can be given as*

$$u \, (t, x) = E f (S \, (\tau, t; x)). \tag{1.53}$$

Here $S (\tau, t; x)$ *is the random evolution family generated by the solution of Equation* (1.2).

Proof. It follows from the above results that the only fact still to be proved is the uniqueness of the solution. So to complete the proof of the theorem, we must prove it. Let $\Phi (t \, , x)$ be a classical solution of (1.43), (1.44). Consider a function

$$\eta \, (t) = \Phi \, (t, \xi \, (t)) \, ,$$

where $\xi (t)$ is a solution of (1.2) and compute its stochastic differential.

In accordance with Ito's formula (see Chapter 4), we obtain the equality

$$\mathrm{d}\, n = \left[\frac{\partial \, \Phi}{\partial \, t} + \mathcal{A}_u \, (t) \, \Phi \right] \mathrm{d}\, t + \Phi' \, (t, \xi \, (t)) \, A_{\xi(t)} \, (t) \; \mathrm{d}\, w,$$

which yields

$$E \; \mathrm{d}\, n = 0 \, ,$$

since $\Phi (t, x)$ satisfies (1.43). But this means that

$$\begin{aligned} E_{t,x} \, n \, (\tau) \; &= \Phi \, (t, x) = E_{t,x} \, n \, (\tau) \\ &= E_{t,x} f \, (\xi \, (\tau)), \end{aligned}$$

i.e., an arbitrary solution $\Phi (t, x)$ of the problem (1.43), (1.44) may be represented in the form (1.53) and, thus, the solution is unique due to the uniqueness of the solution of equation (1.32).

The next theorem, as was mentioned above, is a consequence of Theorem 1.3, though it contains the latter as a special case of $\mathcal{E} = X \times R^1$.

THEOREM 1.4. *Let the fields* $a_X (t)$, $A_X (t)$, Γ_X^X *meet the requirements of Theorem*

1.3 *and the fields*

$$c_x(t) \in L(\mathcal{E}_x), \quad C_x(t) \in L\left(\mathcal{E}_x, L_{12}(H, \mathcal{E}_x)\right), \; \Gamma_x^X$$

be C_k-smooth. Then, for $\varphi \in C_k(x, B, \mathcal{E}^*)$, the problem (1.51) *has a unique solution in the class $C_k(x, B, \mathcal{E}^*)$ which may be represented in the form*

$$\langle y, v_x(\tau)\rangle = E_{t,x}\langle \eta(\tau), \varphi(\xi(\tau))\rangle, \tag{1.54}$$

where $\zeta(\tau) = (\xi(\tau), \eta(\tau))$ is the solution of (1.39) *with $\xi(t) = x$, $\eta(t) = y$.*

THEOREM 1.5. *Let the assumptions of Theorem* 1.3 *be fulfilled for $k = 3$; then the function $g_x(t) = \nabla u_x(t)$ solves the Cauchy problem*

$$\frac{\partial g}{\partial t} + \frac{1}{2}\operatorname{Tr} \nabla^{T\tau}_{A_x(t)} \nabla^{T\tau}_{A_x(t)} g - \frac{1}{2}\operatorname{Tr} \nabla^{T\tau}_{\nabla_{A_x(t)} A_x(t)} g +$$

$$+ \nabla^{T\tau}_{a_x(t)} g + \operatorname{Tr}(\nabla A_x(t))^* \nabla^{T\tau}_{A_x(t)} g +$$

$$+ (\nabla a_x(t))^* g = 0, \quad g_x(\tau) = \nabla f(x), \tag{1.55}$$

where $\nabla^{T\tau}$ is the covariant derivative corresponding to the special connection on TX described in Chapter 2. The function $g_x(t) = \nabla u_x(t)$ may be represented in the form

$$\nabla_h u_x(t) = E\langle \xi'(\tau), f'\xi(\tau))\rangle,$$

where $\xi(\tau)$ is the solution of (1.2) *with $\xi(t) = x$ and $\xi'(t) = h$.*

Proof. Using the connection Γ^{TX} of the form (2.2.26) on TX and the corresponding covariant derivative ∇^τ on X, we may rewrite (1.55) in the form

$$\frac{\partial g}{\partial t} + \frac{1}{2}\operatorname{Tr} \nabla^{\tau}_{A_x(t)} \nabla^{\tau}_{A_x(t)} g - \frac{1}{2}\operatorname{Tr} \nabla^{\tau}_{\nabla_{A^\tau_x(t)} A_x(t)} g +$$

$$+ \nabla^{\tau}_{a_x(t)} g + \operatorname{Tr}\left(\nabla^{\tau}_{A_x(t)}\right)^* \nabla^{\tau}_{A_x(t)} g +$$

$$+ \left(\nabla^{\tau}_{a_x(t)}\right)^* g - \operatorname{Tr} R(A_x(t), g, A_x(t)) = 0,$$

with R the curvature tensor.

2. Quasilinear Parabolic Equations

2.1. GENERAL APPROACH

Let X be a smooth Banach manifold with the τ_2 class model space B equipped with an affine connection. The probabilistic approach developed above permits us to investigate the Cauchy problem for quasilinear parabolic equations on X

$$\frac{\partial u}{\partial t} + \mathfrak{A}_u(t)\, u = 0, \quad u(\tau, x) = \varphi(x). \tag{2.1}$$

where

$$\mathfrak{A}_u(t) = \frac{1}{2}\operatorname{Tr}\left\{\nabla_{A_X(t,u)}\nabla_{A_X(t,u)} - \nabla_{\nabla_{A_X(t,u)}A_X(t,u)}\right\} + \nabla_{a_X(t,u)} \tag{2.2}$$

is a differential operator with coefficients depending on a functional parameter u from a certain functional set θ.

A natural way to solve this problem based on the results of the previous section may be described as follows.

Consider the linear Cauchy problem

$$\frac{\partial f}{\partial t} + \mathfrak{A}_u(t)f = 0, \quad f(\tau, x) = \varphi(x) \tag{2.3}$$

with coefficients depending on a parameter $u \in \Theta$. Due to the results of Section 1, the solution of this problem may be represented in the form

$$f(t, x) = E\,\varphi(\xi(\tau)), \tag{2.4}$$

where $\xi(t)$ is a random process in X satisfying the equation

$$d\,\xi = \exp^X_{\xi(t)}\left(a_{\xi(t)}(t, u(t))\, d\,t + A_{\xi(t)}(t, u(t))\, d\,w\right). \tag{2.5}$$

Putting

$$u = f, \tag{2.6}$$

we obtain (2.1) instead of (2.3). In this way, we may construct a solution of (2.1) by solving the system (2.4)–(2.6). Under some additional assumptions which guarantee that the solution of (2.1) is smooth, the converse holds as well and, thus, under some conditions, (2.1) is equivalent to the system (2.4)–(2.6).

Suppose that both the coefficients and the initial function $\varphi(x)$ in (2.4)–(2.6) satisfying the usual assumptions (see Chapter 4). Then one may construct a solution of (2.4)–(2.6) by the successive approximation method, setting

$$u^0(t, x) = \varphi(x), \tag{2.7}$$

$$d\,\xi^n = \exp^X_{\xi^n(t)}\left(a_{\xi^n(t)}(t, u^n(t))\, d\,t + A_{\xi^n(t)}(t, u^n(t))\, d\,w\right), \tag{2.8}$$

$$u^n(t, x) = E\,\varphi(\xi^{n-1}(\tau)). \tag{2.9}$$

Notice that to be sure the successive approximation method is used correctly, we need to check that the function u^{n+1} of the form (2.9) belongs to the chosen functional class Θ if $u^n \in \Theta$. Here u^n is a function on which the solution $\xi^n(\tau)$ of Equation (2.8) depends. Indeed, if $u^n \in \Theta$ implies $u^{n+1} \in \Theta$, then the mapping Φ defined by (2.9) acts from Θ into itself. Below, we shall describe a situation in which this mapping comes to be a contraction. Finally, due to the contraction mapping theorem, we may state that there exists a unique solution of (2.9) being interpreted as a functional equation of the form

$$U(t, x) = \Phi(t, x\,u) = E\,\varphi(\xi(\tau)). \tag{2.10}$$

The last result implies the existence of a unique solution of the system (2.4)–(2.6).

Due to the results of Section 1, a sequence of functions $u^n(t, x)$ of the form (2.9) satisfies

$$\frac{\partial u^n}{\partial t} + \mathfrak{A}_{u^{n-1}}(t)\, u^n = 0\,, \tag{2.11}$$

and the function

$$u(t, x) = E\,\varphi(\xi(\tau)) \tag{2.12}$$

solves (2.1) (possibly in a weak sense). Thus, we have proved that (2.1) may be solved by the successive approximation method. Generally speaking, (2.12) gives a weak solution of (2.1), until we prove that it is smooth. Recall that to construct a solution of (2.8), we need only the Lipschitz property of $u(t, x)$.

To prove that the function $u(t, x)$ of the form (2.12) is smooth, we may use a technique developed in the previous section. Notice that we may apply this technique with the framework of the above scheme of the construction of the solution of the quasi-linear parabolic equation. Namely, along with Equation (2.5) on the given manifold X, we need to deal with differential extensions of this equation, i.e., with stochastic equations of iterated tangent bundles $T^k X$, $k = 1, 2$. These equations may be easily derived by a formal differentiation procedure applied to the initial equation (see Chapter 4). Stochastic differential equations on manifolds $T^k X$ derived in this way are of the same type as the initial stochastic differential equation. Thus, they may be investigated by the above methods as well. As a result, we prove that if the coefficients and the Cauchy data in (2.1) are smooth enough, then (2.12) gives a classical solution of (2.1).

It is worth mentioning that the main difficulty in the realization of the scheme described above consisted of the correct choice of the functional class Θ adapted to additional structures of X and in the investigation of properties of the mapping $\Phi : \Theta \to \Theta$.

2.2. LOCAL MANIFOLDS

Let $X = B$ be a local manifold (a Banach space belonging to the τ_2 class).

Let Θ be the subspace of $C_0(X, R^1)$ consisting of uniformly Lipschitz functions on X. Denote $|||f|||_\Theta$ as the norm of $f \in \Theta$

$$|||f|||_\Theta = \sup_{x \in X} |f(x)|.$$

Let

$$a_x(t, \upsilon) \in B, \quad A_x(t, \upsilon) \in L_{12}(H, B),$$

$x \in X$, $t \in [0, T]$, $\upsilon \in \Theta$ and the following estimates hold

$$\| a_x(t, \upsilon) \|^2 + \sigma_2^2(A_x(t, \upsilon)) \le C\left(1 + |||\upsilon|||^{2k} + \| x \|^2\right), \tag{2.13}$$

$$\| a_x(t, \upsilon_1) - a_x(t, \upsilon) \|^2 + \sigma_2^2\left(A_x(t, \upsilon_1) - A_x(t, \upsilon_1)\right)$$

$$\le K(\tau)\, |||\upsilon_1 - \upsilon|||^2, \tag{2.14}$$

$$\| a_x(t, \upsilon) - a_y(t, \upsilon) \| + \sigma_2^2 (A_x(t, \upsilon) - A_y(t, \upsilon))$$

$$\le C_\upsilon(\tau) \| x - y \|^2, \tag{2.15}$$

$$x, y \in X, \quad \upsilon, \upsilon_1 \in \Theta, \quad ||| \upsilon ||| \le \tau, \quad ||| \upsilon_1 ||| \le \tau, \quad K(\tau) > 0.$$

Consider the system

$$d\,\xi^u(t) = a_{\xi^u(t)}(t, u(t))\; d t + A_{\xi^u(t)}(t, u(t))\; d w, \tag{2.16}$$

$$u(s, x) = E\,\varphi\left(\xi^u_{s,x}(\tau)\right), \tag{2.17}$$

where $\xi^u_{s,x}(t)$ is a solution of (2.16) such that

$$\xi^u_{s,x}(t) = x. \tag{2.18}$$

The relation (2.17) may be treated as a functional equation of the form

$$u(s, x) = \Phi(s, x, u). \tag{2.19}$$

To prove that it may be solved, we need some auxiliary assertions.

PROPOSITION 2.1. *Let the coefficients of the equation*

$$d\,\xi^\upsilon(t) = a_{\xi^\upsilon(t)}(t, \upsilon(t))\; d t + A_{\xi^\upsilon(t)}(t, \upsilon(t))\; d w, \tag{2.20}$$

satisfy (2.13) – (2.15). *Then there exists a unique solution of this equation such that*

$$E \| \xi^\upsilon_{s,x}(t) \|^2$$

$$\le \left\{ \| x \|^2 + 3 C_1 \int_s^t ||| \upsilon(\tau) |||^2 \, d\tau \right\} \times$$

$$\times \exp\{3 C_1 (t - s)\}, \tag{2.21}$$

$$E \| \xi^\upsilon_{s,x}(t) - \xi^\upsilon_{s,y}(t) \|^2$$

$$\le \| x - y \|^2 \exp\left\{ K \int_s^t [1 + C_\upsilon(\tau)] \, d\tau \right\} \tag{2.22}$$

and

$$E \| \xi^\upsilon_{s,x}(t) - \xi^{\upsilon_1}_{s,y}(t) \|^2$$

$$\le K \left[\int_s^t ||| \upsilon(\tau) - \upsilon_1(\tau) |||^2 \, d\tau \right] \exp K(t - s). \tag{2.23}$$

Proof. The results of Chapter 3 imply that under the conditions of the proposition there exists a unique solution of (2.20). Estimates (2.21)–(2.23) may be proved in a similar way and, therefore, we shall prove only (2.22).

Let us estimate the difference

$$\delta(t) = \| \xi^{\upsilon}_{s,x}(t) - \xi^{\upsilon}_{s,y}(t) \|^2 .$$

Ito's formula ensures that

$$E\,\delta(t)$$
$$\leq \| x - y \|^2 + 2E \int_s^t \| a_{\xi^{\upsilon}_{s,x}(\tau)}(\tau, \upsilon(\tau)) - a_{\xi^{\upsilon}_{s,y}(\tau)}(\tau, \upsilon(\tau)) \|^2 \, d\tau +$$
$$+ E \int_s^t \sigma_2^2 \left(A_{\xi^{\upsilon}_{s,x}(\tau)}(\tau, \upsilon(\tau)) - A_{\xi^{\upsilon}_{s,y}(\tau)}(\tau, \upsilon(\tau)) \right) d\tau$$

and due to (2.13) – (2.15), we obtain

$$E\,\delta(t) \leq \| x - y \|^2 + 3E \int_s^t \delta(\tau)\,[1 + C_\upsilon(\tau)]\, d\tau.$$

Finally, the Gronwall lemma grants that (2.22) holds.

Denote $L_\upsilon(t)$ the Lipschitz constant of a function $\upsilon(t, x)$

$$\| \upsilon(t, x) - \upsilon(t, y) \| \leq L_\upsilon(t) \| x - y \| .$$

PROPOSITION 2.2. *Let* $\varphi, \upsilon \in \Theta$ *and the assumption of Proposition* 2.1 *hold. Then there exists an interval* $\Delta = [s_1, \tau]$ *such that for all* $s \in [s_1, \tau]$ *the function*

$$u(s, x) = E\,\varphi\left(\xi^{\upsilon}_{s,x}(\tau)\right) \tag{2.24}$$

belongs to Θ *as well.*

Proof. Denote $\alpha(s)$ as a positive real function satisfying the inequality $L_\upsilon(s) \leq \alpha(s)$ for each $s \in [0, \tau]$. Let us prove that there exists an interval Δ such that $L_u(s) \leq \alpha(s)$ as well for all $s \in \Delta$.

It results from Proposition 2.1 estimates that the function $u(s, x)$ of the form (2.24) satisfies the estimate

$$| u(s, x) - u(s, y) |^2$$
$$\leq L_f^2 \, \| x - y \|^2 \exp K \int_s^\tau [1 + C_\upsilon(\tau)]\, d\tau. \tag{2.25}$$

Denote $\beta(s)$ as the minimal constant such that

$$| u(s, x) - u(s, y) | \leq \beta(s) \; \| x - y \|$$

for each $s \in [0, T]$. It results from (2.25) that

$$\beta_s \leq L_f^2 \exp K \int_s^t [1 + C_\upsilon(\tau_1)]\, d\tau_1.$$

Let $\gamma(s)$ be a solution of the functional equation

$$\gamma(s) = \gamma(\tau) \exp K \int_s^t [1 + \gamma(\tau_1)]\, d\tau_1$$

satisfying the condition

$$\gamma(\tau) = L_f^2 .$$

The function $\gamma(\tau)$ is a solution of the differential equation

$$-\frac{d\gamma}{ds} = \gamma(s) K [1 + \gamma(s)]$$

as well and, as may be easily verified, has the following representation

$$\gamma(s) = \frac{e^{K(t-s)} L_f^2}{1 - L_f^2 e^{K(t-s)} + L_f^2} . \tag{2.26}$$

The fact that $L_\upsilon(s) \le \gamma(s)$ implies that $L_u(s) \le \gamma(s)$ is due to the above arguments.

It follows from (2.26) that $\gamma(s)$ is bounded over an interval $\Delta = [s_1, \tau]$ whose length $|\tau - s_1|$ satisfies the estimate

$$|\tau - s_1| \le \frac{1}{K} \ln\left(1 + \frac{1}{L_f^2}\right) . \tag{2.27}$$

Thus, we have constructed the interval Δ and the function $\gamma(s)$ such that $L_u(s) \le \gamma(s)$ if $L_\upsilon(s) \le \gamma(s)$ for each $s \in [s_1, T]$. In other words, we have proved that $u(s, x) \in \Theta$ for each $s \in \Delta$, if $\upsilon(s, x) \in \Theta$.

Let $u^n(s, x)$ be a family of successive approximations of the solution of Equation (2.1).

THEOREM 2.1. *Assume that the conditions of Proposition* 2.2 *hold. Then there exists a unique solution* $u(s, x) \in \Theta$ *of equation* (2.19) *for each* $s \in \Delta$, *where* Δ *is the interval whose length satisfies the estimate* (2.27).

Proof. As follows from Proposition 2.2, the mapping Φ acts in Θ. Let us estimate the difference

$$\alpha^n(s, x) = | u^{n+1}(s, x) - u^n(s, x) |.$$

The estimates of Proposition 2.1 result in

$$\alpha^n(s, x) \le L_f^2 \int_s^\tau |||\, u^n(\tau_1) - u^{n-1}(\tau_1) \,|||^2 \, d\tau_1 \times \exp K(\tau - s)$$

and, as a consequence, we obtain that

$$\beta_n(\tau) = \sup_{x \in B} \alpha^n(\tau, x)$$

satisfies the estimate

$$\beta_n(s) \le \delta^n \int_s^\tau \ldots \int_s^{\tau_2} |||\, u^1(\tau_1) - f \,|||^2 \, d\tau_1 \, d\tau_2 \ldots d\tau_n$$

where $\delta = K L_f^2 \exp K(t - s)$.

Since, due to (2.9), all u^n are uniformly bounded, we obtain that

$$|||\, u^1(\tau) - f \,||| \le \text{const} < \infty$$

and, therefore

$$|||\, u^{n+1}(s) - u^n(s) \,||| \le \frac{m^n}{n\,!}\ \text{const},$$

where $m = \delta(\tau - s)$.

In this way, we have proved that for each $s \in \Delta$ the family $u^n(s,x)$ converges to a limiting function $u(s,x)$ in the uniform topology

$$u(s,x) = \lim_{n \to \infty} u^n(s,x). \tag{2.28}$$

Finally, we shall prove that the limiting function $u(s,x)$ satisfies the Lipschitz condition. Indeed, it follows from Proposition 2.2 that for each $s \in \Delta$

$$|\, u^n(s,x) - u^n(s,y) \,| \le \gamma(s)\ \| x - y \|,$$

where $\gamma(s)$ had been defined by (2.26) and this estimate is uniform with respect to n.

To prove that the solution of (2.19) constructed above is unique in Θ, assume first that there are two solutions $u(s,x)$ and $\upsilon(s,x)$ to (2.19) such that

$$u(\tau,x) = \upsilon(\tau,x) = f(x).$$

Using Proposition 2.1 estimates, we may check that

$$|||\, u(s) - \upsilon(s) \,|||^2 \le \text{const} \int_s^\tau |||\, u(\tau_1) - \upsilon(\tau_1) \,|||^2\ d\tau_1$$

and, hence

$$|||\, u(s) - \upsilon(s) \,||| = 0,$$

which implies that the solution is unique.

Consider next the stochastic equation (2.16). If $u(s,x)$ is a bounded Lipschitz function, then it results from (2.13) – (2.15) (see Chapter 3) that there exists a unique Markov process $\xi^u_{s,x}$ solving (2.16). In this case, the function

$$\upsilon(s,x) = E f\big(\xi^u_{s,x}(\tau)\big)$$

for $s \in \Delta$, coincides with $u(s,x)$ due to the uniqueness of the solution of Equation (2.19). The result may be reformulated in the following way.

THEOREM 2.2. *Let* $\varphi \in \Theta$ *and the coefficients* $a_x(s,u)$, $A_x(s,u)$ *satisfy* (2.13), (2.15). *Then there exists a unique solution of* (2.16)–(2.17), *formed by the bounded Lipschitz function* $u(s,x)$ *and the Markov process* $\xi^u_{s,x}(t)$ *in* B.

Applying the results of the previous section, we obtain the following statement.

THEOREM 2.3. *Let the assumption of Theorem* 2.2 *hold. Then the function* $u(s,x)$ *given by* (2.17) *is a weak solution of the Cauchy problem*

$$\frac{\partial u}{\partial t} + \mathfrak{A}_u(t)\, u = 0, \quad u(\tau, x) = f(x), \tag{2.29}$$

where

$$\mathfrak{A}_u(t) = \frac{1}{2}\left(A_x(t, u(t))\, A_x^*(t, u(t))\, D,\ D\right) + $$
$$+ \left(a_x(t, u(t)), D\right). \tag{2.30}$$

Remark 2.1. It is natural to take the function $u(s, x)$ of the form (2.17) as a weak solution of (2.29), since, as will be proved below, it coincides with the classical solution of this problem if we succeed in proving its smoothness.

Assume that X is a product of two Banach spaces of τ_2 class, $X = B \times B_1$. Choose as Θ the subset of $C_0(X, R^1)$ consisting of real functions on X which are linear over B_1, that is, functions of the form

$$\Psi(z) = \langle y, \varphi(x)\rangle, \tag{2.31}$$

where $\varphi \in C_0(B, B_1^*)$ is a Lipschitz function. Denote $|||\Psi|||_\Theta$ the norm in Θ

$$|||\Psi|||_\Theta = \sup_{\|y\|=1} \sup_{x \in B} |\langle y, \varphi(x)\rangle| . \tag{2.32}$$

Along with Θ, we shall need the functional class Θ_1 consisting of those functions in $C_0(B, B_1^*)$ which satisfy the Lipschitz condition. Let $|||\varphi|||_{\Theta_1}$ be the norm in Θ_1

$$|||\varphi|||_{\Theta_1} = \sup_{x \in B} \|\varphi(x)\| . \tag{2.33}$$

Notice that $|||\varphi|||_\Theta = |||\varphi|||_{\Theta_1}$.

Let us state the condition we shall need below on coefficients of the Cauchy problem

$$\frac{\partial \Psi}{\partial s} + \mathfrak{A}_\Psi(s)\, \Psi = 0, \quad \Psi(\tau, x) = \Gamma(z), \tag{2.34}$$

where $z = (x, y)$,

$$\Gamma(z) = \langle y, \varphi(x)\rangle , \tag{2.35}$$

which is similar to (2.1).

Let

$$z \in (x, y) \in B \times B_1, \quad \Psi \in \Theta, \quad g \in \Theta_1,$$
$$a_{1z}(t, \Psi) = (a_x(t, g),\ c_x(t, g)\, y), \tag{2.36}$$
$$A_{1z}(t, \Psi) = (A_x(t, g),\ C_x(t, g)\, y).$$

In this case, the problem (2.34) is equivalent to the problem

$$\frac{\partial g}{\partial s} + \mathrm{Tr}\,\left\langle A_x(s, g(s))\, A_x^*(s, g(s))\, D, D\right\rangle g + $$
$$+ \left\langle a_x(s, g(s)), D\right\rangle g + $$

$$+ \operatorname{Tr} C_x^* (s, g(s)) \left(A_x^* (s, g(s)) D, g\right) +$$

$$+ c_x^* (s, g(s))\, g = 0, \quad g(\tau, x) = \varphi(x). \tag{2.37}$$

Assume that $a_x(s, \upsilon)$, $A_x(s, \upsilon)$ satisfy inequalities (2.13)–2.15) and

$$c_x(s, \upsilon) \in L(B_1), \quad C_x(s, \upsilon) \in L\left(B_1, L_{12}(H, B_1)\right),$$

satisfy the following conditions.

1. *Boundedness*

$$\sigma_2^2 \left(C_x(s, \upsilon)\, y\right) \le \rho \left[1 + \| \upsilon \|^{2k}\right] \| y \|^2. \tag{2.38}$$

2. *Dissipation*

$$\langle c_x(s, \upsilon)\, y, J(y) \rangle \le \left[\rho_0 + \rho \| \upsilon \|^{2k}\right] \| y \|^2, \tag{2.39}$$

where

$$J \in L\left(B_1, B_1^*\right), \quad \| J_y \| = \| y \|, \quad \rho > 0, \quad \rho_0 \le \rho.$$

3. *Lipschitz conditions*

$$\sigma_2^2 \left(C_x(s, \upsilon)\, y - C_{x_1}(s, \upsilon)\, y\right) + \| c_x(s, \upsilon)\, y - c_{x_1}(s, \upsilon)\, y \|^2$$

$$\le L_\upsilon(s) \| x - x_1 \|^2 \| y \|^2,$$

$$\sigma_2^2 \left(C_x(s, \upsilon)\, y - C_x(s, \upsilon_1)\, y\right) + \| c_x(s, \upsilon)\, y - c_x(s, \upsilon_1)\, y \|^2$$

$$\le K(r) \| \upsilon - \upsilon_1 \|^2 \| y \|^2,$$

$$\| \upsilon \| \le r, \quad \| \upsilon_1 \| \le r, \quad K(r) > 0. \tag{2.40}$$

Remark 2.2. The dissipation condition looks like

$$(c_x(s, \upsilon)\, y, y) \le \left[\rho_0 + \rho \| \upsilon \|^{2k}\right] \| y \|^2 \tag{2.41}$$

if $B_1 = H$ is a Hilbert space. Notice that the constant ρ_0 may be negative, while all the other constants in (2.38)–(2.41) are positive.

Consider the system

$$\xi(\tau) = x + \int_s^\tau a_{\xi(t)}(t, g(t))\, dt +$$

$$+ \int_s^\tau A_{\xi(t)}(t, g(t))\, dw(t), \tag{2.42}$$

$$\eta(\tau) = y + \int_s^\tau c_{\xi(t)}(t, g(t))\, \eta(t)\, dt +$$

$$+ \int_s^\tau C_{\xi(t)} (t, g(t)) (\eta(t), \, dt), \tag{2.43}$$

$$\langle y, g(s,x) \rangle = E \langle \eta(\tau), \varphi(\xi(\tau)) \rangle . \tag{2.44}$$

Using (2.36), we may write (2.42)–(2.44) in the form

$$\zeta(\tau) = z + \int_s^\tau a_{1\zeta(t)} (t, \Psi(t)) \, dt +$$

$$+ \int_s^\tau A_{1\zeta(t)} (t, \Psi(t)) \, dw, \tag{2.45}$$

$$\Psi(s, z) = E \, \Gamma(\zeta(\tau)) \tag{2.46}$$

with $z = (x, y)$, $\Gamma(z) = \langle y, \varphi(x) \rangle$.

The system (2.45)–(2.46) is a system of the same type as (2.16)–(2.17) considered above. This permits us to apply the above arguments in order to investigate it.

Nevertheless, some additional difficulties arise when we try to prove that the solutions of linearized equations belong to the chosen functional class. Namely, the trivial proof of the boundedness of solution is now changed for special investigation similar to that developed previously in order to prove the Lipschitz property of the solution.

PROPOSITION 2.3. *Let* (2.13)–(2.14), (2.38)–(2.41) *hold and* $\varphi \in \Theta_1$. *Then there exists an interval* Δ_1 *such that the function*

$$\Psi(s, z) = E \, \Gamma(\zeta^\Phi(\tau)) , \tag{2.47}$$

where $\zeta^\Phi(t)$ *is a solution of the stochastic equation*

$$d\zeta^\Phi = a_{1\zeta^\Phi} (t, \Phi(t)) \, dt + A_{1\zeta^\Phi} (t, \Phi(t)) \, dw \tag{2.48}$$

belongs to Θ *or, what is the same, that the function* $q(s,x)$ *given by*

$$\langle y, q(s,x) \rangle = E \langle \eta^g(\tau), \varphi(\xi^g(\tau)) \rangle , \tag{2.49}$$

where $(\xi^g(\tau), \eta^g(\tau))$ *is a solution of the system*

$$d\xi^g(t) = a_{\xi^g(t)} (t, g(t)) \, dt +$$

$$+ A_{\xi^g(t)} (t, g(t)) \, dw, \tag{2.50}$$

$$d\eta^g(t) = c_{\xi^g(t)} (t, g(t)) \, \eta^g(t) \, dt +$$

$$+ C_{\xi^g(t)} (t, g(t)) (\eta^g(t), \, dw), \tag{2.51}$$

belongs to Θ_1 *for* $\Phi \in \Theta$ *or* $g \in \Theta_1$.

Proof. Let $\gamma(s)$ be a positive function such that

$$K_g(\tau) = \sup_{x \in B} \| g(\tau, x) \|_{B_1^*} \le \gamma(\tau)$$

for all $\tau \in [0, T]$. Let us prove that there exists an interval Δ_1 such that $K_q(\tau) \le \gamma(\tau)$ for all $\tau \in \Delta_1$.

It follows from Proposition 2.1 and coefficient estimates that

$$\| q(s, x) \| \le K_\varphi \exp \int_s^\tau [\rho_0 + \rho K_q(t)] \, dt ,$$

for $K_\varphi = \sup_{x \in B} \| \varphi(x) \|_{B_1^*}$. By arguments similar to those used in the proof of Proposition 2.2, we conclude that a function $\gamma(s)$ of the form

$$\gamma(s) = \frac{[2\rho_0 + 3\rho] \exp [2\rho_0 + 3\rho](\tau - s)}{2\rho_0 + 3\rho - 3\rho K\varphi \exp [2\rho_0 + 3\rho](\tau - s) + 3\rho K\varphi}$$

possesses the necessary property. Namely, the inequality $K_g(\tau) \le \gamma(\tau)$ implies the estimate $K_q(\tau) \le \gamma(\tau)$. Hence, for bounded $\gamma(\tau)$, $q(\tau, x)$ is bounded if $g(\tau, x)$ possesses this property. It follows from the explicit form of the function $\gamma(s)$ that it is bounded over the whole interval $[0, T]$ if $2\rho_0 + 3\rho + 3\rho K_\varphi < 0$. If $2\rho_0 + 3\rho + 3\rho K_\varphi > 0$, $\gamma(s)$ is bounded over the interval $\Delta_1 = [s_1, \tau]$ whose length may be estimated as

$$|\tau - s_1| < \frac{1}{2\rho_0 + 3\rho} \ln \left[1 + \frac{2\rho_0 + 3\rho}{3\rho K\varphi} \right] . \tag{2.52}$$

If $2\rho_0 + 3\rho = 0$, then

$$|\tau - s_1| < \frac{1}{3\rho K\varphi} .$$

To prove that $q(s, x)$ satisfies the Lipschitz condition for all $s \in \Delta$, where Δ is an interval depending on the equation coefficients and on the initial function properties, one must use arguments similar to the ones above. Notice that the necessary calculations appear to be much more involved and we omit them because below we will describe another way to prove the desired facts.

Summing up all the above considerations, we state that $q(s, x) \in \Theta_1$ for all $s \in \Delta \cap \Delta_1$ or, what is equivalent, that $\Psi(s, z) \in \Theta$.

Next, Proposition 2.1 estimates adapted to the considered situation along with arguments used in the proof of Theorem 2.1, lead to the conclusion that the family of functions

$$\begin{aligned} \Psi_{n+1}(s, z) &= \langle y, g_{n+1}(s, x) \rangle \\ &= E \langle \eta_n(\tau), \varphi(\xi_n(\tau)) \rangle = E\,\Gamma(\zeta_n(\tau)) \end{aligned} \tag{2.53}$$

converges to a limiting function

$$\Psi(s, z) = E\,\Gamma(\zeta(\tau)) ,$$

which belongs to Θ (or $g_n(s, x)$ converges to a function $g(s, x)$ given by

$$\langle y, g(s,x)\rangle = E\langle \eta(\tau), \varphi(\xi(\tau))\rangle,$$

which belongs to Θ_1). Processes $\zeta(t)$ and $(\xi(t), \eta(t))$ appearing in the representation of Ψ and g, solve, respectively Equations (2.45) and (2.42)–(2.43).

Hence, we have proved the following statement.

THEOREM 2.4. *Let* $\varphi \in \Theta_1$ *and the coefficients of Equations* (2.42)–(2.43) *satisfy* (2.13)–(2.15), (2.38)–(2.41). *Then there exists a unique solution* $(\xi(t), \eta(t), g(s,x))$ *of the system* (2.42)–(2.44) *for all* $t \in \Delta$, *where* Δ *is the interval constructed in the proof of Proposition* 2.3.

It follows from Chapter 3 and the previous section of this chapter, that the following results hold.

PROPOSITION 2.4. *Suppose that the assumptions of Theorem* 2.4 *hold. Then the function* $\Psi(s,z)$ *of the form* (2.46) *comes to be a weak solution of the Cauchy problem* (2.34), *while the function* $g(s,x)$ *of the form* (2.44) *is a weak solution of the Cauchy problem* (2.37). (2.46) *is the unique solution of* (2.34) *in* Θ *while* (2.44) *is the unique solution of* (2.37) *in* Θ_1.

Remark 2.3. All the above results may be also stated for the systems

$$\xi(t) = x + \int_s^t a_{\xi(\tau)}(\tau, u(\tau, \xi(\tau)))\, d\tau + \int_s^t A_{\xi(\tau)}(\tau, u(\tau, \xi(\tau)))\, dw,$$

$$u(s,x) = Ef(\xi(T)),$$

and

$$\xi(t) = x + \int_s^t a_{\xi(\tau)}(\tau, g(\tau, \xi(\tau)))\, d\tau + \int_s^t A_{\xi(\tau)}(\tau, g(\tau, \xi(\tau)))\, dw,$$

$$\eta(t)\, y + \int_s^t c_{\xi(\tau)}(\tau, g(\tau, \xi(\tau)))\, \eta(\tau)\, d\tau + \int_s^t C_{\xi(\tau)}(\tau, g(\tau, \xi(\tau)))\, (\eta(\tau)\, dw),$$

$$\langle y, g(s,x)\rangle = E\langle \eta(T), \varphi(\xi(T))\rangle,$$

if their coefficients obey the above conditions.

The corresponding parabolic problems in this case have the form

$$\frac{\partial u}{\partial s} + \mathfrak{A}_u(s)\, u = 0, \qquad u(T,x) = f(x),$$

where

$$\mathfrak{A}_u(s)\varphi = \frac{1}{2}\operatorname{Tr}\langle A_x(s,u(s,x))\, A_x^*(s,u(s,x))D, D\rangle\varphi + $$
$$+ \langle a_x(s,u(s,x)), D)\rangle\varphi$$

and

$$\frac{\partial g}{\partial s} + \mathfrak{A}_g(s) + \mathfrak{A}_g^1(s)g = 0, \quad g(T,x) = \varphi(x),$$

where

$$\mathfrak{A}_g^1(s)\varphi = \operatorname{Tr} C_x^*(s,g(s,x))\big(A_x^*(s,g(s,x))D, \varphi(x)\big) +$$
$$+ c_x^*(s,g(s,x))\varphi.$$

2.3. SMOOTH SOLUTIONS OF QUASILINEAR PARABOLIC EQUATIONS

As has been mentioned in Section 1, in order to prove that the solutions of the systems (2.15)–(2.16) or (2.42)–(2.44) are smooth, one has to deal with differential extensions of these equations. By investigation we are able to state the conditions which the initial system coefficients must obey in order to have smooth solutions.

Thus, along with (2.42)–(2.44), let us consider the relations satisfied by

$$\alpha(t) = \xi'_x(t)h \quad \text{and} \quad \beta(t) = \eta'_x(t)h, \quad h \in B,$$

$$\alpha(t) = h + \int_s^t a'_{\xi(\tau)}(\tau, g(\tau))\alpha(\tau)\, d\tau +$$
$$+ \int_s^t A'_{\xi(\tau)}(\tau, g(\tau))(\alpha(\tau), dw(\tau)), \tag{2.54}$$

$$\beta(t) = \int_s^t c'_{\xi(\tau)}(\tau, g(\tau))(\alpha(\tau), \eta(\tau))\, d\tau +$$
$$+ \int_s^t C'_{\xi(\tau)}(\tau, g(\tau))(\alpha(\tau), \eta(\tau), dw) +$$
$$+ \int_s^t c_{\xi(\tau)}(\tau, g(\tau))\beta(\tau)\, d\tau +$$
$$+ \int_s^t C_{\xi(\tau)}(\tau, g(\tau))(\beta(\tau), dw), \tag{2.55}$$

$$\langle y, g'(s,x)\rangle = E\langle\beta(\tau), \varphi(\xi(\tau))\rangle + \tag{2.56}$$
$$+ E\langle\eta(\tau), \varphi'(\xi(\tau))\alpha(\tau)\rangle.$$

Recall that the system (2.42)–(2.44), (2.54)–(2.56) is of the same kind as the initial system (2.42)–(2.44). In other words, a formal differentiation does not lead us out of the class of systems which has been studied in the previous section.

Let us assume now that the coefficients $a_x(t, g)$, $A_x(t, g)$, $c_x(t, g)$ and their derivatives up to the second order, meet the requirements of Section 2. Then the results of that section enable us to verify that the extended system possesses a unique solution, which implies that the initial system has a solution smooth up to second order.

Furthermore, the solvability of the system (2.42)–(2.44), (2.54)–(2.56) due to the results of Section 1, implies that there exists a weak solution of the Cauchy problem

$$\frac{\partial G}{\partial s} + \mathcal{B}(s)\, G = 0, \qquad G(T, x) = (\varphi(x), \varphi'(x)),$$

where

$$G(t, x) = (g(t, x),\ g'(t, x)),$$

$$\mathcal{B}\, G = \begin{pmatrix} \mathfrak{A}_g & 0 \\ 0 & \tilde{\mathfrak{A}}_g \end{pmatrix} \begin{pmatrix} g \\ g' \end{pmatrix},$$

$$\begin{aligned} \tilde{\mathfrak{A}}_g\, g' &= \frac{1}{2} \operatorname{Tr} A'_x(s, g(s))\, A^*_x(s, g(s))\, D\, g' + \\ &+ \frac{1}{2} \operatorname{Tr} A_x(s, g(s))\, (A^*_x)'(s, g(s))\, D\, g' + \\ &+ a'_x(s, g(s))\, g' + \operatorname{Tr}\, (C^*_x)'(s, g(s))\, A_x(s, g(s))\, g' + \\ &+ \operatorname{Tr}\, C^*_x(s, g(s))\, A'_x(s, g(s))\, g' + \\ &+ (c^*_x)'(s, g(s))\, g + \mathfrak{A}_g\, g'. \end{aligned}$$

As a consequence, we obtain the existence of the solution of the Cauchy problem (2.37).

We may construct a smooth (up to the second order) solution of (2.37) by successive applications of the above procedure in case its coefficients and Cauchy data are smooth enough.

Summarizing all the above considerations, we obtain the following result.

THEOREM 2.5. *Let* $\varphi(x) \in C_2(B, B^*_1)$ *and coefficients* $a_x(t, g)$, $A_x(t, g)$, $c_x(t, g)$, $C_x(t, g)$ *along with their derivatives up to the second order with respect to* x, *satisfy estimates similar to* (2.12)–(2.15), (2.38)–(2.41). *Then there exists a unique classical solution of* (2.37) *given in the form of* (2.44).

2.4. CAUCHY PROBLEM FOR QUASILINEAR PARABOLIC EQUATIONS OVER MANIFOLD AND VECTOR BUNDLES

Let X be a smooth Banach manifold with a connection which satisfies all the conditions of Chapter 2. Assume, moreover, that its iterated tangent bundles $T^k X$ are normed vector bundles. Recall that the stochastic equations on X have been investigated in Chapter 4, together with their differential extensions which appeared to be stochastic equations on $T^k X$ of the same kind as the initial ones.

In this section, we are going to realize the probabilistic approach to the construction of the solution of the Cauchy problem (2.1) which have been mentioned in Section 1. While developing this approach in a nonlocal situation, we meet the following obstacles.

1. The relation (2.4) for $u = f$ does not admit the localization and, hence, its solution cannot be constructed by patching together the local problem solutions.

2. To state the Lipschitz conditions on functions defined on a nonlocal manifold X, which have been proved to be enough for the existence of solutions of the stochastic equation, one needs an additional metric structure on X. To avoid introducing this structure, we deal only with smooth coefficients having bounded derivatives. This makes it necessary to assume that X has normed tangent bundles $T^k X$. Recall that we already needed this assumption in Chapter 4.

Choose for Θ a function set out of $C_2(X, R^1)$ with norm $|||u||| = \sup_{x \in X} |u(x)|$ and assume that the coefficients $a_x(t, u)$, $A_x(t, u)$ in (2.5) satisfy the estimates

$$\|\nabla_h^k a_x(t, u)\|_x^2 + \sigma_2^2 \left(\nabla_h^k A_x(t, u)\right) \le L_u(t) \, \|h\|_x^2 , \tag{2.57}$$

$$\|D_u^k a_x(t, u)\|_x^2 + \sigma_2^2 \left(D_u^k A_x(t, u)\right) \le C \left[1 + |||u|||^{2(p-k)}\right], \quad k = 1, 2. \tag{2.58}$$

Consider the system of equations

$$d\,\xi^u(t) = \exp_{\xi^u(t)} \left(a_{\xi^u(t)}(t, u(t))\, dt + A_{\xi^u(t)}(t, u(t))\, dw\right), \tag{2.59}$$

$$u(s, x) = E f\left(\xi_{s,x}^u (T)\right), \tag{2.60}$$

where $\xi_{s,x}^u(t)$ is solution of (2.59) such that

$$\xi_{s,x}^u(s) = x . \tag{2.61}$$

The system (2.59)–(2.60) is essentially nonlocal and cannot be localized. Nevertheless, the corresponding linearized system

$$d\,\xi^n(t) = \exp_{\xi^n(t)} \left(a_{\xi^n(t)}(t, u^n(t))\, dt + A_{\xi^n(t)}(t, u^n(t))\, dw\right), \tag{2.62}$$

$$u^n(s, x) = E f\left(\xi_{s,x}^{n-1}(T)\right), \tag{2.63}$$

admits the localization. More precisely, it is easy to see, thanks to the results of Chapter 4, that the solution of (2.62) may be obtained by patching together the localized equation solutions. Notice that the functions $u^n(s, x)$ of the form (2.63) give once again successive approximations of the functional equation

$$u(s, x) = \Phi(s, x, u(s))$$

solution.

To prove that the sequence $u^n(s, x)$ converges, we need some auxiliary results.

Consider a stochastic equation

$$d\,\xi^{\upsilon}(t) = \exp_{\xi^{\upsilon}(t)}\left(a_{\xi^{\upsilon}(t)}(t, \upsilon(t))\, dt + A_{\xi^{\upsilon}(t)}(t, \upsilon(t))\, dw\right) \tag{2.64}$$

with the coefficients depending upon the parameter υ. A random process $\xi_x^{\upsilon}(t)$ which satisfies (2.64) and the condition

$$\xi_x^{\upsilon}(s) = x,$$

depends on two parameters, namely $x \in X$ and $\upsilon \in \Theta$. If (2.57)–(2.58) hold, then due to results of Chapter 4 $\xi_x^{\upsilon}(t)$ depends smoothly on x and υ.

Along with $\xi_x^{\upsilon}(t)$, we shall need some information about the two processes $\alpha(t) = D_x \xi_x^{\upsilon}(t)\, h$, $h \in B$ and $\beta(t) = D_{\upsilon}\, \xi_x^{\upsilon}(t)\, \gamma$, $\gamma \in \Theta$, solving the equations

$$d\,\alpha(t) = \exp^{TX}_{\alpha(t)}\left(a^1_{\alpha(t)}(t, \upsilon(t))\, dt + A^1_{\alpha(t)}(t, \upsilon(t))\, dw\right., \tag{2.65}$$

$$\begin{aligned} d\,\beta(t) &= \exp^{TX}_{\beta(t)}\left[a^2_{\beta(t)}(t, \upsilon(t)) + a^1_{\beta(t)}(t, \upsilon(t))\right] dt + \\ &\quad + \left[A^2_{\beta(t)}(t, \upsilon(t)) + A^1_{\beta(t)}(t, \upsilon(t))\right] dw, \end{aligned} \tag{2.66}$$

where

$$a^1_{x,h}(t, \upsilon) = (a_x(t, \upsilon), \quad a'_x(t, \upsilon)\, h),$$

$$A^1_{x,h}(t, \upsilon) = (A_x(t, \upsilon), \quad A'_x(t, \upsilon)\, h),$$

$$a^2_{x,h}(t, \upsilon)\, \gamma = (D_{\upsilon}\, a_x(t, \upsilon)\, \gamma, \ 0),$$

$$A^2_{x,h}(t, \upsilon) = (D_{\upsilon} A_x(t, \upsilon)\, \gamma, \ 0),$$

$$x \in X, \quad h \in B, \quad \gamma \in \Theta.$$

Let us derive the estimates on processes $\alpha(t)$ and $\beta(t)$ satisfying (2.64) and (2.65). For this purpose we shall use the multiplicative representation of stochastic equation solution (4.2.12) and (4.3.7) and apply the results obtained in the previous section about local equation solutions. Then the standard technique which has been used more than once above, permits us to obtain the following estimates

$$E\, \|\alpha(T)\|^2_{\xi(T)} \le \|h\|^2_x \exp C \int_s^T \left[\rho_0 + \rho\, K_{\upsilon}(\tau)\right] d\tau, \tag{2.67}$$

$$E\, \|\beta(T)\|^2_{\xi(T)} \le C_1\, |T - s| \exp C\,(T - s) \tag{2.68}$$

in case the coefficients a, A satisfy (2.13)–(2.15), while $D_x a_x$, $D_x A_x$ satisfy (2.38)–(2.41).

Let Φ be a mapping defined by the left-hand side of (2.60). In order to verify that Φ acts in Θ, let us construct a positive function $\gamma(s)$ such that the estimate

$$|\langle \upsilon'(s,x),\ h\rangle| \le \gamma(s)\ \| h \|$$

implies the estimate

$$|\langle u'(s,x),\ h\rangle| \le \gamma(s)\ \| h \| \tag{2.69}$$

for the function

$$u(s,x) = E f\,(\xi^{\upsilon}_{s,x}\ (T)). \tag{2.70}$$

Let $f \in \Theta$; then it results from (2.67), (2.68) that

$$| \langle h, u'(s,x) \rangle |^2 \le E\ |||f'|||^2\ \| \alpha(T) \|^2 \le$$

$$\le \| h \|^2\ \exp\ C \int_s^T\ [2\,\rho_0 + 3\,\rho_1\,K_{\upsilon'}(\tau)]\ d\tau\, K_{f'}\ , \tag{2.71}$$

where $K_{\upsilon'}(s)$ is the minimal constant such that

$$|\langle h, \upsilon'(s\,,x) \rangle| \le K_{\upsilon'}(s)\ \| h \|\,.$$

It follows from (2.71) that

$$K_{u'}(s) \le K_{f'}\ \exp\ C \int_s^T\ [2\,\rho_0 + 3\,\rho\,K_{\upsilon'}(\tau)]\ d\tau\,,$$

and the arguments of the previous section show that the function

$$\gamma(s) = \frac{2\rho_0\ \exp\,(2\,\rho_0 + 3\,\rho)\,(T-s)}{2\,\rho_0 + 3\,\rho\,K_{f'}\ \exp\ (2\,\rho_0 + 3\,\rho)\,(T-s) + 3\,\rho\,K_{f'}} \tag{2.72}$$

meet all the necessary requirements. This implies that $u'(s,x)$ is a bounded function over the interval $\Delta = [s, T]$ whose length may be estimated as

$$|T - s| \le \frac{1}{2\,\rho_0}\ \ln\ \left(1 + \frac{2\rho_0}{3\,\rho\,K_{f'}}\right)\,, \tag{2.73}$$

if $2\,\rho_0 + 3\,\rho_1\,K_{f'} > 0$.

Hence, for all $s \in \Delta$, $u(s,x) \in \Theta$, if $\upsilon(s,x) \in \Theta$. In addition, the mapping Φ defined by the relation

$$u(s,x) = \Phi(s,x,u) = E f\big(\xi^{u}_{s,x}\ (T)\big)\,, \tag{2.74}$$

where $\xi^{u}_{s,x}(t)$ is a solution of (2.64) such that $\xi^{u}_{s,x}(s) = x$, acts in Θ.

Denote by $\mathcal{N}$ the set of continuous bounded mapping from Δ into Θ and put

$$\| \alpha \|_{\mathcal{N}} = \sup_{\tau \in \Delta}\ \| \alpha(\tau) \|_{\Theta}\,.$$

The above consideration implies that the mapping Ψ defined by (2.60) acts in $\mathcal{N}$. Let us prove that there exists $N > 0$ such than for any $n > N, \Phi^n$ is a contraction mapping.

Consider

$$\ell_n (s) = \| \Phi^n (s, \upsilon_1 (s)) - \Phi^n (s, \upsilon (s)) \|_\Theta .$$

It follows from (2.74) and (2.67)–(2.68) that

$$| \ell_1 (s, \upsilon) - \ell_1 (s, \upsilon_1) | \leq K_{f'} (T - s) \exp C (T - s) \| \upsilon - \upsilon_1 \|_{\mathcal{N}}$$

and, as a consequence,

$$| \ell_n (s, \upsilon) - \ell_n (s, \upsilon_1) | \leq \frac{K_{f'}^n (T-s)^n C^n}{n!} \| \upsilon - \upsilon_1 \|_{\mathcal{N}} .$$

Finally

$$\| \Psi^n (\upsilon_1) - \psi^n (\upsilon) \|_{\mathcal{N}}$$

$$= \sup_s \| \Phi^n (s, \upsilon_1 (s)) - \Phi^n (s, \upsilon (s)) \|_\Theta \leq$$

$$\leq \frac{K_{f'}^n (T-s)^n C^n}{n!} \times \| \upsilon - \upsilon_1 \|_{\mathcal{N}} .$$

Since given constants $K_{f'}$, C and $\delta = | T - s |$, we can always find a number N such that for $n > N$

$$\frac{K_{f'}^n C^n \delta^n}{n!} < 1 ,$$

we deduce that Ψ^n is a contraction mapping in $\mathcal{N}$, and as a consequence, Ψ is a contraction as well. Now the existence of a unique solution of (2.11) results from the contraction mapping theorem.

Reassuming all the above speculations, we obtain the following statement.

THEOREM 2.5. *Let the coefficients of* (2.59) *satisfy* (2.57)–(2.58) *and* $f \in C_2 (X, R^1)$. *Then there exists a unique solution of the system* (2.59)–(2.60) *which gives a Markov process* $\xi (t) \in X$ *and a smooth bounded function* $u (s, x)$ *on* $\Delta \times X$. *Here* Δ *is an interval whose length satisfies the estimate* (2.73).

It results from Subsection 1.4 of the previous section, that the function $u (s, x)$ of the form (2.60) is a weak solution of the Cauchy problem (2.1).

Consider now a more complicated situation.

Let $\pi : \mathcal{E} \to X$ be a normed vector bundle and TX be a normed tangent bundle over the base X. Denote Γ^π, Γ^X, $\Gamma^{\mathcal{E}}$, respectively, the local connection coefficients of the bundle π and of manifolds X and $\mathcal{E}$.

Consider a stochastic differential equation on $\mathcal{E}$

$$d\gamma = \exp^{\mathcal{E}}_{\gamma(t)} \left(a^3_{\gamma(t)} (t, \Phi (t)) \, dt + A^3_{\gamma(t)} (t, \Phi (t)) \, dw \right), \tag{2.75}$$

and suppose that its coefficients

$$a_z^3 \in T_z\,\mathcal{E}, \quad A_z^3 \in L_{12}\,(H, T_z\,\mathcal{E}), \quad z \in \mathcal{E},$$

have the form

$$a_z^3\,(t, \Phi) = \left(a_x\,(t, g),\ c_x\,(t, g)\,y - \Gamma_x^\pi\,(a_x\,(t, g), y)\right),$$

$$A_z^3\,(t\,, \Phi) = \left(A_x\,(t, g),\ C_x\,(t, g) - \Gamma_x^\pi\,(A_x\,(t, g), y)\right),$$

$$z = (x, y), \quad \Phi = \langle y, g_x \rangle, \quad y \in \mathcal{E}_x, \quad g_x \in \mathcal{E}_x^*, \tag{2.76}$$

for a_z^3, a_z^3 (Notice that $a'y = \nabla_y\, a - \Gamma^\pi\,(y, a)$.)

Denote by Θ the set of functions defined on $\mathcal{E}$ which are linear with respect to the fibre arguments, i.e., functions given in the form

$$G\,(z) = \langle y, g\,(x)\rangle_x\,, \tag{2.77}$$

$$g\,(x) \in \delta\,(\pi^*),$$

where $\delta\,(\pi^*)$ is the set of bounded smooth sections of the cobundle $\pi^* : \mathcal{E}^* \to X$. Put

$$|||\,G\,|||_\Theta = \sup_{\|y\|_x = 1}\ \sup_{x \in X}\ |\langle y, g\,(x)\rangle|$$

and

$$|||\,g\,|||_\delta = \sup_{x \in X}\ \|\,g\,(x)\,\|\,.$$

It is worth noticing that

$$|||\,G\,|||_\Theta = |||\,g\,|||_\delta\,.$$

Let the coefficients in (2.75) satisfy the following conditions:

1. $a_x\,(t, g) \in T_x X, \quad A_x\,(t, g) \in L_{12}\,(H,\ T_x X\,), \quad x \in X,$

and the inequalities (2.57)–(2.58) hold for them.

2. $c_x\,(t\,, g) \in L\,(\mathcal{E}_x), \quad C_x\,(t, g) \in L\left(\mathcal{E}_x\,, L_{12}\,(H,\ \mathcal{E}_x)\right)$

are such that

$$\langle c_x\,(t, g)\,y,\ J\,(y)\rangle_x \le \left[\,\rho_0 + \rho\,\|\,g\,\|^{2p}\right]\ \|\,y\,\|^2, \tag{2.78}$$

$$\sigma_2^2\,(C_x\,(t, g)\,y) \le \rho\,\|\,g\|^{2p}\ \|\,y\,\|^2, \tag{2.79}$$

$$\|\,\nabla_h^k\ c_x\,(t, g)\,y\,\|^2 + \sigma_2^2\ (\nabla_h^k\ C_x\,(t, g)\,y\,) \le L_g\,(t)\,\|\,y\,\|^2\ \|\,h\,\|^2, \tag{2.80}$$

$$\|\,D_g^k\ c_x\,(t, g)\,y\|^2 + \sigma_2^2\ (D_g\ C_x\,(t, g)\,y)$$

$$\leq C\left(1+\|g\|^{2(p-k)}\right)\|y\|^2 . \tag{2.81}$$

One may easily verify that these assumptions imply that the coefficients $a^3_z\,(t,\Phi)$, $A^3_z\,(t,\Phi)$ to meet the requirements of Theorem 2.4 and, hence, that there exists a unique solution of (2.75) such that it satisfies the estimates of type (2.21).

Assuming that the coefficients in (2.75) are smooth, one may deal with differential extensions of this equation. These extensions give differential equations on $T\mathcal{E}$ or $T^2\mathcal{E}$ with respect to $\alpha(t) = D_x\,\xi^{\upsilon}_{s,x}\,(t)$,

$$\beta(t) = D_{\upsilon}\,D_x\,\xi^{\upsilon}_{s,x}\,(t)$$

and

$$\alpha_1(t) = D^2_x\,\xi^{\upsilon}_{s,x}\,(t), \quad \beta_1(t) = D^2_x\,\xi^{\upsilon}_{s,x}\,(t),$$

respectively.

Hence, the above arguments permit us to construct their solutions and to obtain the desired estimates, similar to (2.67)–(2.68), for them. After all those preparations, consider the system

$$\mathrm{d}\gamma = \exp^{\mathcal{E}}_{\gamma(t)}\left(a^3_{\gamma(t)}\,(t, G(t))\,\mathrm{d}t + A^3_{\gamma(t)}\,(t, G(t))\,\mathrm{d}w\right), \tag{2.82}$$

$$G(s,z) = E\,\Gamma\left(\gamma^G_{s,z}\,(T)\right), \tag{2.83}$$

where $z = (x, y) \in \mathcal{E}$,

$$\Gamma(z) = \langle y, g(x)\rangle_x . \tag{2.84}$$

Due to (2.57)–(2.58) and (2.78)–(2.81), we may conclude that the system (2.82)–(2.83) satisfies the conditions of Theorem 2.5, which imply that there exists a unique solution $(\gamma(t),\ G(s,z))$ of this system. Here $\gamma(t) \in \mathcal{E}$ is a Markov process on $\mathcal{E}$ and $G(s,z)$ is a function in Θ.

Applying the above result to the case $\mathcal{E} = TX$, we may prove that the solution of (2.59)–(2.60) is smooth if the coeffficients of the system are smooth.

Indeed, let $\mathcal{E} = TX$ and

$$a^3_z\,(t, G) = \left(a_x(t,u),\ \ a'_x(t,u)\,y\right),$$

$$A^3_z\,(t, G) = \left(A_x(t,u),\ \ A'_x(t,u)\,y\right).$$

In this case, the system (2.82)–(2.83) is a system with respect to the pair $(\gamma(t), G(s,z))$, where the process $\gamma(t)$ has the form

$$\gamma(t) = (\xi(t),\ D_x\xi(t))$$

in a local trivialization, and the function G is of the form $G(s,z) = \langle y, u'(s,x)\rangle$. Now the existence of a solution of this system implies that the function $u(s,x)$ satisfying (2.60) possesses a first-order derivative.

Changing TX to T^2X, we may prove in a similar way that the function

$u(s,x)$ is smooth up to the second order if the coefficients of (2.60) and the function f are smooth enough.

Analogous arguments show that the solution of the system (2.82)–(2.83) possesses derivatives up to the second order for smooth enough coefficients.

Finally, let us mention that the results of the previous section open the way to prove the existence of a smooth solution of the following Cauchy problems

$$\frac{\partial u}{\partial s} + \mathfrak{A}_u(s)\, u = 0, \quad u(T,x) = f(x), \tag{2.85}$$

where

$$\mathfrak{A}_u(s) = \frac{1}{2}\left\{\nabla_{A_x(s,u)} \nabla_{A_x(s,u)} - \nabla_{\nabla_{A_x(s,u)} A_x(s,u)}\right\} - \nabla_{a_x(s,u)}$$

and

$$\frac{\partial g}{\partial s} + \mathfrak{A}_g(s)\, g + \mathfrak{A}^1_g(s)\, g = 0, \quad g(T,x) = \varphi(x), \tag{2.86}$$

where

$$\mathfrak{A}^1_g(s) = \mathrm{Tr}\ C^*_x(s, g(s))\, \nabla^{\pi^*}_{A_x(s,g(s))} + c^*_x(s, g(s)).$$

As a result, we deduce the following assertions.

THEOREM 2.6. *Let the coefficients of* (2.85) *satisfy* (2.57)–(2.58) *for* $k = 2$ *and* $f \in C_2(X, R^1)$. *Then there exists a unique* C_2*-smooth solution of* (2.85) *given in the form*

$$u(s,x) = E f(\xi(T)), \tag{2.87}$$

where $\xi(t)$ *is a solution of* (2.59).

THEOREM 2.7. *Let the coefficients of* (2.86) *satisfy* (2.57)–(2.58), (2.78)–(2.81) *for* $k = 2$ *and* $\varphi \in \sigma_2(\pi^*)$. *Then there exists a unique* C_2*-smooth solution of this problem given in the form* (2.83).

3. Forward Kolmogorov Equations

3.1. EVOLUTION FAMILIES IN THE SPACE OF MEASURES

A random family $S(t,\tau)$ of manifold mappings defined by the solution of the stochastic equation (1.2), gives rise to both an evolution family $U(\tau,t)$ acting in the space of functions on X which was studied in the previous section, and to an adjoint family $U^*(t,\tau)$ acting in the space of the bounded measures on X

$$\int_X U(\tau,t)\,\varphi(x)\,\mu(dx) = \int_X \varphi(x)\,[U^*(t,\tau)\,\mu](dx). \tag{3.1}$$

It immediately follows from the definition that $U^*(t,\tau)$ is an evolution family

$$U^*(t,\tau) = U^*(t,s)\, U^*(s,\tau), \quad U^*(t,t) = I,$$

$$(t > s > \tau),$$

with the generator satisfying the relation

$$\int_X \mathfrak{A}_U(\tau)\,\varphi(x)\,\mu(\mathrm{d}x) = \int_X \varphi(x)\,(\mathfrak{A}_{U*}(\tau)\,\mu)\,(\mathrm{d}x) \tag{3.2}$$

for $\varphi \in \mathcal{D}_{\mathfrak{A}_U}$, $\mu \in \mathcal{D}_{\mathfrak{A}_{U^*}}$. Here

$$\mathfrak{A}_{U*}(\tau)\,\mu = \lim \frac{U^*(\tau + \Delta\tau, \tau) - I}{\Delta\tau}\,\mu = \mathfrak{A}_U^*(\tau)\,\mu\,.$$

Let $\mathcal{D}^*$ be a set of measures invariant under the action of $U^*(t, \tau)$, so that

$$\mathfrak{A}_{U*}(t)\,\mathcal{D}^* \subset \mathcal{D}^* \tag{3.3}$$

for any $t \in [0, T]$.

Given $\mu \in \mathcal{D}^*$ and $\mu_t = U^*(t, \tau)\,\mu$, it is easy to obtain the relation

$$\frac{\partial\,\mu_t}{\partial\,t} = \mathfrak{A}_{U*}(t)\,\mu_t\,, \qquad \mu_0 = \mu\,. \tag{3.4}$$

This equation is called the forward Kolmogorov equation for the random family $S(t, \tau)$.

It follows from (3.1) that $U^*(t, \tau)$ may be represented using $S(t, \tau)$. Let $\mu_t^S = \mu \circ S^{-1}(t, \tau)$. Then

$$\int_X \varphi(x)\,[U^*(\tau, t)\,\mu]\,(\mathrm{d}x)$$

$$= E \int_X \varphi(S(t, \tau) \circ x)\,\mu(\mathrm{d}x)$$

$$= \int_X \varphi(x)\,E\,\mu_t^S\,\mathrm{d}x)$$

and, hence,

$$U^*(t, \tau)\,\mu = E\,\mu \circ S^{-1}(t, \tau)\,. \tag{3.5}$$

Notice that this formula yields the relation

$$(U^*(t, \tau)\,\mu)\,(A) = \int_X \mu(\mathrm{d}x)\,P(\tau, x, t, A)\,,$$

where

$$P(\tau, x, t, A) = P\,\{S(t, \tau) \circ x \in A\}$$

is the transition probability of the considered process.

Recall (see Chapter 4) that given $\varphi \in C_2(X, B, R^1)$, we have

$$\mathfrak{A}_U(\tau)\,\varphi(x) = \Big[\frac{1}{2}(A(\tau, x)\,A^*(\tau, x)\,\nabla, \nabla) + (a(\tau, x), \nabla)\Big]\,\varphi.$$

The operator $\mathfrak{A}_U^*$ which is the operator adjoint to $\mathfrak{A}_U$ had been derived in Chapter 2. Given $\mu \in \mathcal{M}_2(X)$, it looks like

$$\mathfrak{A}_U^*(t)\,\mu = \Big\{\frac{1}{2}((\nabla + \lambda) \otimes (\nabla + \lambda), A \otimes A) - \langle \nabla + \lambda\,, a\rangle\Big\}\,\mu$$

$$= \mathfrak{A}_U^*(t)\,\mu\,. \tag{3.6}$$

Thanks to this, the Kolmogorov equation (3.4) may be represented in the form

$$\frac{\partial\,\mu_t}{\partial\,t} = \left\{\frac{1}{2}\left((D+\lambda)\otimes(D+\lambda),\,A\otimes A\right) - \langle D+\lambda\;a\rangle\right\}\mu_t\,. \tag{3.7}$$

The measure $U^*(t,\tau)\,\mu$ given by (3.5) is a solution of the Cauchy problem (3.4), possibly in the weak sense

$$\frac{\partial}{\partial\,t}\int_X \varphi(x)\,\mu_t(\mathrm{d}\,x) = \int_X \varphi(x)\,\mathfrak{A}_U^*(t)\,\mu_t(\mathrm{d}\,x),$$

$$\int_X \varphi(x)\,\mu_t(\mathrm{d}\,x)|_{t=\tau} = \int_X \varphi(x)\,\mu(\mathrm{d}\,x)\,. \tag{3.8}$$

To answer the question of whether this solution is smooth or not, we need to know if the smoothness property of a measure is invariant under a random mapping or not.

Let $S : X \to X$ be a smooth invertible mapping, then (see Chapter 1) the relation

$$\rho_{\mu^S}(h^S, y)|_{y=S\circ x} = \rho_\mu(h, x) \tag{3.9}$$

holds for $h^S(y) = S'(x)\,h(x)$. This relation is equivalent to

$$\int_X \langle h^S(y),\varphi'(y)\rangle\,\mu^S(\mathrm{d}\,y)$$

$$= \int_X \varphi(S\circ x)\,\rho_\mu(h,x)\,\mu(\mathrm{d}\,x)\,. \tag{3.10}$$

Actually, the considered mapping is smooth only in a rather weak sense: for almost all x there exists a sequence of smooth mappings S_n such that

$$S_n\circ x \to S\circ x, \qquad S'_n(x) \to S'(x)$$

on a set K such that $\mu(K) = 1$ while $S'(x)$ is defined ad hoc and only in the mean square sense. Still under these conditions, we may pass to the limit in (3.10) and so prove (3.9) in the considered situation.

Let $h^S = \eta$ be a vector field and

$$h(x) = \eta^{S^{-1}}(x) = [S'(x)]^{-1}\,\eta(S\circ x)\,.$$

It follows from (3.10) that

$$\int_X \langle\eta(y),\varphi'(y)\rangle\,[U^*(t,\tau)\,\mu]\,(\mathrm{d}\,y)$$

$$= \int_X \varphi(S(t,\tau)\circ x)\,\rho_\mu(\eta^{S^{-1}};x)\,\mu(\mathrm{d}\,x)$$

and the estimate

$$\int_X \langle\eta(y),\varphi'(y)\rangle\,[U^*(t,\tau)\,\mu]\,(\mathrm{d}\,y)$$

$$\le \left\{E\int_X |\varphi(y)|^2\,\mu_t^S\,(\mathrm{d}\,y)\right\}^{1/2}$$

$$\leq \{E \int_X |\rho_\mu (\eta^{s^{-1}} (x))|^2 \mu (d x)\}^{1/2} \tag{3.11}$$

holds.

Under the condition

$$\int_X E |\rho_\mu (\eta^{s^{-1}}; x)| \mu (d x) < \infty , \tag{3.12}$$

the left-hand side of (3.11) is a continuous linear functional in $\mathcal{L}_2 (X, \mu)$ with respect to φ and thus there exists a function $\alpha \in \mathcal{L}_2 (X, U^* \mu)$ such that

$$\int_X \langle \eta (y), \varphi' (y)\rangle (U^*\mu)_t (d y)$$

$$= - \int_X \varphi (y) \alpha (y) (U^*\mu)_t (d y).$$

By definition, this function has the logarithmic derivative

$$\alpha (y) = \rho_{U^*\mu} (\eta, y).$$

Similar arguments may be used to calculate higher derivatives. Now we may state the following assertion.

THEOREM 3.1. *Let both the coefficients of the stochastic Equation* (1.2) *and the connection coefficients* Γ^X *meet the conditions of Theorem* 4.3.2 *which guarantee the existence of the smooth* (*in the mean square sense*) *flow* $S (t, \tau)$. *Then the set* $\mathcal{M}_2 (X)$ *of measures having logarithmic derivatives* $\rho_\mu (x) \in \mathcal{L}_2 (X, \mu)$ *is invariant with respect to the mapping* (3.5)

$$\mu \mapsto U^* (t, \tau) \mu .$$

COROLLARY 3.1. *Under the conditions of Theorem* 3.1 *formula* (3.5) *gives the solution of the Cauchy problem* (3.7) *for the forward Kolmogorov equation belonging to* $\mathcal{M}_2 (X)$.

COROLLARY 3.2. *Equation* (3.7) *permits us to derive an equation which describes the evolution of the logarithmic derivative* $\lambda_t (x)$ *of the measure* μ_t. *Notice that to this aim we need only local calculations.*

Let us differentiate both the left-and right-hand sides of Kolmogorov equation (3.4) along a direction h, and use the relation

$$\mathfrak{A}_U^* (t) \mu = \rho_{\mathfrak{A}^*} (t, x) \mu ,$$

where

$$\rho_{\mathfrak{A}^*} = \frac{1}{2} (\nabla + \lambda,\ \nabla + \lambda,\ \mathrm{Tr}\ A \otimes A) - (\nabla + \lambda, a) .$$

Then

$$\frac{\partial}{\partial t} (h , \nabla) \mu_t = [(h, \nabla) \rho_{\mathfrak{A}^*} (t, x) + (h, \lambda) \rho_{\mathfrak{A}^*} (t, x)] \mu_t$$

and, moreover,

$$\frac{\partial}{\partial t} \langle h, \lambda_t (x)\rangle \mu_t + \langle h, \lambda_t (x)\rangle \frac{\partial \mu_t}{\partial t}$$

$$= \left[(h, \nabla) \rho_{\mathfrak{A}^*} (t, x) + \langle h, \lambda_t (x)\rangle \rho_{\mathfrak{A}^*} (t, x)\right] \mu_t . \tag{3.13}$$

Notice that (3.7) includes only one item depending on

$$\lambda'_t (x) - (\nabla \otimes \lambda, \ \mathrm{Tr}\ A \otimes A) .$$

Using the relation

$$(h, \nabla) (h_2, \lambda) = (h_2, \nabla) (h, \lambda) ,$$

which results from logarithmic derivative definition, we obtain

$$\langle h, \nabla \rangle (\nabla \otimes \lambda, \ A A^*)$$

$$= (\nabla \otimes \nabla, \ A A^*) (\lambda, h) + \delta (\nabla, \lambda),$$

where, in $\delta (\nabla, \lambda)$, we include items depending on λ and λ'.

Hence, we obtain the relation

$$\frac{\partial \lambda}{\partial t} = \frac{1}{2} (\nabla \otimes \nabla, \ \mathrm{Tr}\, A A^*) \lambda + \delta_1 (\nabla, \lambda),$$

where $\delta_1 (\nabla, \lambda)$ is nonlinear differential expression which includes items depending on λ and λ'. We do not write the exact form of $\delta_1 (D, \lambda)$, although it is not too complicated.

The resulting quasilinear equation may be investigated following the approach of the preceding section.

3.2. SMOOTHNESS PROPERTY OF TRANSITION PROBABILITY

The relation (3.5) shows that the transition probability

$$P (\tau, x, t, A) = (U^* (t, \tau) \nu_x) (A)$$

is a result of the action of the evolution family $U^* (t, \tau)$ on a point measure

$$\nu_x (A) = \begin{cases} 1, & \text{if } x \in A \\ 0, & \text{if } x \notin A \end{cases} .$$

Notice that for $t = \tau$ $P (\tau, x, t, A) = \nu_x (A)$.

As a measure ν_x is in no case smooth, the above arguments do not imply that the transition probability is smooth.

We now describe another approach which permits us to prove the smoothness of $P (\tau, x, t, A)$ for $t > \tau$ under some additional assumptions.

The stochastic equation (1.2) gives rise to mapping

$$F_x (t, \tau) : \kappa \mapsto \xi_{\tau, x} (t)$$

from the white noise space (i.e., the Hilbert space $\mathcal{H}_-$ described in Chapter 3) onto the manifold X. The image of the canonical (with respect to $\mathcal{H}_-$) Gaussian measure μ under this mapping is just the measure we are interested in

$$P(\tau, x, t, A) = \mu \circ F_x^{-1}(t, \tau)(A).$$

That is why we may state that the transition probability smoothness is a consequence of the smoothness of the Gaussian measure μ, in case the mapping $F_x(t, \tau)$ possesses the corresponding smoothness property.

First consider Equation (1.2) in a linear space $X = B$. Let $h \in \mathcal{H}_0$. By substituting $\kappa + \alpha h$ in (1.2) we obtain the relation

$$\xi(t, \alpha) = x + \int_\tau^t a(s, \xi(s, \alpha))\, ds + \int_\tau^t A(s, \xi(s, \alpha))\, dw + \alpha \int_\tau^t A(s, \xi(s, \alpha))\, h(s)\, ds$$

for the random variable

$$\xi(t, \alpha) = F_x(t, \tau)(\kappa + \alpha h).$$

We may differentiate this relation (in the mean square sense) with respect to the parameter α and calculate the derivative for $\alpha = 0$. As a result, we obtain the linear equation

$$\eta_{\tau,h}(t) = h + \int_\tau^t a'(s, \xi(s))\, \eta_{\tau,h}(s)\, ds + \int_\tau^t A'(s, \xi(s))(\eta_{\tau,h}(s), dw(s)) + \int_\tau^t A(s, \xi(s))\, h(s)\, ds \tag{3.14}$$

for the derivative

$$\xi'_\alpha(t, 0) = F'_x(t, \tau)(\kappa) \circ h = \eta_{\tau,h}(t). \tag{3.15}$$

Let $V(t, \tau)$ be an operator multiplicative functional of the process $\xi(t)$ corresponding to a linear equation

$$d\eta = a'(t, \xi(t))\, \eta(t)\, dt + A'(t, \xi(t))(\eta(t), dw). \tag{3.16}$$

One may give a solution of the nonuniform equation (3.15) in the form

$$\eta_{\tau,h}(t) = F'_\kappa(t, \tau)\, h = \int_\tau^t V(t, s; \xi(\cdot))\, A(s, \xi(s))\, h(s)\, ds. \tag{3.17}$$

Consider now a vector field $z(y)$ over X and a $\mathcal{H}_-$-valued vector field $\upsilon(\kappa)$ over $\mathcal{H}_-$, $F_x(t, \tau)$-compatible with z. The $F_x(t, \tau)$ compatibility condition for vector fields υ and z, which reads

$$z(y)\big|_{y = F_x(t,\tau) \circ \kappa} = F'_x(t, \tau)(\kappa)\, \upsilon, \tag{3.18}$$

leads to (3.15).

As it follows from the results of Chapter 2, the measure

$$P(t, x, \tau, A) = P\{\omega : \xi_{\tau,x}(t) \in A\}$$

giving the $F_x(t, \tau)$-image of the Gaussian measure μ, possesses a square integrable logarithmic derivative if the estimate

$$|||\upsilon|||^2 = E\{\|\upsilon\|^2 + \sigma_2^2(\upsilon')\}$$

$$= \int_{\mathcal{H}_-} \{\|\upsilon(\kappa)\|^2_{\mathcal{H}_0} + \sigma_2^2(\upsilon'(\kappa))\} \mu(dx) < \infty \tag{3.19}$$

holds.

So to prove the smoothness property of $P(\tau, x, t, A)$, we ought to state the conditions which guarantee that, given $z(y)$, the vector field υ which solves (3.18), satisfies the estimate (3.19).

Let us look for a solution of (3.18) in the form

$$\upsilon = A^*(\tau, \xi(\tau)) V^*(\tau, t, \xi(\cdot)) q.$$

Then given $q \in \mathcal{H}$, we obtain the equation

$$R_q = z, \tag{3.20}$$

where

$$R = \int_\tau^t V(t, s; \xi(\cdot)) A(s, \xi(s)) A^*(s, \xi(s)) V^*(s, t, \xi(\cdot)) \, ds. \tag{3.21}$$

If, moreover, $q = R^{-1} z$ exists a.s., then

$$\upsilon(s, \kappa) = A^*(s, \xi(s)) V^*(s, t, \xi(\cdot)) R^{-1} z \, (F_x(s, \tau) \circ \kappa). \tag{3.22}$$

Now we may state an intermediate assertion.

PROPOSITION 3.1. *Assume that on the space $\mathcal{H}_-$ endowed with the Gaussian measure μ, there exists a vector field υ of the form* (3.22) *which satisfies* (3.19). *Then the transition probability $P(\tau, x, t, A)$ is differentiable along the vector field* (3.18).

Let us consider conditions to guarantee the solvability of (3.20).

It follows from the results of Chapter 3 that the multiplicative functional V is (a.e.) invertible

$$V^{-1}(\tau, t, \xi(\cdot)) = V(t, \tau; \xi(\cdot)) \quad (t > \tau).$$

The right-hand side random linear operator solves an inverted (with respect to the time variable) linear stochastic equation in the space of the Hilbert–Schmidt operators and, under the conditions of Theorem 3.5.1, it satisfies estimates of the type

$$E \|V(t, \tau; \xi(\cdot))\|^{2k} \le C < \infty. \tag{3.23}$$

Using the relation

$$V(t, \tau; \xi(\cdot)) = V(t, s; \xi(\cdot)) V(s, \tau; \xi(\cdot)),$$

we change (3.22) into

$$\upsilon(t,\kappa) = A^*(t,\ \xi(t))\ V^*(\tau, t;\xi(\cdot))\ R_0^{-1} z_0\,. \tag{3.24}$$

where

$$z_0 = V(\tau, t; \xi(\cdot))\, z\,(F_x(t,\tau)\circ\kappa), \tag{3.25}$$

$$R_0 = V(\tau, t; \xi(\cdot))\, R\, V^*(t,\tau; \xi(\cdot)) \tag{3.26}$$

$$= \int_\tau^t (s,\tau;\xi(\cdot))\, A(s,\xi(s))\, A^*(s,\xi(s))\, V^*(\tau, s; \xi(\cdot))\ ds\,.$$

Now suppose that X is equipped with a Hilbert–Schmidt structure $(X, \mathcal{H}, i)$ and the coefficients $a(t,x)$, $A(t,x)$ of Equation (1.2) possesses the properties

$$A(t,x) \in L_{12}(H_0, \mathcal{H}),$$

$$A'(t,x) \in L_{22}(H_0 \times \mathcal{H}, \mathcal{H}),$$

$$a(t,x) \in L_{12}(\mathcal{H}). \tag{3.27}$$

In addition, assume that all the above functions are bounded in corresponding norms uniformly over $[0, t] \times X$.

The representation

$$A(t,x) = A_0(t) + A_1(t,x) \tag{3.28}$$

now follows from (3.23), where

$$A_0(t) \in L(H_0, \mathcal{H}) \text{ and } A_1(t,x) \in L_{12}(H_0, \mathcal{H}),\ A'(t,x) \equiv A'_1(t,x).$$

Due to properties of the embedding operator $i \in L(\mathcal{H}, X)$ the stochastic equation (1.2) may be solved in the space X. On the other hand, the operator multiplicative functional $V(t,\tau)$ defined by Equation (3.16) acts on $\mathcal{H}$ and $V(t,\tau) - I \in L_{12}(\mathcal{H})$ (a.s.)

Given a vector field $z(x) \in \mathcal{H}$, $x \in X$, we may consider the operator R as an element of the space $L(H)$ and, thus, omit the embedding operator.

Assume now that $A(t,x)$ is an invertible operator in a neighborhood $S_{x,\mathcal{E}} = \{y : \| y - x \| < \mathcal{E}\}$ of a certain point $x \notin X$ and that

$$\| [A(t,x)\, A^*(t,x)]^{-1} \| \le \alpha\,. \tag{3.29}$$

Consider a random interval $[0, \tilde{\tau}]$ for which the relations

$$\xi_{\tau,x}(t) \in S_{x,\mathcal{E}}, \quad t \in [0, \tilde{\tau}], \tag{3.30}$$

$$\sigma_2^2\,[V(t,\tau) - I] \le r < 1 \tag{3.31}$$

hold, and estimate the norm of the operator R_0^{-1}.

Given $\varphi \in \mathcal{H}$, we have

$$(R_0\,\varphi, \varphi) \ge \int_\tau^t (A(s,\xi(s))\, A^*(s,\xi(s))\, V^*(\tau, s; \xi(\cdot))\, \varphi,$$

$$V^* (\tau, s; \xi(\cdot))\, \varphi)\, d s$$

$$\geq \frac{1}{\alpha} \int_\tau^t \| V^* (\tau, s; \xi(\cdot)) \|^2 d s$$

$$\geq \frac{\| \varphi \|^2}{\alpha} |\tilde{\tau} - \tau|$$

and thus,

$$\| R_0^{-1} \| \leq \frac{\alpha}{|\tilde{\tau} - \tau|} . \tag{3.32}$$

This estimate implies that the solution $q \in \mathcal{H}$ of Equation (3.20) does exist μ-a.e. and, hence, that the desired vector field υ over the space $\mathcal{H}_-$ does exist as well.

To check (3.19), we need the following statement.

PROPOSITION 3.2. *Let* (3.23) *and* (3.25) *be fulfilled. Then the length of the interval* $[\tau, \tilde{\tau}]$ *for which* (3.30) *and* (3.31) *hold satisfies the estimate*

$$E \frac{1}{|\tilde{\tau} - \tau|^m} < \infty \quad (m = 1, 2) . \tag{3.33}$$

Proof. The random process $\gamma(s) = (\xi(s), V(s, \tau))$ solves a stochastic equation with linear (with respect to the second argument) coefficients in the Hilbert space $\mathfrak{H} = \mathcal{H} \oplus L_{12}(\mathcal{H})$. Without loss of generality, we may assume that the coefficients are bounded over the whole space, as we may change them out of a ball with a radius $r_1 > r$ without changing the solution inside the considered region. So we have to estimate the first exit time for the process

$$\gamma(t) - \gamma(0) = \int_0^t a_1(s)\, d s + \int_0^t A_1(s)\, d w$$

from a fixed neighborhood of zero assuming that

$$\| a_1 \|_{\mathcal{H}} \leq C, \quad \sigma_2(A_1) \leq C, \quad \gamma(0) = (x, 0) .$$

First, notice that for $|t| > r/2C$

$$\left\| \int_0^t a_1(s)\, d s \right\| \leq \| a_1 \| t \leq \frac{r}{2} .$$

Due to this, for small ℓ, we have

$$P\{\tau \leq \ell\} = P\left\{ \sup_{0 \leq t \leq \ell} \| \gamma(t) - \gamma(0) \| > \mathcal{E} \right\}$$

$$\leq P\left\{ \sup_{0 \leq t \leq \ell} \left\| \int_0^t A_1(s)\, d w(s) \right\| > \frac{\mathcal{E}}{2} \right\} = o\left(\frac{\ell^m}{\mathcal{E}^{2m}}\right) .$$

The last estimate implies the convergence of the integral in

$$E\left(\frac{1}{|\tilde{\tau}-\tau|^m}\right) \le \int_0^\infty \frac{1}{s^m}\, dP(\tau < s) = m \int_0^\infty \frac{P(\tau < s)}{s^{m+1}}\, ds < \infty .$$

COROLLARY. *Under conditions* (3.23)–(3.25), *the estimate*

$$E \| \upsilon \|^{2k} < \infty , \quad (k = 1, 2)$$

is valid.

Let us investigate the second item in the right-hand side of (3.19). Since, by differentiation of (3.22) with respect to $\kappa \in \mathcal{H}_-$, we obtain rather cumbersome expressions, we prefer to discuss only the specific features of their components.

First, the differentiation of the operator R_0^{-1} gives factors like R_0^{-k} in the correspondent products. Those factors may be estimated on the basis of (3.28) and (3.29). The derivative of the evolution operator V may be computed by differentiating both parts of the equation which this operator obeys in a way similar to that in deriving (3.17). This leads to dealing with equations whose coefficients are derivatives of the initial equation coefficients. Factors of the type $E \| V(t, \tau) \|^k$ which are present in all the expressions, may be estimated as solutions of corresponding stochastic equations.

Finally, each item of the resulting expression include, along with the above-mentioned factors, another factor defined by the linear map

$$h \mapsto F'_x(t, \tau)\, h .$$

This map, due to (3.17), is a linear integral operator acting from $\mathcal{H}_0 = L_2([0, T], H_0)$ into B and, hence, presents a Hilbert–Schmidt operator. In this way, we may estimate

$$E\, \sigma_2^2(\upsilon'(x)) < \infty .$$

The final result may be stated as follows.

THEOREM 3.2. *Let* $\xi(t)$ *be a solution of* (1.21) *in a space* B *equipped with a Hilbert–Schmidt structure* (B, H, i). *Assume that the coefficients of* (1.21) *satisfy* (3.27), $a \in C_k(X)$, $A \in C_k(X)$ *and* $A(t, x)$ *is degenerate in a neighborhood of the point* x_0 *(i.e.* (3.29) *holds). Then the transition probability* $P(\tau, x_0, t, A)$ *of the process* $\xi(t)$ *which solves* (1.21), *is differentiable along the vector fields out of* $\sigma_k(\tau)$, *and possesses a square integrable logarithmic derivative.*

Remark 3.1. The arguments which have been discussed before stating Theorem 3.2, show that, under theorem conditions, the transition probability is as smooth as the equation coefficients a, A. If those coefficients are C_k-smooth and their derivatives give Hilbert–Schmidt mappings, then the transition probability possesses a C_k-smooth (along $\mathcal{H}$) logarithmic derivative.

Now we will show that a slight modification of the above arguments enables us to extend them to smooth manifold valued stochastic equations, assuming that the manifold has a normed (for example a Hilbert) structure.

Consider a stochastic equation

$$d\,\xi(t) = \exp^{X}_{\xi(t)}\left(a_{\xi(t)}(t)\,d\,t + A_{\xi(t)}(t)\,d\,w\right) \tag{3.34}$$

with smooth coefficients a, A along with its differential extension

$$d\,\gamma(t) = \exp^{TX}_{\gamma(t)}\left(a^{1}_{\gamma(t)}(t)\,d\,t + A^{1}_{\gamma(t)}(t)\,d\,w\right). \tag{3.35}$$

The last equation generates a multiplicative functional in the tangent bundle (see Chapter 4)

$$Y(t,s)\circ Y(s,\tau) = Y(t,\tau) \qquad \text{(a.s.)}$$

which is an invertible mapping (a.e.).

Assume that the manifold linear connection is compatible with its Hilbert–Schmidt structure and the coefficients a, A satisfy conditions like (3.27)

$$A(t,x) \in L(H_x, \mathcal{H}_x),$$

$$A'(t,x) \in L_{22}(H_x \times \mathcal{H}_x, \mathcal{H}_x),$$

$$a'(t,x) \in L_{12}(\mathcal{H}_x).$$

Then the mapping

$$Y(t,\tau) : T_{\xi(\tau)}X \to T_{\xi(t)}X$$

is compatible with this structure as well

$$Y(t,\tau) : \mathcal{H}_{\xi(\tau)} \to \mathcal{H}_{\xi(t)}$$

and satisfies the estimate

$$E\,\|Y(t,\tau)\|^2 \le \exp\{C(t-\tau)\}.$$

PROPOSITION 3.3. *The derivative* $F'_x(t,\tau)$ *of the mapping* (3.14) *along* $h \in \mathcal{H}$ *is given by*

$$F'_x(t,\tau)\,h = \int_\tau^t Y(t,s;\xi(\cdot))\,A(s,\xi(s))\,h(s)\,d\,s \in \mathcal{H}_{\xi(t)}. \tag{3.36}$$

Proof. Let us use the multiplicative representation of

$$F_{x,q}(t,\tau) = \tilde{F}_x(t,t_{n-1})\circ\tilde{F}(t_{n-1},t_{n-2})\circ\ldots\circ\tilde{F}(t_1,\tau)\circ x$$

and compute its derivative with respect to x along $h \in \mathcal{H}$. Since each factor in this representation is local, its derivative may be computed according to (3.17). As a result, we obtain that

$$F'_{x,q}(t,\tau)\,h = \sum_k Y_q(t_{k+1},t_k)$$

which leads to (3.36) after passing to the limit.

All further constructions may be done like those in the local case.

Consider the equation

$$z_{F_x(t,\tau)} = \int_\tau^t Y(t, s; \xi(\cdot))\, A(s, \xi(s))\, \upsilon(s, \kappa)\, ds \tag{3.37}$$

with respect to a vector field υ on the space $\mathcal{H}_-$ connected with a vector field z on X.

To find the solution of this equation, we write υ in a form similar to (3.24),

$$\upsilon(s, \kappa) = A^*(s, \xi(s))\, Y^*(\tau, s; \xi(\cdot))\, R_0^{-1} z_0, \tag{3.38}$$

where $z_0 = Y(s, \tau, \xi(\cdot))\, z$ is a vector field on X being a section of the Hilbert bundle and

$$R_0 = \int_\tau^t Y(t, s; \xi(\cdot))\, A(s, \xi(s))\, A^*(s, \xi(s))\, Y^*(s, t; \xi(\cdot))\, ds$$

is an operator in a fixed Hilbert space $\mathcal{H}_x$, $x = \xi(\tau)$.

Notice that the estimate of R_0^{-1}, as well as the proof of the proposition connected with this estimate, are of a local nature and, thus, may be extended without any change to the considered case. All the other arguments necessary for the estimate of the vector field $\upsilon(s, \kappa)$ derivative are connected with linear space mappings. Like in the local situation, we omit detailed computations which come out to be even more complicated but are of the same nature and restrict ourselves to stating the final result.

THEOREM 3.3. *Let X be a C_k-smooth manifold with a normed Hilbert–Schmidt structure (for example, a Riemannian manifold). Assume that Ito's field (a_x, A_x) possesses the properties*

$$a_x \in \mathcal{H}_x, \quad A_x \in L_{12}(H_x, \mathcal{H}_x), \quad A'_x \in L_{22}(H_x \times \mathcal{H}_x, \mathcal{H}_x).$$

Then to guarantee the existence of the square integrable logarithmic derivative of the transition probability $P(\tau, x, t, A)$, it is enough that there exists a solution υ of (3.37) *satisfying* (3.19). *This solution does exist if A is nondegenerate*

$$\| A(t, x)\, A^*(t, x) \|^{-1} \le C$$

in a neighborhood of an initial point $x \in X$.

Remark. Similar to the case of linear space, the transition probability $P(\tau, x, t, A)$ is as smooth as the coefficients of (3.34) up to infinite order of smoothness.

Let us mark the important case when $\gamma \subset \tau_x$ is a trivialized bundle. Assuming that it is trivial one may choose a trivial connection (since another connection will only lead to a change of the vector field). Then all the speculations necessary for the investigation of (3.34) are similar to those in the linear case.

CHAPTER 6

Diffusion Processes on Lie Groups and Principal Fibre Bundles

Let X be a smooth Banach manifold and G a group of smooth transformations of X. Sometimes we shall denote such a pair by (X, G). In this chapter, we shall deal with stochastic processes on X compatible in a certain sense with G-actions. Particular attention will be paid to the case $X = G$ and $X = \mathcal{P}$, where $\mathcal{P}$ is a principal fibre bundle over a certain manifold Y with G the structural group of $p : \mathcal{P} \to Y$. The most interesting in those two cases are equations with invariant (under actions of the group G) coefficients.

In the end, we shall consider smooth measures on manifolds generated by solutions of stochastic equation and state sufficient conditions under which those measures are absolutely continuous. As a consequence, a way to construct measures equivalent to a given one will be obtained.

1. Lie Groups and Smooth Bundles

1.1. LIE GROUPS AND LIE ALGEBRAS

In what follows, the term smooth manifold (mapping) will stand for a C^∞-smooth manifold (mapping). The exceptions will be expressly mentioned. Moreover, we always assume that a smooth partition of unity does exist on the considered manifold. Notice that these assumptions are valid for Riemannian manifolds.

A smooth Banach manifold G is called a Lie group if its elements form a group and the group operation is compatible with the smooth structure of the manifold. This means that the mapping

$$(g, h) \mapsto g\, h^{-1}$$

is smooth.

In what follows, G is supposed to be connected as well.

Lie group elements give rise to important diffeomorphisms of the manifold G:

a left shift	$L_g : x \mapsto g\, x, \quad x \in G,$
a right shift	$R_g : x \mapsto x\, g,$
an inner automorphism	$\alpha_g : x \mapsto g\, x\, g^{-1}.$

The corresponding tangent mappings which act in the tangent bundle TG are denoted

$$T L_g = \mathrm{d} L_g : T_x G \to T_{gx} G,$$

$$T R_g = \mathrm{d} R_g : T_x G \to T_{gx} G,$$

$$T \alpha_g = \mathrm{d} \alpha_g : T_x G \to T_{gxg^{-1}} G.$$

Since the unity e of the group G is invariant with respect to the mapping α_g for any $g \in G$, the tangent space $T_e G$ is easily proved to be invariant under the tangent mapping $\mathrm{d}\,\alpha_g$, that is

$$A\,\mathrm{d}\, g = \mathrm{d}\,\alpha_g : T_e G \to T_e G.$$

A linear space $\mathcal{L}$ is called a Lie algebra if a bilinear mapping $\mathcal{L} \times \mathcal{L} \to \mathcal{L}$ (called a bracket and denoted $[\cdot, \cdot]$) is defined and possesses the following properties

1. $[A, B] = -[B, A]$ for any $A, B \in \mathcal{L}$.
2. The Jacobi identity

$$\big[A, [B, C]\big] + \big[B, [C, A]\big] + \big[C, [A, B]\big] = 0$$

is satisfied for any $A, B, C \in \mathcal{L}$.

Smooth vector fields over a smooth manifold X form a Lie algebra which is denoted $\mathcal{X}(X)$. The bracket in $\mathcal{X}(X)$ has the form

$$[\xi, \eta] = T\,\xi \circ \eta - T\,\eta \circ \xi.$$

In a local trivialization the principal part of the vector field $[\xi, \eta]$ looks like

$$[\xi, \eta]_x = \xi'(x)\,\eta(x) - \eta'(x)\,\xi(x).$$

Recall that we denote by the same letter both a vector field and its principal part if it does not lead to misunderstandings.

Let $X = G$ be a Lie group. Then there is a Lie subalgebra $\mathcal{G}$ in the Lie algebra $\mathcal{X}(G)$ consisting of left invariant vector fields. Recall that a vector field ξ over G is left invariant if the equality $\xi = \mathrm{d}\,L_g\,\xi$ holds. In other words, ξ is called invariant if the diagram

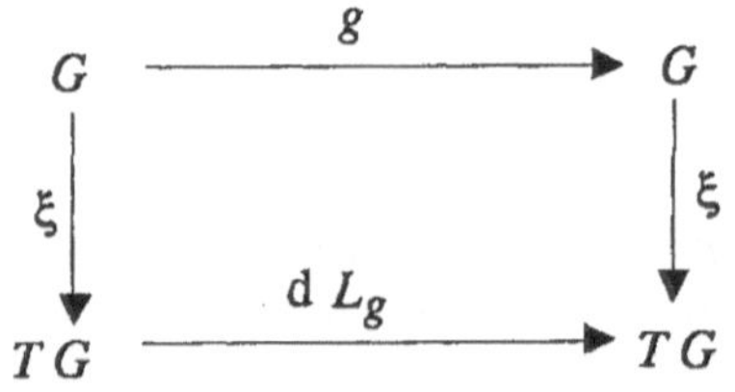

is commutative. Since $\xi_g = \mathrm{d}\,L_g\,\xi_e$, the left invariant vector field ξ is uniquely determined by its value ξ_e. That is why the Lie algebra $\mathcal{G}$ may be identified with the space $T_e G$ if one introduces the bracket

$$[A, B] = [\xi^A, \xi^B]$$

in $T_e G$. Here $\xi_g^A = \mathrm{d} L_g A$ denotes the left invariant field induced by an element $A = \xi_e^A \in \mathrm{d} T_e G$. The Lie algebra $\mathcal{G}$ is called the Lie algebra of the Lie group G.

Let $GL(B)$ be the group of automorphisms of a certain linear space B. A homomorphism

$$\rho : g \in G \mapsto \rho(g) \in GL(B)$$

from G into $GL(B)$ is called a (linear) representation of the group G while B is called the representation space.

The mapping $\mathrm{Ad} : G \to GL(\mathcal{G})$ defined above gives a linear representation of G, while $\mathcal{G}$ is the representation space. Ad is called the adjoint representation of the Lie group G.

A homomorphism from a Lie algebra $\mathcal{A}$ into a Lie algebra $\mathcal{B}$ is called a representation of $\mathcal{A}$ by $\mathcal{B}$. In particular, if $\mathcal{B}$ is a Lie algebra of linear operators acting in a certain linear space $\mathcal{L}$ (with the bracket $[a, b] = a\,b - b\,a,\ a, b \in \mathcal{L}$) then a homomorphism $\varphi : \mathcal{A} \to \mathcal{B}$ is called a linear representation of $\mathcal{A}$ in the space $\mathcal{B}$. Notice that the word linear will be omitted, since we shall deal only with linear representations.

The linear mapping

$$\mathrm{ad} = \mathrm{d}\,(\mathrm{Ad}\, g)\big|_{g=e} : T_e G \to T_e G ,$$

which is tangent to the adjoint representation of a Lie group, satisfies the relation

$$(\mathrm{ad}\, A)\, B = [A, B] ,$$

the Jacobi identity yields

$$\mathrm{ad}\,[A, B] = \mathrm{ad}\, A\ \mathrm{ad}\, B - \mathrm{ad}\, B\ \mathrm{ad}\, A$$

and implies that ad gives a representation of the Lie algebra $\mathcal{G}$ by the Lie algebra of linear operators acting in $\mathcal{G}$. This representation is called the adjoint representation of the Lie algebra $\mathcal{G}$.

Consider now a flow of Lie group G diffeomorphisms induced by a left invariant vector field. Notice that this flow is generated by a solution of the equation

$$\frac{\mathrm{d}\, g(t)}{\mathrm{d}\, t} = \mathrm{d} L_{g(t)} A_e = A_{g(t)} \tag{1.1}$$

such that

$$g(0) = e. \tag{1.2}$$

For any $g_1 \in G$, we obtain

$$\frac{\mathrm{d} L_{g_1} g(t)}{\mathrm{d}\, t} = \mathrm{d} L_{g_1} \frac{\mathrm{d}\, g(t)}{\mathrm{d}\, t} = \mathrm{d} L_{g_1}\, \mathrm{d} L_{g(t)} A = \mathrm{d} L_{g_1 g(t)} A ,$$

and the equality

$$g_1 g(t, e) = g(t, g_1)$$

follows from the uniqueness of the solution $g(t, e)$ of the Cauchy problem (1.1)–(1.2). Hence

$$g(\tau, e)\, g(t, e) = g(t, g(\tau, e)) = g(t + \tau, e) \tag{1.3}$$

and, consequently, there exists a global integral path of a left invariant vector field, namely, a mapping $t \mapsto g(t, e)$ is defined for all $t \in R^1$. The relation (1.3) shows that the path $g_A(t) = g(t, e)$ gives a global one-parameter subgroup of the group G.

A mapping $\exp : \mathcal{G} \to G$ given by

$$\exp A = \gamma_A(1)$$

(where $\gamma_A(t)$ is the integral path of the field A starting at e) is called an exponential mapping. It is easy to verify that

$$\gamma_A(t) = \exp tA.$$

The last relation implies that an exponential mapping derivative at the zero point gives an identity mapping in $\mathcal{G}$: $d \exp A|_0 = I$. Finally, the implicit function theorem permits us to state that

$$\exp : U_0 \to V_e$$

is an invertible smooth mapping from a neighborhood $U_0 \subset \mathcal{G}$ of the zero into a neighborhood V_e of the unity $e \in G$.

Taking $\mathcal{G} = T_e G$ as the model space of the manifold, we obtain a chart $(V_e, \varphi_e = \exp^{-1})$ which is called a canonical chart. Obviously, one may construct a left invariant atlas by making use of left shifts, setting

$$\varphi_g = \exp^{-1}) \circ L_{g^{-1}} : V_g \to U_0 .$$

Assume G is equipped with a left invariant affine connection and denote Γ^G as its local coefficients. Let $\mathrm{Exp} : TG \to G$ be the exponential mapping induced by the given affine connection in the way described in Chapter 2. In order to understand the relation between this mapping and the mapping exp described above in this chapter, let us notice first that we can associate a left invariant vector field A to vector ξ in a unique way, be setting $\xi = A_{\tau_G(\xi)}$. In other words, each tangent vector at any point $g \in G$ may be extended in a unique way to a left invariant vector field over G. Let the vector field η over TG be a horizontal lift of the vector field A over G. If $\gamma_A(t)$ is an integral path of the field A, then a path $\gamma_A'(t)$ is an integral path of the field TA. Hence, for any $t \in R^1$, $\gamma_A''(t) = TA$ and $\tau_G(\gamma_A'(t)) = A_{\gamma_A(t)}$. Next, if $(TA)_\xi = \xi$ holds, we obtain $\gamma_A''(t) = \eta_\xi$. Recall (see Chapter 2) that the principal part of the field η, the horizontal lift of the field A, in local coordinates looks like

$$(A_x - \Gamma_x^G (A_x, A_x)),$$

and the integral paths of the field A satisfy the relations

$$\gamma_A'(t) = A_{\gamma_A(t)},$$

$$\gamma_A''(t) = -\Gamma^G_{\gamma_A(t)}\left(\gamma_A'(t), \gamma_A'(t)\right). \tag{1.4}$$

These relations imply that $\gamma_A(t)$ is geodesic curve over G and, hence, the mappings Exp and exp coincide over $\mathcal{Y}$ if $\Gamma^G\ (A, A) = -A^2$.

To conclude this section, consider a two-parameter family of group diffeomorphisms generated by a vector field $A(t)$ over G, i.e. by the equation

$$\frac{d\varphi_e(t,\tau)}{dt} = dL_{\varphi_e(t,\tau)}\alpha(t), \quad \varphi_e(t,\tau) = e. \tag{1.5}$$

Here $\alpha : [0, T] \to \mathcal{Y}$ is a curve in $\mathcal{Y}$ such that $A_e(t) = \alpha(t)$.

Given $\tau \in [0, T]$ $\varphi_\ell(t, \tau)$ is a curve in G which is called a left multiplicative integral. Denote

$$\varphi_\ell(t,\tau) = \overset{\frown}{\exp} \int_\tau^t \alpha(s)\ ds.$$

If follows from the uniqueness of the Cauchy problem (1.5) that the relation

$$\varphi_\ell(t,s) \circ \varphi\ (s,\tau) = \varphi_\ell(t,s)$$

holds. It means that

$$\overset{\frown}{\exp} \int_\tau^t \alpha(s)\ ds$$

is an evolution family.

Thus, a left multiplicative integral is a two-parameter evolution family of transformations induced by a nonautonomous left invariant vector field.

In the same way, one may define a right multiplicative integral induced by a nonautonomous right invariant vector field denoted

$$\varphi_r(\tau, t) = \overset{\frown}{\exp} \int_\tau^t \alpha(s)\ ds.$$

1.2. TRANSFORMATION GROUPS AND FIBRE BUNDLES

Consider a Lie group G and a smooth manifold X. The group G is said to act on X if the homorphisms

$$L : g \mapsto \ell_g, \qquad \ell_{gh} = \ell_g \circ \ell_h,$$

from G into the group of the manifold diffeomorphisms are defined and the mapping

$$(g, x) \mapsto \ell_g x$$

is a smooth mapping from $G \times X$ into X.

The action of G on X induces a mapping

$$\lambda : \mathcal{Y} \to \mathcal{X}(X),$$

from the Lie algebra $\mathcal{Y}$ into the Lie algebra of vector fields over the manifold X. Indeed, given $A \in \mathcal{Y}$, set

$$\lambda(A) = \frac{\mathrm{d}}{\mathrm{d}\,t} \; \ell_{\exp tA}\, x\,|_{t=0}\,.$$

The mapping λ is a homomorphism of Lie algebras.

Let X and $\mathcal{F}$ be smooth manifolds and G be a Lie group acting on $\mathcal{F}$.

Let us now introduce the notion of a smooth fibre bundle with typical fibre $\mathcal{F}$, structural group G and total space $\mathcal{E}$. Acting in a way analogous to that of Chapter 2, let us consider first a local smooth fibre bundle. Namely, consider the direct product $X \times \mathcal{F}$, which is a local fibre bundle, and define local smooth fibre bundle morphisms

$$\Phi = (\varphi, g) : X \times \mathcal{F} \to X \times \mathcal{F}$$

by

$$\Phi = (x, h) \mapsto ((\varphi, (x), \; \ell_{g(x)}\, h).$$

Here $\varphi : X \to X$ and $g : X \to G$ are manifold morphisms.

A manifold $\mathcal{E}$ is said to be equipped with a smooth fibre bundle structure if a one-to-one smooth mapping

$$F : \mathcal{E} \to X \times \mathcal{F}$$

is given. Suppose $F_1 : \mathcal{E} \to X \times \mathcal{F}$ is another mapping of this kind. The mappings F and F_1 are said to determine the same smooth fibre bundle structure in $\mathcal{E}$ if

$$\Phi = F \circ F_1^{-1} : X \times \mathcal{F} \to X \times \mathcal{F}$$

is a morphism of the type described above.

The mapping F gives a way of introducing a projection $\pi : \mathcal{E} \to X$ by setting $\pi = \mathrm{pr}_X \circ F$ where pr_X is the projection onto the first component. It also permits us to define a fibre $\mathcal{E}_x = \pi^{-1}(x) = F^{-1}\{(x, h), h \in \mathcal{F}\}$. Notice that two mappings F and F_1 which are compatible in the above sense, give rise to the same projection π.

Let us come to consider, at last, global objects. Consider a smooth manifold mapping

$$\pi : \mathcal{E} \to X$$

which is assumed to be locally trivial. It means that for each U_α belonging to a covering $\{U_\alpha\}$ of the manifold X one may introduce in $Y_\alpha = \pi^{-1}(U_\alpha)$ the structure of a local smooth fibre bundle with a fibre $\mathcal{F}$ and the structural group G. Moreover, these introduced structures coincide over the intersections $U_\alpha \cap U_\beta$ and define a projection π.

In this case, we say that a smooth fibre bundle $(\mathcal{E}, X, \pi, \mathcal{F}, G)$ is given.

It follows immediately from the above construction that for each intersecting pair $U_\alpha \cap U_\beta \neq \emptyset$, a mapping $g_{\alpha\beta} : U_\alpha \cap U_\beta \to G$ is determined in such a way as to define localization morphisms

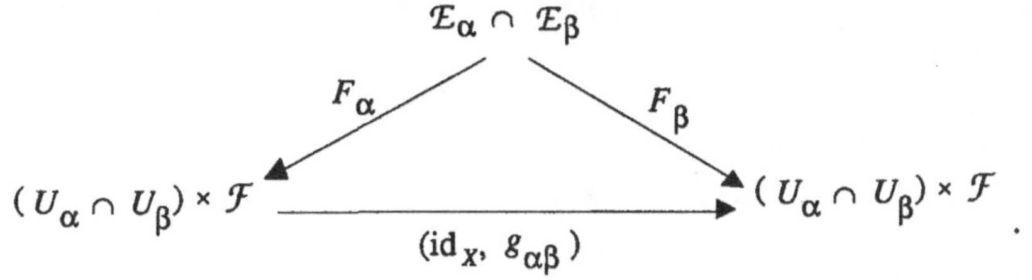

The mapping $g_{\alpha\beta}$ is called a transition function. It may be easily verified that

$$g_{\alpha\beta} = g_{\beta\alpha}^{-1}, \quad g_{\gamma\beta}(x)\, g_{\beta\alpha}(x) = g_{\gamma\alpha}(x), \quad x \in U_\alpha \cap U_\beta \cap U_\gamma .$$

Vice versa, given manifolds $X, \mathcal{F}$ and a family of transition functions $g_{\alpha\beta}$ possessing the above properties, we may define a smooth fibre bundle.

A fibre bundle whose typical fibre coincides with its structural group is called a principal fibre bundle. There is a unique principal fibre bundle $(\mathcal{P}, X, p, G)$ corresponding to a smooth fibre bundle $(\mathcal{E}, X, \pi, \mathcal{F}, G)$. The latter bundle is said to be a fibre bundle associated with $\mathcal{P}$.

Let $\mathcal{P}$ be a representation of the group G with a linear space $\mathcal{L}$ as the representation space and p^ρ be a bundle associated with the principal fibre bundle p whose typical fibre is $\mathcal{L}$ and structural group is $\rho(G)$. A vector bundle $p^\rho : \mathcal{L}_X \to X$ over X with a typical fibre $\mathcal{L}$ and a structural group $\rho(G)$, is called a bundle associated with p by the representation ρ. In particular, if $\mathcal{L} = \mathcal{G}$ is the Lie algebra of the group G and $\rho = \mathrm{Ad}$, then the bundle $p^{\mathrm{Ad}} : \mathcal{G}_X \to X$ is called the adjoint bundle.

1.3. CONNECTIONS ON PRINCIPAL FIBRE BUNDLES

Let $\mathcal{P} = (\mathcal{P}, X, p, G)$ be a principal fibre bundle over a manifold X. Defining a connection on the bundle P means giving a way of constructing a decomposition of the tangent bundle $T\mathcal{P}$ (or, to be more precise, a decomposition of the tangent bundle total space) and to decompose it as a direct sum of its horizontal and vertical components. In Chapter 2, we described the construction of a connection on a vector bundle. We shall apply the scheme used there to construct a connection on a principal fibre bundle. This scheme gives a natural way of constructing a global connection on p.

According to the scheme mentioned above, we shall start with local descriptions of the necessary constructions, keeping in mind that all constructions must be invariant under bundle morphisms.

Consider a trivial principal fibre bundle $p : \mathcal{P} = X \times G \to X$ and a bundle morphism

$$\Phi = (\varphi, \ell_{g(x)}) : X \times G \to X \times G, \tag{1.6}$$

where $\varphi : X \to X$ and $g : X \to G$. The tangent bundle $T\mathcal{P}$ over $\mathcal{P}$ looks like

$$Tp : T(X \times G) \to X \times G.$$

Below, we omit the description of the charts of the manifold X identifying, if necessary, a neighborhood U_x of a point $x \in X$ with a neighborhood of zero in the model space B.

Applying the functor T to (1.6), we obtain a transformation rule for the trivialization of the considered bundle

$$T\Phi\,(x, a, y, \xi) \mapsto (\varphi\,(x),\ g\,(x)\,a),$$

$$\varphi'\,(x)\,y,\ \mathrm{d}\ \ell_{g\,(x)}\,\xi + \mathrm{d}\,r_a\ g'\,(x)\,y)\,. \tag{1.7}$$

It follows from (1.7) that vertical tangent components (corresponding to $y = 0$) are transformed in simple way, namely

$$T\Phi\,(x, a, 0, \xi) = (\varphi\,(x),\ g\,(x)\,a, 0, \mathrm{d}\ \ell_{g\,(x)}\,\xi)\,. \tag{1.8}$$

Hence, we may introduce a vertical tangent bundle

$$T_\upsilon\,\mathcal{P} : V\,T\,\mathcal{P} \to \mathcal{P}$$

in a way similar to that used in Chapter 2 for a vector bundle. The vertical bundle is a subbundle of the bundle Tp and vertical tangent trivializations have the form

$$T_\upsilon\,p : X \times T\,G \to X \times G\,.$$

Notice that the tangent spaces of the group G, i.e., fibres $\mathcal{Y}_g$ of the tangent space $T\,G$, are isomorphic to $\mathcal{Y} = T_e\,G$

$$\mathrm{d}\,L_a : \mathcal{Y} \to T_a\,G = \mathcal{Y}_a\,, \qquad a \in G.$$

Let us represent an element of a vertical tangent bundle fibre $T_{(x,a)}\,\mathcal{P}$ in the form $(x, a, 0, \mathrm{d}\,r_a\ A)$ with $A \in \mathcal{Y}$ and transform (1.8) into

$$T\Phi : (x, a, 0, \mathrm{d}\,r_a\ A) \mapsto$$

$$\mapsto (\varphi\,(x),\ g\,(x)\,a\,, 0, \mathrm{d}\,r_{ga}\,(\mathrm{Ad}\,g)\,A)\,. \tag{1.9}$$

Let $\mathrm{Ad}\,p : \mathcal{Y}_x \to X$ be the adjoint vector bundle with morphisms

$$\mathrm{Ad}\,\Phi : (x\,, A) \mapsto \big(\varphi\,(x), [\mathrm{Ad}\,g\,(x)]\,A\big)\,. \tag{1.10}$$

The above result permits us to associate $T_V\,p$ with $\mathrm{Ad}\,p$ in the following way. Consider the inverse image of $\mathrm{Ad}\,p : \mathcal{Y}_x \to X$ along p

$$p^*\,(\mathrm{Ad}\,p) : p^*\,(\mathcal{Y}_x) \to X \times G$$

which clearly is a bundle isomorphic to the vertical tangent bundle. Denote by υ the correspondent isomorphism

$$\upsilon : (x, a, 0, \mathrm{d}\ r_a A) \mapsto (x, a, A)\,. \tag{1.11}$$

Notice that all the above constructions are invariant with respect to bundle morphisms and, hence, may be extended to nonlocal bundles. In particular, the isomorphism υ maps $V\,T\,\mathcal{P}$ into $p^*\,(\mathcal{Y}_x)$

$$\upsilon : V\,T\,\mathcal{P} \to p^*\,(\mathcal{Y}_x)\,.$$

A mapping Γ^p

$$\Gamma^p : x \mapsto \Gamma^p_x \in L\,(T_x X, \mathcal{Y}\,), \quad x \in X$$

is called a local connection coefficient if it transforms according to the rule

$$\begin{aligned} &\Gamma^{p\,\Phi}_{\varphi(x)}\ (\varphi'(x)\,y) \\ &\qquad = \mathrm{Ad}\,(g\,(x))\,\Gamma^p_x\,(y) - \mathrm{d}\,r_{g^{-1}(x)}\,g'(x)\,y, \\ &\qquad y \in B,\ g'(x) : B \to \mathcal{Y} \end{aligned} \tag{1.12}$$

under the action of a morphism Φ.

Given a connection coefficient, we may construct a connection map

$$K^p : (x, a, y, \mathrm{d}\ r_a\ A) \mapsto \left(x, A + \Gamma^p_x\ (y)\right). \tag{1.13}$$

It is easy to check by direct calculations that the diagram

$$\begin{array}{ccc} T(X \times G) & \xrightarrow{\quad T\,\Phi \quad} & T(X \times G) \\ \Big\downarrow K_p & & \Big\downarrow K\Phi_p \\ X \times \mathcal{Y} & \xrightarrow[\quad \mathrm{Ad}\,\Phi \quad]{} & X \times \mathcal{Y} \end{array}$$

is commutative. This implies that K^p may be regarded as a vector bundle morphism

$$K^p : T\,\mathcal{P} \to \mathcal{Y}_x\,.$$

Hence, we may change local constructions to global ones in a way similar to that described above, in Chapter 2, for the case of vector bundles.

In particular, we obtain the possibility of constructing a global connection in p (assuming that there exists a smooth partition of unity on X). This construction may be used to prove the existence of a global connection on $\mathcal{P}$.

The kernel of the connection map

$$H\,T_{(x,a)}\,\mathcal{P} = \mathrm{Ker}\ K^p = \left\{(x, y, a, \xi)\ \ \xi = -\,\mathrm{d}\,r_a\ \Gamma^p_x\,(y)\right\}$$

is called the horizontal tangent space at the point (x, a). Thus, the horizontal tangent space is obviously defined in an invariant way and possesses the property

$$H\,T_{(x,\,ga)}\,\mathcal{P} = \mathrm{d}\ r_g\,H\,T_{(x,a)}\,\mathcal{P}\,.$$

The relation

$$(x, a, y, \xi)\ =\ (x, a, y,\ -\mathrm{d}\ r_a\ \Gamma^p_x\,(y)) + (x, a, 0, \xi\ + \mathrm{d}\ r_a\ \Gamma^p_x\,(y))$$

defined an invariant (under the action of G) decomposition into vertical and horizontal components for an element of the fibre of the vector bundle $T\mathcal{P}$ as the Whitney sum

$$T\mathcal{P} = H\,T\mathcal{P} \oplus V\,T\mathcal{P}.$$

Given a connection K^P, we obtain a way to construct lifts of vector fields.

Let η be a vector field over X with principal part $\eta(x) \in B$ in a given local trivialization. A horizontal lift of the field η gives a horizontal vector field h over $\mathcal{P}$ with the principal part given by

$$h : (x, a) \mapsto \big(x, a, \eta(x) - \mathrm{d}\, r_a\, \Gamma_x^P(\eta(x))\big).$$

Integral paths of this field may be described as solutions of the system of differential equations

$$\frac{\mathrm{d}\,x}{\mathrm{d}\,t} = \eta(x), \tag{1.14}$$

$$\frac{\mathrm{d}\,a}{\mathrm{d}\,t} = -\mathrm{d}\, r_a\, \Gamma_x^P(\eta(x)). \tag{1.15}$$

The above arguments permit us to prove that this system is invariant with respect to bundle morphisms. Under natural assumptions, like the condition of uniform boundedness of all the considered fields, the above system determines an integral flow of parallel displacements along integral paths of the vector field η in fibres of the principal fibre bundle.

Indeed, Equation (1.15) is a linear differential equation on the group G generated by a nonautonomous left invariant vector field. Equations of this type have been dealt with in the previous section. The results obtained there show that a solution $a(t)$ of (1.15) such that $a(0) = e$ may be given in the form

$$a(t) = \overset{\curvearrowright}{\exp} \int_0^t \Gamma_x^P(\eta(s))\, \mathrm{d}\, s.$$

Notice that this multiplicative integral, which is sometimes called a multiplicative exponent, is a linear operator of parallel displacement of a fibre along the path $x(t)$ for the point $x(0)$ to the point $x(t)$.

Remark 1.1. Given the connection coefficient Γ^P, we may construct a connection in the adjoint bundle. The connection coefficient of the adjoint bundle is a bilinear mapping

$$\Gamma_x^{\mathrm{Ad}\,P} : B \times \mathcal{Y} \to \mathcal{Y},$$

which transforms under bundle morphisms according to the rules described in Chapter 2,

$$\Gamma_{\varphi(x)}^{\mathrm{Ad}\,P\,\Phi}\ (\varphi'(x)\,y,\ \mathrm{Ad}\,g(x)\,a)$$

$$= \mathrm{Ad}\,(g(x))\,\Gamma_x^{\mathrm{Ad}\,P}\,(y, a) + (g'(x)\,y\ \mathrm{d}\,r_{g^{-1}})\,a. \tag{1.16}$$

It is easily verified that

$$\Gamma_x^{\mathrm{Ad}\,p}(y, a) = \left[\Gamma_x^p(y), a\right] \tag{1.17}$$

satisfies the above conditions.

The connection map in this case is given by

$$K^{\mathrm{Ad}\,p}(x, a, y, b) = b + \left[\Gamma_x^p(y), a\right]. \tag{1.18}$$

Remark 1.2. Given the connection coefficient Γ_x^p, we may define a connection on an arbitrary vector bundle associated with the given principal fibre bundle p by means of a representation ρ.

If $\mathcal{L}$ is a representation space, then the mapping

$$\Gamma_x^{\rho p} : B \times \mathcal{L} \to \mathcal{L}$$

is a connection coefficient. This mapping obeys the transformation rule

$$\Gamma_{\varphi(x)}^{\rho p \Phi}(\varphi'(x)\, y, \rho(g(x))\, a)$$

$$= \rho(g(x))\, \Gamma_x^{\rho p}(y, a) + \mathrm{d}\, r_{\rho(g^{-1})}\, \rho(g(x))'\, y\, a$$

under bundle morphism actions.

1.4. LINEAR CONNECTION ON THE TOTAL SPACE OF A FIBRE BUNDLE

Let $p : T\mathcal{P} \to X$ be a principal fibre bundle over a manifold X, and let Γ^x, Γ^p be, respectively, the connection coefficients of the manifold X and the bundle p. Construct a linear connection on the manifold $\mathcal{P}$ which is the total space of the fibre bundle p, i.e. a connection map

$$K^{\mathcal{P}} : T^2\mathcal{P} \to T\mathcal{P}.$$

This construction will be performed within the framework of the scheme used in Chapter 2 while constructing a linear connection on the total space of a vector bundle.

Recall that to construct a map $K^{\mathcal{P}}$, it is enough to construct its vertical and horizontal components. Given the projection $p : \mathcal{P} \to X$, consider the iterated tangent map $T^2 p : T^2\mathcal{P} \to T^2 X$. For a connection mapping $K^X : T^2 X \to TX$, the composition $K^X \circ Tp$ acts from $T^2\mathcal{P}$ into TX, $K^X \circ T^2 p : T^2\mathcal{P} \to TX$. The required element of $HT\mathcal{P}$ may be constructed by a lift of the vector field obtained above.

In local coordinates, we have the following chain of mappings

$$p : (x, a) \mapsto x,$$

$$Tp : (x, a, y, A) \mapsto (x, y),$$

$$T^2 p : (x, a, y, A, z, B, u, C) \mapsto (x, y, z, u),$$

$$K^X \circ T^2 p : (x, a, y, A, z, B, u, C) \mapsto \left(x, u + \Gamma_x^X(y, z)\right),$$

$$H K^{\mathcal{P}} : (x, a, y, A, z, B, u, C) \mapsto$$

$$\mapsto \left(x, a, u + \Gamma^X_x (y, z), -\Gamma^p_x (u + \Gamma^X_x (y, z))\right). \qquad (1.19)$$

Notice that omitting the map $d r_a$ in the last expression means changing a chart $U_x \times V_a \times B \times \mathcal{Y}_a$, $V_a \subset G$ of the manifold $T\mathcal{P}$ to the chart $U_x \times V_a \times B \times \mathcal{Y}$. In order to compute the vertical component

$$V K^{\mathcal{P}} : T^2 \mathcal{P} \to V T \mathcal{P},$$

let us make use of the mappings $K^{\mathcal{P}} : T^2 \mathcal{P} \to \mathcal{Y}_X$, and $T K^{\mathcal{P}} : T^2 \mathcal{P} \to T\, \mathcal{Y}_X$. If we add to the above chain the connection mapping of the adjoint bundle

$$K^{\mathrm{Ad}\, p} : T\, \mathcal{Y}_X \to \mathcal{Y}_X$$

and finally the mapping υ^{-1}, the resulting mapping gives the vertical component.

In a coordinate representation, the resulting chain of mappings acts as follows

$$T K^p : (x, a, y, A, z, B, u, C) \mapsto$$

$$\mapsto \left(x, A + \Gamma^p_x (y), z, C + (\Gamma^p_x)' (y, z) + \Gamma^p_x (u)\right),$$

$$K^p : \left(x, A + \Gamma^p_x (y), z, C + (\Gamma^p_x)' (y, z) + \Gamma^p_x (u)\right) \mapsto$$

$$\mapsto \left(x, C + (\Gamma^p_x)' (y, z) + \Gamma^p_x (u) + [\Gamma^p_x (z), A] + \right.$$

$$\left. + [\Gamma^p_x (z), \Gamma^p_x (y)]\right)$$

and finally

$$V K^p : (x, a, y, A, z, B, u, C) \mapsto$$

$$\mapsto \left(x, a, 0, C + (\Gamma^p_x)' (y, z) + \Gamma^p_x (u) + \right.$$

$$\left. + [\Gamma^p_x (z), A + \Gamma^p_x (y)]\right). \qquad (1.20)$$

The expressions obtained in (1.19) and (1.20) give a decomposition of $T\mathcal{P}$ in the form of a direct sum $H T \mathcal{P} \oplus V T \mathcal{P}$ and, thus, their sum is nothing but a connection map $K^{\mathcal{P}}$ acting as

$$K^{\mathcal{P}} : \left(\begin{pmatrix} x \\ a \end{pmatrix}, \begin{pmatrix} y \\ A \end{pmatrix}, \begin{pmatrix} z \\ B \end{pmatrix}, \begin{pmatrix} u \\ C \end{pmatrix}\right) \mapsto$$

$$\mapsto \left(\begin{pmatrix} x \\ a \end{pmatrix}, \begin{pmatrix} u + \Gamma^X_x (y,z) \\ C + (\Gamma^p_x)' (y,z) + [\Gamma^p_x (y), \Gamma^p_x (z)] + [\Gamma^p_x (z), A] - \Gamma^p_x \left(\Gamma^X_x (y,z)\right) \end{pmatrix}\right).$$

The last relation implies that the linear connection coefficients of the manifold $\mathcal{P}$ may be given in the form

$$\Gamma^{\mathcal{P}}_{\binom{x}{a}} : B \times \mathcal{G} \times B \times \mathcal{G} \to B \times \mathcal{G},$$

$$\Gamma^{\mathcal{P}}_{\binom{x}{a}} \left(\binom{y}{A}, \binom{z}{B} \right)$$

$$= \begin{pmatrix} \Gamma^X_x (y,z) \\ (\Gamma^p_x)' (y,z) + \left[\Gamma^p_x (y), \Gamma^p_x (z) \right] - \Gamma^p_x \left(\Gamma^X_x (y,z) \right) + \left[\Gamma^p_x (z), A \right] \end{pmatrix} .$$

Changing the above special chart of $\mathcal{P}$ to an arbitrary one, we obtain the linear connection coefficients written in the form

$$\Gamma^{\mathcal{P}}_{\binom{x}{a}} \left(\binom{y}{A_a}, \binom{z}{B_a} \right)$$

$$= \begin{pmatrix} \Gamma^X_x (y,z) \\ M \ (x, a, y, A_a, z, B_a) \end{pmatrix} , \tag{1.21}$$

where $A_a \in T_a G, B_a \in T_a G,$

$$M \ (x, a, y, A_a, z, B_a)$$

$$= \mathrm{d} L_a \left\{ (\Gamma^p_x)' (y, z)) - \Gamma^p_x \left(\Gamma^X_x (y \, , z) \right) + \left[\Gamma^p_x (y), \Gamma^p_x (z) \right] \right\} +$$

$$+ \left[\mathrm{d} L_a \ \Gamma^p_x (y), A_a \right] + \ \left[\mathrm{d} L_a \ \Gamma^p_x (z), B_a \right] .$$

Remark 1.3. The vertical tangent bundle over $\mathcal{P}$ admits the following description. Consider a left action of G on $\mathcal{P}$ and notice that it induces an isomorphism between the Lie algebra $\mathcal{G}$ of the group G and the Lie algebra $\tilde{\mathcal{G}}$ of vector fields over $\mathcal{P}$. Define this isomorphism $\lambda : \mathcal{G} \to \tilde{\mathcal{G}}$ by the relation

$$\lambda (A) (z) = \frac{\mathrm{d}}{\mathrm{d} t} \ \exp (t A) (z) \big|_{t=0} = \tilde{A} (z) \ ,$$

$$A \in \mathcal{G} , \quad z \in \mathcal{P} .$$

Elements $\lambda (A)$ are called fundamental vector fields. It is easy to verify that fundamental vector fields are vertical and, moreover, at each point z they generate the space $V T_z \mathcal{P}$.

For a fundamental vector field $\lambda (A)$, the relation

$$\mathrm{d} L_g \lambda (A) = \lambda (\mathrm{Ad} \, g \, A), \quad g \in G$$

may be derived by direct calculations.

Recall that, given a linear connection on th manifold, we may introduce a covariant differentiation, and (see Chapter 2) the covariant derivative of the vector field ψ over $\mathcal{P}$ along the vector field Φ has in local coordinates the form

$$\nabla_{\Phi}^{\mathcal{P}} \psi = \psi' \Phi + \Gamma^{\mathcal{P}} (\Phi , \psi) .$$

Let $\mathrm{Ad}\,(p) : \mathcal{Y}_x \to X$ be the adjoint bundle over X. The covariant derivative $\nabla^{\mathrm{Ad}\,p}$ of a bundle $\mathrm{Ad}\,p$ section u along the vector field υ has the following representation

$$\nabla_{\upsilon}^{\mathrm{Ad}\,p}\ u = u' \upsilon + [\, \Gamma_x^p \upsilon, u\,] .$$

Below, we shall also need some covariant expressions including derivatives of the second order, namely

$$\begin{aligned} \nabla_A^{\mathrm{Ad}\,p}\ \nabla_B^{\mathrm{Ad}\,p}\ u \ &= u'' (B, A) + u' B' A + \\ &+ \left[(\Gamma^p)' (B, A), u\right] + [\Gamma^p B' \ A), u] + \\ &+ [\Gamma^p B, u' \ A\,] + [\Gamma^p A\ , u' \ B] + \left[\Gamma^p B,\ [\Gamma^p B, u\,]\right] , \end{aligned}$$

and

$$\begin{aligned} &\nabla_A^{\mathrm{Ad}\,p}\ \nabla_B^{\mathrm{Ad}\,p}\ u - \nabla_{\nabla_A B}^{\mathrm{Ad}\,p}\ u \\ &= u'' (B, A) + \left[(\Gamma^p)' (B, A)\ , u\right] + [\Gamma^p B, u' \ A\,] + \\ &+ [\Gamma^p A, u' \ B\,] + \left[\Gamma^p A,\ [\Gamma^p B, u\,]\right] - u' \ \Gamma^{\tau} (B, A) - \\ &- [\Gamma^p \Gamma^{\tau} (B, A), u] . \end{aligned} \tag{1.22}$$

2. Invariant Stochastic Equations

2.1. EQUATIONS INVARIANT UNDER ONE-PARAMETER GROUP ACTIONS

Consider a stochastic equation

$$\mathrm{d}\,\xi = \exp_{\xi(t)} (a_{\xi(t)}\ (t)\ \mathrm{d}\,t + A_{\xi(t)}\ (t)\ \mathrm{d}\,w) \tag{2.1}$$

on a Banach manifold X, corresponding to a time-dependent Ito field

$$\mathbb{A}\ (t) = (a_x\ (t),\ A_x\ (t)) .$$

Without special mention, we shall assume below that the manifold X, its connection, and the Ito field meet all the requirements of Chapter 4. Recall that those requirements are sufficient to guarantee the existence and uniqueness of the solution of Equation (2.1) satisfying the condition $\xi\,(\tau) = x$. As a consequence, those assumptions grant the existence of an evolution random family $S\,(t, \tau; x)$ generated by the above-mentioned solution: $\xi\,(t) = S\,(t, \tau; x)$.

Let f be a smooth mapping adapted to the connection on X in the sense that the diagram

$$\begin{array}{ccc} X & \xrightarrow{\quad f \quad} & X \\ {\scriptstyle \exp}\downarrow & & \downarrow{\scriptstyle \exp} \\ TX & \xrightarrow[\quad Tf \quad]{} & TX \end{array} \tag{2.2}$$

is commutative. Consider a smooth mapping $f(t): X \to X$ depending on a parameter $t \in [0, t]$ and smooth with respect to t

$$\frac{\partial f_x(t)}{\partial t} \in T_{f_x(t)} X.$$

Assume next that $f(t)$ is adapted to the connection on X. The morphism $f(t)$ transforms $\mathfrak{A}(t)$ as

$$\mathfrak{A}(t) \mapsto Tf(t)\, \mathfrak{A}(t) = \left\{ a^f_{f_x(t)},\ A^f_{f_x(t)} \right\}, \tag{2.3}$$

where

$$a^f_{f_x(t)}(t) = Tf_x(t)\, a_x(t) + \frac{\partial f_x(t)}{\partial t},$$

$$A^f_{f_x(t)}(t) = Tf_x(t)\, A_x(t).$$

If $a_x(t)$, $A_x(t)$ are invariant with respect to $f_x(t)$, i.e.

$$a^f_{f_x(t)}(t) = a_x(t), \quad A^f_{f_x(t)}(t) = A_x(t), \tag{2.4}$$

then the transformed Ito field coincides with the initial one. The uniqueness of the solution of the Cauchy problem (2.1) implies that

$$S(t, \tau, f_x(\tau)) = f_{S(t,\tau;x)}(t) \tag{2.5}$$

and, correspondingly, that the transition probability

$$P(\tau, x, t, Q) = P\{S(t, \tau; x) \in Q\}, \quad Q \in \mathcal{Z}_x$$

of the process $\xi(t)$ satisfies the relation

$$P(\tau, f_x(\tau), t, f_Q(t)) = P(\tau, x, t, Q). \tag{2.6}$$

We call (2.5), (2.6) the invariance relations with respect to the mapping $f_x(t)$ for stochastic equation solutions and their probabilities.

If the mapping $f: X \to X$ does not depend on the time parameter, then (2.3) is transformed to

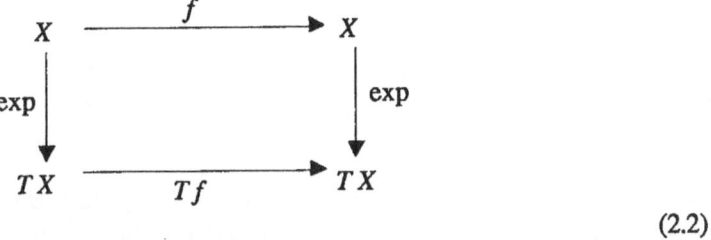

$$\begin{array}{ccc} X & \xrightarrow{\quad f \quad} & X \\ {\scriptstyle z}\downarrow & & \downarrow{\scriptstyle z^f} \\ TX & \xrightarrow[\quad Tf \quad]{} & TX \end{array} \tag{2.7}$$

which means that the vector field $z = \{a(t),\ A(t)\,h\},\ (h \in H)$ is f-compatible with $z^f = \{a^f(t),\ A^f(t)\,h\}$. In this case the invariance relations look like

$$S(t, \tau; f_x) = f_{S(t,\tau;x)}, \tag{2.8}$$

$$P(\tau, f_x, t, f_Q) = P(\tau, x, t, Q).$$

Let $f(\alpha),\ (\alpha \in R^1)$ be a one-parameter transformation group acting on X

$$f_x(\alpha + \beta) = f_{f_x(\beta)}(\alpha).$$

This family solves the differential equation

$$\frac{\partial f_x(\alpha)}{\partial \alpha} = F(f_x(\alpha)),\ f_x(\alpha)\big|_{\alpha=0} = x, \tag{2.9}$$

where the vector field F is given by

$$F(x) = \frac{\partial f_x(\alpha)}{\partial \alpha}\bigg|_{\alpha=0}.$$

The invariance relations (2.8) now imply that the vector fields F and $z = \{a(t),\ A(t)\,h\}$ do commute. Indeed, differentiating the relation

$$\{T f_x(\alpha)\}\, z(x) = z(f_x(\alpha))$$

with respect to α and computing the derivative for $\alpha = 0$, we obtain, due to (2.9), that

$$[z, F] = 0, \quad z = \{a(t), A(t)\,h,\ h \in H\}.$$

Conversely, (2.9) yields the invariance relation

$$S(t, \tau; f_x(\alpha)) = f_{S(t,\tau;x)}(\alpha). \tag{2.11}$$

Let us state the result obtained as a rigorous assertion.

THEOREM 2.1. *The solutions of Equation* (2.1) *and their distributions are invariant under the action of the one-parameter group* $f_x(\alpha)$ *if and only if* (2.10) *holds.*

COROLLARY 2.1. *Let* G *be a Lie group acting on* X. *The invariance relations* (2.8) *hold if and only if the vector fields in* (2.11) *commute with the Lie algebra of the vector fields generated by elements of the Lie algebra* $\mathcal{G}$.

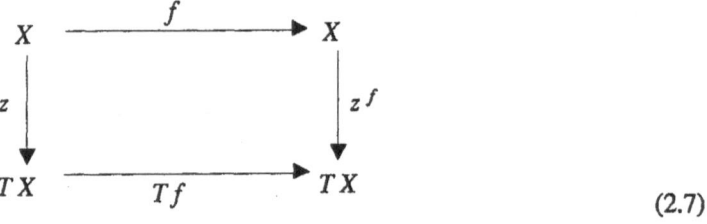

2.2. STOCHASTIC EQUATIONS ON A LIE GROUP

Consider a manifold $X = G$ equipped with a Lie group structure and assume the connection generated by the Exp mapping to be invariant with respect to left shifts.

Let $\mathcal{G}$ by the Lie algebra of the group G and exp be the exponential of the group. Recall that

$$\mathrm{Exp}_g \circ \mathrm{d}\, L_g = L_g \exp . \tag{2.12}$$

Consider a stochastic equation on a Lie group with left invariant coefficients

$$a_x(t) = \mathrm{d}\, L_x\, a(t) , \quad A_x(t) = \mathrm{d}\, L_x\, A(t) , \tag{2.13}$$

where

$$a(t) \in \mathcal{G}, \quad A(t) \in L_{12}(H, \mathcal{G}), \quad x \in G, \quad t \in [0, T].$$

In this case, the stochastic equation (2.1) may be written in the form

$$\mathrm{d}\,\xi = \mathrm{Exp}_{\xi(t)} \left(a_{\xi(t)}(t)\ \mathrm{d}\,t + A_{\xi(t)}(t)\ \mathrm{d}\,w \right) .$$

It follows from (2.12) and (2.13) that the last relation may be rewritten as

$$\mathrm{d}\,\xi = \xi(t) \circ \exp\ \{a(t)\ \mathrm{d}\,t + A(t)\ \mathrm{d}\,w\} . \tag{2.14}$$

The solutions of equation (2.14) generate an evolution family which is left invariant due to previous section results, i.e.

$$S(t, \tau; g\, x) = g\, S(t, \tau; x) , \tag{2.15}$$

$$\xi(t) = S(t, \tau; x), \quad x, g \in G.$$

In particular, for $x = e$, we obtain

$$S(t, \tau; g) = g\, S(t, \tau; e) . \tag{2.16}$$

Therefore, in the considered case, the action of the evolution family is nothing more than a multiplication from the right by the group element

$$\delta(t, \tau) = S(t, \tau; e).$$

The evolution property

$$S(t, \tau; e) = S(t, s; S(s, \tau, e))$$

leads to the relation

$$\delta(t, \tau) = \delta(t, s)\, \delta(s, \tau) ,$$

and, in general, for any partition

$$\tau \le t_1 < \ldots \le t_n = t ,$$

we obtain

$$\delta(t, \tau) = \delta(t, t_{n-1}) \ldots \delta(t_1, \tau) .$$

Let us call $\delta(t\,\tau)$ a stochastic multiplicative integral and denote

$$\delta(t,\tau) = \overset{\frown}{\exp}\left\{\int_\tau^t a(s)\,ds + \int_\tau^t A(s)\,dw(s)\right\}. \tag{2.17}$$

The usual multiplicative integral properties come to be a natural consequence of the definition itself.

Remark 2.1. As follows from (2.16), the tangent map satisfies the relation

$$T\,S(t,\tau;x) = d\,R_{S(t,\tau;x)}\,.$$

In a similar way, dealing with right invariant fields one obtains the right multiplicative integral

$$\overset{\frown}{\exp}\left\{\int_\tau^t a(s)\,ds + \int_\tau^t A(s)\,dw(s)\right\}.$$

Now let $\varphi(t)$ be a path in $T_e G$ and

$$\psi(t,\tau) = \overset{\frown}{\exp}\int_\tau^t \varphi(s)\,ds\,.$$

PROPOSITION 2.1. *The random process*

$$\eta(t) = \overset{\frown}{\exp}\int_\tau^t \varphi(s)\,ds\;\overset{\frown}{\exp}\left\{\int_\tau^t a(s)\,ds + \int_\tau^t A(s)\,dw(s)\right\} \tag{2.18}$$

solves the following stochastic equation

$$d\eta = \mathrm{Exp}_{\eta(t)}\left\{d\,L_{\eta(t)}\,a(t) + b_{\eta(t)}(t))\,dt + d\,L_{\eta(t)}A(t)\,dw\right\}$$

$$= L_{\eta(t)}\,\exp\left\{a(t) + d\,L_{\eta^{-1}(t)}\,b_{\eta(t)}(t))\,dt + A(t)\,dw\right\}\,. \tag{2.19}$$

Proof. A transformation $f_x = L_{\overset{\frown}{\exp}\int_\tau^t \varphi(s)\,ds}$ of the manifold G induces a corresponding transformation of the Ito field (a, A) which may be computed by using (2.13). First notice that the commutative diagram (2.7) applies if G is a Lie group and f is a Lie group morphism. Moreover, the action of Tf comes to be merely the action of the mapping $d\,L_{\overset{\frown}{\exp}\int_\tau^t \varphi(s)\,ds}$ and, therefore,

$$A_y^f(t) = d\,L_{\overset{\frown}{\exp}\int_\tau^t \varphi(s)\,ds}\;A\;d\,L^{-1}{}_{\overset{\frown}{\exp}\int_\tau^t \varphi(s)\,ds}\,,$$

which means that the diffusion coefficient does not change.

The drift coefficient is transformed in a similar way though its expression includes an extra term

$$b(t) = \frac{\partial f_x(\alpha)}{\partial \alpha}\bigg|_{x=L^{-1}\overset{\frown}{\exp}\int_\tau^t \varphi(s)\,ds\,y} = d\,L_{\overset{\frown}{\exp}\int_\tau^t \varphi(s)\,ds}\,y\,.$$

Hence, the process $\eta(t)$ has a stochastic differential defined by Ito's field

$$(\mathrm{d} L_x a(t) + b_x(t),\ \mathrm{d} L_x A(t))$$

and, as a consequence, solves Equation (2.19).

Consider a stochastic equation with left invariant coefficients on a Lie group G and assume that the coefficients depend, in addition, on a parameter $x \in X$ with X being a smooth Banach manifold. Let

$$\mathrm{d}\eta = L_{\eta(t)} \exp\left\{c(t, \xi(t))\ \mathrm{d}t + C(t, \xi(t))\ \mathrm{d}w\right\},$$

where

$$c(t, x) \in \mathcal{Y},\quad C(t, x) \in L_{12}(H, \mathcal{Y}),\quad x \in X$$

and $\xi(t)$ is a diffusion process on X.

A stochastic equation

$$\mathrm{d}\eta = \mathrm{Exp}_{\eta(t)}\left(c_{\eta(t)}(t, \xi(t))\ \mathrm{d}t + C_{\eta(t)}(t, \xi(t))\ \mathrm{d}w\right)$$

generates a G-valued multiplicative functional of the Markov process $\xi(t)$ given by the relation

$$\delta(t, \tau) = \widehat{\exp}\left\{\int_\tau^t c(s, \xi(s))\ \mathrm{d}s + \int_\tau^t C(s, \xi(s))\ \mathrm{d}w\right\},$$

where

$$c_g(t, x) = \mathrm{d} L_g c(t, x),$$

$$C_g(t, x) = \mathrm{d} L_g C(t, x).$$

Notice that in local coordinates we have

$$\mathrm{d}\delta = \delta\left\{c(t, \xi(t))\ \mathrm{d}t + \frac{1}{2}\operatorname{Tr} C^2(t, \xi(t))\ \mathrm{d}t + C(t, \xi(t))\ \mathrm{d}w\right\}.$$

2.3. STOCHASTIC EQUATIONS ON PRINCIPAL FIBRE BUNDLES

Let $p : \mathcal{P} \to X$ be the principal fibre bundle considered in the previous section. In subsection 3 of section 1, we showed that, given a connection on the base X and a connection of the principal fibre bundle, we may define a connection of the fibre bundle total space $\mathcal{P}$. Naturally, due to the results of Chapter 4, we may consider a stochastic differential equation on the manifold $\mathcal{P}$. Nevertheless, as in the vector bundle case, we may expect that a stochastic equation solution would have certain special properties if the equation coefficients are compatible with the principal fibre bundle structure.

In accordance with the notation of section 1 let Γ^x, Γ^p be connection coefficient of, respectively, the base X and the principal fibre bundle p, while $\Gamma^{\mathcal{P}}$ denotes the connection coefficient of the principal fibre bundle total space $\mathcal{P}$. Denote by $\exp^{\mathcal{P}}$ the exponential mapping corresponding to $\Gamma^{\mathcal{P}}$.

A stochastic equation with respect to a random process valued on the manifold $\mathcal{P}$ looks like

$$\mathrm{d}\,\gamma = \exp^{\mathcal{P}}_{\gamma(t)} \{c_{\gamma(t)}(t)\ \mathrm{d}\,t + C_{\gamma(t)}(t)\ \mathrm{d}\,w\}, \tag{2.20}$$

where (c_z, C_z) is an Ito field on $\mathcal{P}$, $z \in \mathcal{P}$,

$$c : \mathcal{P} \to T\,\mathcal{P}, \quad C : \mathcal{P} \to L_{12}\ (H_X, T\,\mathcal{P})\,.$$

It follows from the results of Chapter 4 that under corresponding conditions, there exists a random evolution family $\mathcal{U}(t, \tau)$ of manifold $\mathcal{P}$ mappings generated by Equation (2.20). In order to investigate the structure of this family, first consider a local principal fibre bundle $\mathcal{P} = X \times G$.

Let the principal parts of the fields c, C have the form

$$c_{\binom{x}{g}}(t) = (a_x(t),\ \mathrm{d}\,L_g\,b_x(t)),$$

$$C_{\binom{x}{g}}(t) = (A_x(t),\ \mathrm{d}\,L_g\,B_x(t)),$$

where

$$a_x(t) \in B, \quad b_x(t) \in \mathcal{Y}, \quad A_x(t) \in L_{12}\ (H, B),$$

$$B_x(t) \in L_{12}\ (H, \mathcal{Y}), \quad x \in X.$$

Notice that, under a local bundle morphism

$$\Phi = (f, g_1) : (x, g) \mapsto (f(x),\ g_1\,g),$$

the coefficient

$$c_{\binom{x}{g}}(t) = (a_x,\ \mathrm{d}\,L_g\,b_x) = (a_x, g\,b_x),$$

transforms according to

$$c_{\binom{x}{e}} \mapsto (f'(x)\,a_x,\ g_1\,b_x + g'_1\,a_x).$$

This shows that the linear character of the dependence of b on g is invariant under bundle morphisms. To grant the coefficient linear dependence on the group element, invariant under bundle morphisms, define c as follows

$$c_{\binom{x}{g}}(t) = \mathrm{d}\,L_g\,\{b_x(t) - \Gamma^p_x\ (a_x(t))\}\,,$$

where $b_x(t)$ is the principal part of a vector bundle $\mathrm{Ad}\,p$ section.

Due to arguments similar to those used to construct the component c and changing the local considerations to global ones, we obtain the following result.

PROPOSITION 2.2 *An Ito field (c, C) over the total space of a principal fibre bundle $p : \mathcal{P} \to X$ compatible with the principal fibre bundle structure may be defined by giving fields a_x, b_x, A_x, B_x over X which are, respectively, sections of τ_X,* $\mathrm{Ad}\,p$

$L_{12}\,(\theta, \tau_X)$, $L_{12}\,(\theta, \operatorname{Ad} p)$ *bundles.*

For a local bundle $\mathcal{P} = X \times G$ we obtain

$$c_{\binom{x}{g}} = \left(a_x,\ \mathrm{d} L_g\,(b_x - \Gamma_x^p\,(a_x))\right),$$

$$C_{\binom{x}{g}}(t) = \left(A_x,\ \mathrm{d} L_g\,(B_x - \Gamma_x^p\,(A_x))\right). \tag{2.21}$$

The formulas obtained above show how one can construct a local stochastic equation for $\gamma = (\xi, g)$ invariant under bundle morphisms

$$\mathrm{d}\,\xi = a_{\xi(t)}(t)\ \mathrm{d}t - \frac{1}{2} \operatorname{Tr} \Gamma^X_{\xi(t)} (A_{\xi(t)}(t) \cdot,\ A_{\xi(t)}(t) \cdot)\ \mathrm{d}t + \\ + A_{\xi(t)}(t)\ \mathrm{d}w, \tag{2.22}$$

$$\mathrm{d}\,g = \mathrm{d} L_g \Big\{ \frac{1}{2} \operatorname{Tr} \big\{ (\Gamma^p_{\xi(t)})' (A_{\xi(t)}(t) \cdot,\ A_{\xi(t)}(t) \cdot) - \\ - \Gamma^p_{\xi(t)} (\Gamma^X_{\xi(t)} (A_{\xi(t)}(t) \cdot,\ A_{\xi(t)}(t) \cdot)) \big\}\ \mathrm{d}t + \\ + [\,\Gamma^p_{\xi(t)} (A_{\xi(t)}(t) \cdot),\ B_{\xi(t)}\,(t) \cdot\,]\ \mathrm{d}t + B_{\xi(t)}(t)\ \mathrm{d}w + \\ + b_{\xi(t)}(t)\ \mathrm{d}t - \Gamma^p_{\xi(t)} (a_{\xi(t)}(t))\ \mathrm{d}t - \Gamma^p_{\xi(t)} (A_{\xi(t)}(t)\ \mathrm{d}w) \Big\}. \tag{2.23}$$

As had to be expected, the first equation in this system is merely a stochastic equation on the base X of the bundle p and its solution is an X-valued stochastic process $\xi(t) = p(\gamma(t))$. The second equation may be turned, by substituting $\xi(t)$, into a linear stochastic equation on the structural group G which is a fibre of the bundle $p : \mathcal{P} \to X$ corresponding to a chosen trivialization. It is obvious that this equation defines an operator multiplicative functional $Y(t, \tau)$ of the Markov process $\xi(t)$.

Thus, a solution of a stochastic equation on a local principal fibre bundle generates a random evolution family of fibre bundle total space mappings $Y(t, \tau) : \gamma(\tau) \mapsto \gamma(t)$ which possesses, almost surely, the following properties.

1. The projection of $Y(t, \tau)$ onto the base coincides with the corresponding random family $S(t, \tau)$ generated by the projection $\xi(t) = p(\gamma(t))$

$$p \circ Y(t, \tau) = S(t, \tau) \circ p. \tag{2.24}$$

2. The mapping $Y(t, \tau)$ acts from the fibre through the point $\xi(\tau)$ to the fibre $\mathcal{P}_{\xi(t)}$ through the point $\xi(t)$ and is generated by a solution of a linear stochastic equation on the group G.

DEFINITION 2.1. A random evolution family of a principal fibre bundle total space $\mathcal{P}$ mappings possessing the above properties is called a p-multiplicative functional of the

Markov process $\xi(t)$.

According to this definition, a solution of a local stochastic equation on a principal fibre bundle with coefficients of type (2.21) generates a p-multiplicative functional of its projection onto the base.

As in the vector bundle case, we may change local considerations to nonlocal ones. due to results of Theorem 2.1 of Chapter 4. Notice that the application of this result is justified by the left invariance property of coefficient fields. Indeed, coefficient properties permit us to state that the exit of the process out of the domain of a certain chart depends on the behaviour of its projection onto the base only. Therefore, it follows from Theorem 2.1 of Chapter 4 that an evolution family $\mathcal{V}(t, \tau)$ exists such that

$$\gamma(t) = \mathcal{V}(t, \tau)\, \gamma$$

is the unique solution of Equation (2.20) with coefficients of type (2.21) corresponding to an $\mathcal{F}_\tau$-measurable initial condition γ.

The family $\mathcal{V}(t, \tau)$ may be given in the form of a local random evolution family composition, and stabilization takes place in this representation, as it does in the vector bundle case.

Consider finally a stochastic equation

$$d\gamma = \exp^{\mathcal{P}}_{\gamma(t)} \left(\alpha_{\gamma(t)}(t)\, dt + \beta_{\gamma(t)}(t)\, dw\right) \tag{2.25}$$

with coefficients

$$\alpha_{\binom{x}{g}} = \left(a_x - dL_g\, \Gamma^p_x(a_x)\right),$$

$$\beta_{\binom{x}{g}} = \left(A_x - dL_g\, \Gamma^p_x(A_x)\right).$$

In local coordinates, Equation (2.25) is a system which coincides with (2.22), (2.23) if $b_x \equiv 0$ and $B_x \equiv 0$ in (2.23).

Equation (2.25) describes the stochastic parallel displacement in a principal fibre bundle along trajectories of the diffusion process $\xi(t) = p(\gamma(t))$.

2.4. EVOLUTION FAMILIES IN SECTIONS OF PRINCIPAL AND ASSOCIATED BUNDLES

Let $p : \mathcal{P} \to X$ be a principal fibre bundle with a structural group G and $\rho(p) : \mathcal{L}_X \to X$ be an associated vector bundle with the structural group $\rho(G)$ acting in a representation space $\mathcal{L}$, which is the typical fibre of the bundle $\mathcal{L}_X$.

A bounded function ψ on $\mathcal{P}$ valued in $\mathcal{L}$ is called an equivariant function if the relation

$$\psi(g z) = \rho(g)\, \psi(z)$$

holds.

A bundle morphism $\psi : \mathcal{P} \to \mathcal{L}_X$ is called equivariant if the diagram

$$\begin{array}{ccc} \mathcal{P} & \xrightarrow{\quad g \quad} & \mathcal{P} \\ \psi \downarrow & & \downarrow \psi \\ \mathcal{L}_X & \xrightarrow[\quad \rho(g) \quad]{} & \mathcal{L}_X \end{array}$$

is commutative.

The random flow of manifold $\mathcal{P}$ mappings considered in the previous section, may be mapped in a natural way to $\mathcal{L}_X$. Indeed, in each trivialization it is described by a pair of equations, namely by the stochastic equation (2.22) on the base X and the linear equation (2.23) on the structural group G with the coefficients depending on the solution of the equation on the base.

The representation ρ permits us to transform Equation (2.23) into a linear equation on the group $GL(\mathcal{L})$ of transformations of the typical fibre $\mathcal{L}$ of the bundle $\mathcal{L}_X$. Hence, it generates a linear stochastic equation in the bundle $\mathcal{L}_X$. We may develop this procedure for local equations verifying its invariance. As a consequence, due to the localization principle described in Chapters 4 and 5, it is also possible to deal with global objects.

Recall (See Chapter 5) that a random flow $\mathcal{V}(t, \tau)$ generated by the solution of the stochastic equation (2.20) defines an evolution family acting in the space of functions on $\mathcal{P}$. The above procedure, which gives a way to map a random flow on $\mathcal{P}$ into a random flow on $\mathcal{L}_X$, permits us to define an evolution family of mappings acting on the space of functions on $\mathcal{L}_X$ and, hence, an evolution family acting in sections of the bundle $\rho(p)$.

Evolution families constructed in this way, define the solutions of the Cauchy problem for second-order parabolic equations with respect to sections of the bundle $\rho(p)$.

Notice that we may derive those equations by a special trick.

Assume that we deal with local objects and consider the trivial bundles

$$p : X \times G \to X \,, \quad \rho(p) : \mathcal{L}_X = X \times \mathcal{L} \to X \,.$$

A random evolution family $\mathcal{V}(t, \tau)$ acting on $\mathcal{P}$ which is generated by the solution of system (2.22), (2.23) may be given in the form

$$\mathcal{V}(t, \tau) : (x, g) \mapsto (S(t, \tau; x) \,, \ \sigma_x(t, \tau)\, g) \,,$$

where $\sigma_x(t, \tau)$ is a G-valued multiplicative functional of the process $\xi(t) = p(\gamma(t)) = S(t, \tau; x)$.

Let $\psi : \mathcal{P} \to \mathcal{L}_X$ be an equivariant bundle morphism and $\varphi : X \to \mathcal{L}$ be the principal part of the bundle $\rho(p)$ section satisfying the relation $\psi(x, g) = \rho(p)\, \varphi(x)$.

Assume that $\varphi \in \sigma_2(\rho(p))$ and set

$$\mathcal{N}(\tau, t)\, \varphi(x) = E\, \rho(\sigma(t, \tau))\, \varphi(S(t, \tau; x)) \,.$$

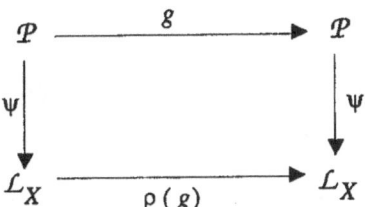

The family $\mathcal{N}(\tau, t)$ possesses the evolution property

$$\mathcal{N}(\tau, t)\, \varphi(x)$$

$$= E\, E_\theta\, \rho\, (\sigma(t, \theta)\, \sigma(\theta, t)) \times \varphi\, (S(t, \theta)\, S(\theta, \tau) \circ x)$$

$$= E\, \rho\, (\sigma(t, \theta)) \times (\mathcal{N}(\tau, \theta)\, \varphi)\, (S(t, \theta; x))$$

$$= \mathcal{N}(\tau, \theta)\, \mathcal{N}(\theta, t)\, \varphi\, (x)\,.$$

It may be proved that $\mathcal{N}(\tau, t)$ satisfies a parabolic equation of the second order.

To make our treatment more precise assume for the moment that $p : \mathcal{P} \to X$ is a principal bundle with the right action of the structural group equipped with connection. Denote $K^p : T\mathcal{P} \to \mathcal{Y}_X$, $K^{\mathcal{P}} : T^2\mathcal{P} \to T\mathcal{P}$ as the corresponding connection maps for the principal bundle p and $K^{\mathrm{Ad}\,p} : T\mathcal{Y}_X \to \mathcal{Y}_X$, $K^{\mathcal{Y}_X} : T^2\mathcal{Y}_X \to T\mathcal{Y}_X$ for the adjoint bundle $\mathrm{Ad}\,p : \mathcal{Y}_X \to X$. In local trivializations, those maps act as follows

$$K^p\,(x, g, y, \tilde{B}) = (x, \mathrm{Ad}\,(g)\,(B + \Gamma_x^p\, y))\,, \tag{2.26}$$

$$K^{\mathcal{P}}\,(x, g, y, \tilde{A}, z, \tilde{B}, u, C) = \Big(x, g, u + \Gamma_x^X\,(z, y),$$

$$C + \big\{\, \mathrm{Ad}\,(g)\, \{(\Gamma_x^p)'\,(z, y) + [\Gamma_x^p\, z,\ \Gamma_x^p\, y] -$$

$$- \Gamma_x^p\, \Gamma_x^X\,(z, y) + [\Gamma_x^p\, z,\ A] + [\Gamma_x^p\, z,\ B]\,\big\}\, g\Big), \tag{2.27}$$

$$K^{\mathrm{Ad}\,p}\,(x, A, y, B) = \big(x, B + [\Gamma_x^p\, y,\ A]\big), \tag{2.28}$$

$$K^{\mathcal{Y}_X}\,(x, A, y, B, z, C, u, \mathcal{D}) = \Big(x, A, u + \Gamma_x^X\,(z, y),$$

$$\mathcal{D} + [(\Gamma_x^p)'\,(z, y), A] + \big[\Gamma_x^p\, z, [\Gamma_x^p\, y,\ A]\big] + [\Gamma_x^p\, y,\ C] +$$

$$+ [\Gamma_x^p\, z,\ B] - [\Gamma_x^p\, \Gamma_x^X\,(z, y), A]\Big)\,. \tag{2.29}$$

(Recall that $B\,g = \mathrm{d}\ R_g\ B,\ g \in G,\ B \in \mathcal{Y}$.)

Next, define covariant derivatives ∇^p and $\nabla^{\mathrm{Ad}\,p}$ corresponding to K^p and $K^{\mathrm{Ad}\,p}$, respectively. For φ being a section of p, $\nabla^p \varphi : TX \to \mathcal{Y}_X$ and for Ψ being a section of $\mathrm{Ad}\,p$, $\nabla^{\mathrm{Ad}\,p}\, \Psi : TX \to \mathcal{Y}_X$. In local coordinates, the action of corresponding covariant derivatives looks like the following

$$\nabla_y^p\ \varphi = \varphi'\,(x)\, y + \varphi\,(x)\, \Gamma_x^p\, y, \tag{2.30}$$

$$\nabla_z^{\mathrm{Ad}\,p}\ \Psi = \Psi'\,(x)\, z + [\Gamma_x^p\, z, \Psi\,(x)]\,. \tag{2.31}$$

Below, we shall also need second-order invariant operators such as

$$\mathfrak{A}^z\ \varphi\,(x) = \nabla_z^{\mathrm{Ad}\,p}\ \nabla_z^p\ \varphi\,(x) - \nabla_{\nabla_z z}^p\ \varphi\,(x) \tag{2.32}$$

with local representation

$$\mathfrak{A}^{z,y}\ \varphi(x) = \varphi''(x)(z,y) + \varphi(x)\left\{(\Gamma^p_x)'(z,y) + \right.$$

$$+ [\Gamma^p_x y, \varphi^{-1}(x)\varphi'(x) z] + [\Gamma^p_x z, \varphi^{-1}(x)\varphi'(x) y] +$$

$$\left. + [\Gamma^p_x z, \Gamma^p_x y] - \Gamma^p_x\, \Gamma^X_x(z,y)\right\} -$$

$$- \varphi'(x)\, \Gamma^X_x(z,y), \tag{2.33}$$

acting in sections of principal bundle p and

$$\mathfrak{A}^z_1 \Psi(x) = \nabla^{\mathrm{Ad}^* p}_z\ \nabla^{\mathrm{Ad}^* p}_z\ \psi(x) - \nabla^{\mathrm{Ad}^* p}_{\nabla_z z}\ \psi(x) \tag{2.34}$$

with local representation (1.22) acting in sections of the bundle $\mathrm{Ad}^* p : \mathcal{G}^*_X \to X$.

We now aim at demonstrating the probabilistic approach to solving the following Cauchy problems

$$\frac{\partial u}{\partial s} + \frac{1}{2}\ \mathrm{Tr}\,\mathfrak{A}^A u + \nabla^p_a u +$$

$$+ u\ \mathrm{Tr}\,[u^{-1}\nabla^p_A u,\ B\,\cdot\,] + u\, b = 0, \tag{2.35}$$

$$u(t,x) = \varphi(x)\,, \tag{2.36}$$

and

$$\frac{\partial \upsilon}{\partial s} + \frac{1}{2}\ \mathrm{Tr}\,\mathfrak{A}^A_1\ \upsilon + \nabla^{\mathrm{Ad}^* p}_a\ \upsilon +$$

$$+ C^*(\nabla^{\mathrm{Ad}^* p}_A\ \upsilon, \cdot\,) + c^*\upsilon = 0, \tag{2.37}$$

$$\upsilon(t,x) = \Psi(x)\,. \tag{2.38}$$

To deal with the Cauchy problem (2.35), (2.36), consider the stochastic equation

$$\mathrm{d}\,\gamma = \exp^{\mathcal{P}}_{\gamma(t)}\left(q_{\gamma(t)}\ \mathrm{d}\,t + Q_{\gamma(t)}\ \mathrm{d}\,w\right). \tag{2.39}$$

In local trivialization, Equation (2.39) may be presented as a system of stochastic equations

$$\mathrm{d}\,\xi = a_{\xi(t)}\ \mathrm{d}\,t - \frac{1}{2}\ \mathrm{Tr}\,\Gamma^X_{\xi(t)}\left(A_{\xi(t)}\,\cdot,\ A_{\xi(t)}\,\cdot\right)\ \mathrm{d}\,t + A_{\xi(t)}\ \mathrm{d}\,w, \tag{2.40}$$

$$\mathrm{d}\,g = g(t)\left\{b_{\xi(t)}\ \mathrm{d}\,t - \Gamma^p_{\xi(t)}\, a_{\xi(t)}\ \mathrm{d}\,t + B_{\xi(t)}\ \mathrm{d}\,w - \right.$$

$$- \Gamma^p_{\xi(t)}\, A_{\xi(t)}\ \mathrm{d}\,t + \frac{1}{2}\ \mathrm{Tr}\left\{(\Gamma^p_{\xi(t)})'\left(A_{\xi(t)}\,\cdot,\ A_{\xi(t)}\,\cdot\right) - \right.$$

$$\left.\left. - \Gamma^p_{\xi(t)}\,\Gamma^X_{\xi(t)}\left(A_{\xi(t)}\,\cdot,\ A_{\xi(t)}\,\cdot\right)\right\}\ \mathrm{d}\,t + \mathrm{Tr}\left[\Gamma^p_{\xi(t)}\, A_{\xi(t)}\,\cdot,\ B_{\xi(t)}\right]\ \mathrm{d}\,t\right\}. \tag{2.41}$$

Here $\exp^{\mathcal{P}}$ corresponds to the connection map (2.27),

$$a_x \in T_x X, \quad A_x \in L_{12}(H, T_x X), \quad b_x \in \mathcal{Y}, \quad B_x \in L_{12}(H,$$

$$q_\gamma = (a_x,\ b_x - \Gamma_x^p\, a_x), \quad \mathcal{Q}_\gamma = (A_x,\ B_x - \Gamma_x^p\, A_x).$$

Due to results of the previous section, we may state that if the coefficients of (2.39) are smooth enough, then there exists a unique solution of (2.39) (or (2.40), (2.41) which satisfies the condition

$$\gamma(s) = (\xi(s), g(s)) = (x, e). \tag{2.42}$$

Consider next a curve in G determined by the relation

$$\eta(t) = \varphi(\xi(t))\, g(t), \tag{2.43}$$

where $\gamma(t) = (\xi(t), g(t))$ is the solution of (2.39), (2.42) and φ is a smooth section of p. By direct computations and taking into account the results of previous sections, one may prove that

$$E\left\{d\eta\,\eta^{-1}(t) - \varphi(\xi(t))\ dg\,g^{-1}(t)\,\varphi^{-1}(\xi(t))\ d\varphi(\xi(t))\,\varphi^{-1}(\xi(t))\right\}\Big|_{t=s}$$

$$= \mathrm{Ad}\,(\varphi(x))\left\{\frac{1}{2}\ \mathrm{Tr}\left\{\varphi^{-1}\,\mathrm{Tr}\,\varphi''(A_x\cdot,\ A_x) -\right.\right.$$

$$\left. - \varphi^{-1}\varphi'\ \Gamma_x^X(A_x\cdot,\ A_x\cdot)\right\} +$$

$$+ \varphi^{-1}\varphi' a_x + b_x - \Gamma_x^p\, a_x + \frac{1}{2}\ \mathrm{Tr}\left\{(\Gamma_x^p)'(A_x\cdot,\ A_x\cdot) -\right.$$

$$\left. - \Gamma_x^p\,\Gamma_x^X(A_x\cdot,\ A_x\cdot)\right\} + \mathrm{Tr}\left[\Gamma_x^p\, A_x\cdot,\ B_x\cdot\right] +$$

$$\left. + \mathrm{Tr}\left[\Gamma_x^p\, A_x\cdot,\ \varphi^{-1}\varphi'(x)\, A_x\cdot\right]\right\} = \mathrm{Ad}\,(\varphi(x))\ \mathcal{B}\varphi(x). \tag{2.44}$$

Comparing the resulting expression with (2.31), one may see that

$$\mathrm{Ad}\,(\varphi(x))\ \mathcal{B}\varphi(x) = \left\{\frac{1}{2}\ \mathrm{Tr}\,\mathfrak{A}^A\,\varphi(x) + \nabla_a^P\ \varphi(x) +\right.$$

$$\left. + \varphi(x)\,\mathrm{Tr}\left[\varphi^{-1}\,\nabla_{A\cdot}^P\ \varphi(x),\ B\cdot\right] + \varphi(x)\, b\right\}\varphi^{-1}(x). \tag{2.45}$$

Now define an evolution family acting on the set of principal bundle sections with $\mathcal{B}$ being its generator. Put

$$U(t, s)\,\varphi(x) = \varphi(x)\ \exp\ \mathcal{B}\ (t - s). \tag{2.46}$$

It follows from standard computations that

$$u(s, x) = U(t, s)\,\varphi(x) \tag{2.47}$$

satisfies Equations (2.35) and (2.36).

Now we may state the following assertion.

THEOREM 2.2. *Let* $a_x \in T_x X$, $A_x \in L_{12}(H, T_x X)$, $b_x \in \mathcal{Y}$, $B_x \in L_{12}(H, \mathcal{Y})$ *be smooth fields along with local connection coefficients. Then there exists a unique smooth solution of the Cauchy problem* (2.35), (2.36) *which may be presented in the form* (2.46) *with generator* $\mathcal{B}$ *having probabilistic representation* (2.44).

Consider next the stochastic equation in the total space of the bundle Ad $p : \mathcal{Y}_X \to X$.

$$d\beta(t) = \exp_{\beta(t)}^{\mathcal{Y}_X} \left(m_{\beta(t)}(t)\, dt + M_{\beta(t)}(t)\, dw\right) \tag{2.48}$$

which looks like the system of equations in local trivialization

$$d\xi = a_{\xi(t)}(t)\, dt - \frac{1}{2} \operatorname{Tr} \Gamma^X_{\xi(t)} \left(A_{\xi(t)}(t)\,\cdot, A_{\xi(t)}(t)\,\cdot\right) dt + \\ + A_{\xi(t)}(t)\, dw, \tag{2.49}$$

$$\begin{aligned} d\eta(t) = {} & c_{\xi(t)}(t)\,\eta(t)\, dt - \left[\Gamma^p_{\xi(t)}\, a_{\xi(t)}(t), \eta(t)\right] dt + \\ & + C_{\xi(t)}(t)(\eta(t), dw) - \left[\Gamma^p_{\xi(t)}\, A_{\xi(t)}(t)\, dw, \eta(t)\right] - \\ & - \frac{1}{2} \operatorname{Tr}\Big\{ \left[(\Gamma^p_{\xi(t)})'\left(A_{\xi(t)}(t)\cdot, A_{\xi(t)}(t)\cdot\right), \eta(t)\right] + \\ & + \Big[\Gamma^p_{\xi(t)}\, A_{\xi(t)}(t)\cdot, \left[\Gamma^p_{\xi(t)}\, A_{\xi(t)}(t)\cdot, \eta(t)\right]\Big] - \\ & - \left[\Gamma^p_{\xi(t)}\, \Gamma^X_{\xi(t)}\left(A_{\xi(t)}(t)\,\cdot, A_{\xi(t)}(t)\,\cdot\right)\eta(t)\right]\Big\} + \\ & + \operatorname{Tr}\left[\Gamma^p_{\xi(t)}\, A_{\xi(t)}(t)\cdot, C_{\xi(t)}(t)\,\eta(t)\cdot\right] dt. \end{aligned} \tag{2.50}$$

Here, $\exp^{\mathcal{Y}_X} : T\mathcal{Y}_X \to \mathcal{Y}_X$ stands for the exponential mapping on $\mathcal{Y}_X$ corresponding to the connection map (2.29),

$$a_x(t) \in T_x X, \qquad A_x(t) \in L_{12}(H, T_x X),$$

$$c_x(t) \in L(\mathcal{Y}), \qquad C_x(t) \in L_{12}(H, T(\mathcal{Y})), \qquad x \in X.$$

Assuming Equation (2.48) coefficients to be smooth and bounded, we may deduce from results of Chapter 4 that a unique solution of Equation (2.48) exists satisfying the condition

$$\beta(s) = (\xi(s), \eta(s)) = (x, h)\,. \tag{2.51}$$

Notice, in addition, that (2.50) defines a multiplicative operator functional of the process

$\xi(t) = \mathrm{Ad}\, p\,(\xi(t))$.

Consider the coadjoint vector bundle $\mathrm{Ad}^*\, p : \mathcal{G}^*_X \to X$ with a typical fibre isomorphic to coalgebra $\mathcal{G}^*$. Let Ψ be a bounded smooth section of $\mathrm{Ad}^*\, p$. Due to the results of the previous chapter, one may easily prove the following statement.

THEOREM 2.3. *Let*

$$a_x(t) \in T_x X, \qquad A_x(t) \in L_{12}(H, T_x X),$$

$$c_x(t) \in L(\mathcal{G}), \Gamma^p_x, \qquad C_x(t) \in L_{12}\ (H, T(\mathcal{G})), \Gamma^{\mathcal{P}}_x$$

be smooth bounded fields. Then a unique solution exists for the Cauchy problem (2.37), (2.38) *which may be presented in the form*

$$\langle h, u(s, x)\rangle = E\langle \eta(t), \Psi(\xi(t))\rangle, \tag{2.52}$$

where $\beta(t) = (\xi(t), \eta(t))$ *is the solution of the stochastic Cauchy problem* (2.48), (2.51).

3. Stochastic Equations on Manifolds and their Solution Distribution Properties

3.1. FORWARD AND BACKWARD DERIVATIVES AND THEIR CONNECTIONS

Let X be a smooth manifold, f_α be a one-parameter group of transformations of X, $\alpha \in R^1$. Consider an X-valued random process $\xi(t)$ solving (2.1) and invariant with respect to f_α. As has been shown in Section 1, its transition probability satisfies the relation

$$\int_X \varphi(f_\alpha(y))\, P(\tau, x, t, \mathrm{d}y) = \int_X \varphi(y)\, P(\tau, f_\alpha(x), t, \mathrm{d}y)\,. \tag{3.1}$$

Let $P(\tau, x, t, Q)$ depend smoothly on x and be a smooth function of the set Q along the vector field $F(x)$ generating the group f_α. The conditions which guarantee the smoothness of the transition probability has been discussed in Chapter 5. Here we assume that those conditions are fulfilled.

Let $\varphi \in C_1(X, R^1)$. If we compute the derivative of both parts of (3.1) with respect to α at $\alpha = 0$, we obtain the relation

$$\int_X \langle F(y), \varphi'(y)\rangle\, P(\tau, x, t, \mathrm{d}y) = \int_X \varphi(y)\langle F(x), P'_x(\tau, x, t, \mathrm{d}y)\rangle\,.$$

Integration by parts changes this relation to

$$\int_X \varphi(y)\left\{\nabla_F\, P(\tau, x, t, \mathrm{d}y) + \langle F(x), P'_x(\tau, x, t, \mathrm{d}y)\rangle\right\} = 0. \tag{3.2}$$

Therefore, it the vector field F commutes with stochastic equation coefficients (see (2.10)), then the transition probability derivatives with respect to forward and backward variables are connected by the relation

$$\nabla_F P(\tau, x, t, \cdot) = -\langle F(x), P'_x(\tau, x, t, \cdot)\rangle\,.$$

3.2. STOCHASTIC EQUATIONS ON LIE GROUPS AND THE SMOOTHNESS PROPERTY OF THEIR SOLUTION DISTRIBUTIONS

Let $X = G$ be a Lie group and $\xi(t)$ be a G-valued stochastic process which solves (2.14). To investigate this process distribution behaviour under a group action, let us follow the approach suggested in Chapter 5.

Let H be a Hilbert space densely imbedded in $\mathcal{Y} = T_e G$, $i : H \to \mathcal{Y}$ be the embedding operator and the triple $(\mathcal{Y}, H, i)$ give a Hilbert–Schmidt structure. Consider a subbundle κ of the tangent bundle τ_G generated by the subset H of the Lie algebra $\mathcal{Y}$, $\kappa_g^{-1} = \mathrm{d} L_g H$, and assume that $w(t)$ is the Wiener process associated in the canonic way with the Hilbert space H.

If the coefficients of Equation (2.1) are left invariant then, as has been proved in Section 2, the equation itself may be written in the form

$$\mathrm{d}\xi = L_\xi \,\{a(t)\,\mathrm{d}t + A(t)\,\mathrm{d}w\}, \tag{3.3}$$

$$a(t) \in H,\; A(t) \in L_{12}(H), \quad t \in [0, T],$$

while the multiplicative integral

$$\sigma(t, \tau) = \overset{\curvearrowright}{\exp}\left\{\int_\tau^t b(s)\,\mathrm{d}s + \int_\tau^t A(s)\,\mathrm{d}w(s)\right\},$$

$$b(s) = a(s) - \frac{1}{2}\,\mathrm{Tr}\; A(s) A^*(s)$$

provides a representation of its solution.

Due to results of Chapter 5 (Section 3) in order to verify that the transition probability $P(\tau, x, t, Q)$ of the process $\xi(t) = \sigma(t, \tau) \circ x$ is differentiable along trajectories of κ sections, we have to check the invariance of $\mathrm{d}\xi$ with respect to the tangent map

$$T S(t, \tau; x) = \mathrm{d} R_{S(t, \tau; x)}$$

(where $S(t, \tau; x) = \sigma(t, \tau) \circ x$). This relation means that

$$(\mathrm{Ad}\;\sigma(t, \tau))\, H \subset H. \tag{3.4}$$

Consider, in particular, the simplest case $a \equiv 0$, $A(t) \equiv A$. According to Theorem 3.2 of Chapter 5, the relation

$$(\mathrm{Ad}\, g) A h - A h \in L_{12}(H) \tag{3.5}$$

is sufficient to grant that (3.4) holds.

The last relation shows that the difference between the mapping $\mathrm{Ad}\, g$ and the identity mapping is a Hilbert–Schmidt operator. Notice that a verification of (3.5) would be connected with the investigation of concrete infinite-dimensional groups, which is beyond the scope of this book.

To conclude this section, we point out another approach to the above problem, connected with restrictions of another type.

The relation (3.4) is proved to be valid if

$$[a(t), H] \subset H, \quad \Big[\int_0^t A(s)\, \mathrm{d}w(s)\, H \Big] \subset H.$$

The first inclusion takes place if H is a Lie subalgebra of the Lie algebra $\mathcal{G}$ and $a(t) \in H$. Nevertheless, since the values of the stochastic integral $\int_0^t A(s)\ \mathrm{d}w(s)$ may leave H, in order to be sure the second inclusion holds, one needs to assume that H is an ideal of the Lie algebra $\mathcal{G}$.

Under these conditions, the transition probability $P(\tau, x, t, Q)$ of the process $\xi(t) = \sigma(t, \tau) \circ x$ is differentiable along shifts on elements of the invariant subgroup generated by H.

3.3. ABSOLUTELY CONTINUOUS SMOOTH MEASURES

Smoothness is a natural property of a Borel measure on a smooth manifold.

Throughout all the exposition, we mean by a smooth measure a measure with smooth logarithmic derivatives along a large enough number of vector fields.

The quasi-invariance of the measure μ is another property which usually may be of interest if it is the quasi-invariance with respect to a large enough collection of mappings.

It is easy to show that for a measure μ smoothness in the classical sense and quasi-invariance with respect to shifts are not equivalent. Indeed, a measure μ on X having a smooth density with respect to a Lebesgue measure, cannot be quasi-invariant if it has a finite support. It should be noticed that, in this case, the measure logarithmic derivative has singularities at boundary points of the support.

We shall now use simple arguments to show that if a measure is smooth in the above sense, then it is quasi-invariant as well.

Consider a family of nonnegative measures $\mu(t, \cdot)$ depending on a parameter $t \in [0, T]$ and defined on a measurable space $(X, \mathcal{Z})$. Assume that this family is differentiable in a weak sense with respect to t. It means that there exists a real valued measure $\nu(t, \cdot)$ such that for any bounded measurable function φ, the relation

$$\frac{\mathrm{d}}{\mathrm{d}t} \int_X \varphi(x)\, \mu(t, \mathrm{d}x) = \int_X \varphi(x)\, \nu(t, \mathrm{d}x)$$

holds. If in addition $\nu(t, \cdot)$ is absolutely continuous with respect to $\mu(t, \cdot)$, then the density

$$\rho(t, x) = \frac{\nu(t, \mathrm{d}x)}{\mu(t, \mathrm{d}x)}$$

is called the logarithmic derivative of the family $\mu(t, \cdot)$.

In case the family $\mu(t, \cdot)$ consists of equivalent measures

$$\frac{\mu(t, \mathrm{d}x)}{\mu(0, \mathrm{d}x)} = \lambda(t, x) > 0 \quad \text{(a.e.)}$$

and λ is a differentiable function of t, we obtain that $\nu(t, \mathrm{d}x) = \lambda'_t(t, x)\, \mu(0, \mathrm{d}x)$ and

$$\rho(t, x) = \frac{\nu(t, dx)}{\mu(t, dx)} = \frac{d}{dt} \ln \lambda(t, x).$$

Therefore

$$\lambda(t, x) = \exp \int_0^t \rho(s, x) \, ds$$

at those points for which the right-hand side is finite.

The converse is valid as well.

PROPOSITION 3.1. *Let a family of measures* $\mu(t, \cdot)$, $(0 \le t \le T)$ *defined on a measurable space* $(X, \mathcal{Z})$ *possess the logarithmic derivative* $\rho(t, x)$ *and the following conditions hold.*

(i) $\rho(t, x)$ *is a continuous function of the argument* $t \; \mu(t, \cdot)$ a.e.

(ii) $\rho(t, x)$ *is a* σ*-bounded function in the sense that there exists a decomposition* $X = \cup_{k=1}^{\infty} X_k$ *such that* $\rho(t, x)$ *is bounded over each set*

$$[t_1, t_2] \times X_k, \quad (0 < t_1 < t_2 < T).$$

Then the measure family consists of mutually equivalent measures and

$$\ln \frac{\mu(t, dx)}{\mu(0, dx)} = \int_0^t \rho(s, x) \, ds.$$

Proof. Consider a bounded measurable function $\varphi(x)$ which is equal to zero outside of a certain set X_k and put

$$\alpha(t) = \int_X \varphi(x) \exp\left\{ - \int_0^t \rho(s, x) \, ds \right\} \mu(t, dx).$$

Under the conditions of the proposition, one may differentiate the function $\alpha(t)$ with respect to t, which gives

$$\alpha'(t) = - \int_X \varphi(x) \rho(t, x) \exp\left\{ - \int_0^t \rho(s, x) \, ds \right\} \times$$

$$\times \mu(t, dx) +$$

$$+ \int_X \varphi(x) \rho(t, x) \exp\left\{ - \int_0^t \rho(s, x) \, ds \right\} \mu'_t(t, dx) \equiv 0,$$

which implies that $\alpha(t) = \alpha(0)$. Since $\varphi(x)$ is an arbitrary function, we obtain

$$\mu(t, dx) = \exp\left\{ - \int_0^t \rho(s, x) \, ds \right\} \mu(0, dx).$$

Consider now a measure μ defined on a smooth manifold X and a vector field $z(x)$ with integral flow $S(t; x)$

$$\frac{d S(t; x)}{dt} = z(S(t; x)), \quad S(0; x) = x.$$

The relation $\mu(t,\cdot) = \mu \circ S^{-1}(t)$ defines a measure $\mu(t,\cdot)$ on X depending on t which inherits the smoothness property of the original measure μ. If μ is differentiable along z and has logarithmic derivative $\rho_\mu(z;x)$, then the measure $\mu(t,\cdot)$ possesses the same property and

$$\rho_{\mu(t,\cdot)}(z;x) = \rho_\mu(z; S(-t;x)).$$

Notice next that

$$\frac{d}{dt}\,\mu(t,\cdot) = \nabla_z\,\mu(t,\cdot) = \rho_{\mu(t,\cdot)}(z;x)\,\mu(t,\cdot),$$

and hence the logarithmic derivative of the measure $\mu(t,\cdot)$ with respect to t coincides with its logarithmic derivative along the vector field z.

Assume the function $\rho_\mu(z;x)$ to be continuous over the manifold X and to be locally bounded in the sense that it is bounded over the union of an arbitrary finite number of charts. Let us mention that in concrete situations, one may take $\rho_\mu(z;x)$ to be bounded by changing the field z out of a certain domain. Notice that inside the domain, the logarithmic derivative is not changed in this way. It is easily deduced from Proposition 3.1 that the following assertion holds.

THEOREM 3.1. *Let μ be a measure on a manifold X having a continuous bounded logarithmic derivative along a vector field z. Then its shifts along this field trajectories are equivalent and the relation*

$$\ln \frac{\mu(t, dx)}{\mu(0, dx)} = \int_0^t \rho_\mu(z; S(-\tau;x))\,d\tau$$

holds.

COROLLARY 3.1. *Let X be a manifold with a given Hilbert–Schmidt structure (B, H, i) and the measure μ defined on X possesses a continuous vector logarithmic derivative $\lambda(x)$. Then, for any vector field $z(x)$, such that the function*

$$\rho_\mu(z;x) = \langle \lambda(x), z(x)\rangle + \operatorname{div} z(x)$$

is locally bounded, the measure μ is equivalent to its shifts along the vector field $z(x)$.

Hence, in this case, the measure μ has a large enough collection of equivalent shifts.

Remark 3.1. Under the conditions of Chapter 5, the above results may be applied to investigate diffusion process transition probabilities. In particular, in the way mentioned in Section 2, one may construct measures on Lie groups having a large enough collection of equivalent group shifts.

3.4. ABSOLUTELY CONTINUOUS STOCHASTIC EQUATION SOLUTION DISTRIBUTIONS

In this section, we shall deal with another approach to the investigation of absolutely continuous measures generated by stochastic equation solutions. The crucial point of this approach is the use of the parabolic differential equations which stochastic equation solution functional expectations obey, due to the results of Chapter 5.

Consider a pair of stochastic equations on a Banach manifold

$$d\xi = \exp^X_{\xi(t)} \left(a_{\xi(t)}(t)\, dt + A_{\xi(t)}(t)\, dw\right), \tag{3.6}$$

$$d\eta = \exp^X_{\xi(t)} \left(b_{\eta(t)}(t)\, dt + A_{\eta(t)}(t)\, dw\right), \tag{3.7}$$

and suppose that their coefficients are connected by the relation

$$b_x(t) = a_x(t) + A_x(t)\,\alpha_x(t)\,. \tag{3.8}$$

Assume that the coefficients a, b, A meet all the requirements which are necessary in order to guarantee the existence and uniqueness of smooth solutions of (3.6) and (3.7). Due to results of Chapter 5, we know that solutions of (3.6) and (3.7) generate evolution families of linear operators acting in $C_2(X, R^1)$.

Moreover, if $\eta_{\tau,x}(t)$ is a solution of (3.7) satisfying $\eta_{\tau,x}(t) = x$ and $M_{\tau,x}$ is the space of X-valued trajectories $y(s)$, $(s > \tau)$ starting from the point $y(\tau) = x$, then the function

$$\begin{aligned} u(\tau, x) &= U_1(\tau, t) f(x) = E f(\eta_{\tau,x}(t)) \\ &= \int_X f(y)\, P^{\eta}(\tau, x, t, dy) \\ &= \int_{M_{\tau,x}} f(y(s))\, \mu^{\eta}_{\tau,x}(dy). \end{aligned} \tag{3.9}$$

solves the Cauchy problem

$$\begin{aligned} &\frac{\partial u}{\partial \tau} + \frac{1}{2} \langle A_x(t) A_x^* \nabla, \nabla \rangle u + \\ &\quad + \langle b_x(t), \nabla \rangle u = 0, \qquad u(t, x) = f(x)\,. \end{aligned} \tag{3.10}$$

On the other hand, the unique solution of this problem may be represented as well in terms of the process $\xi(t)$. Comparing the resulting representations, we may easily deduce the relation between the distributions μ^{ξ} and μ^{η} of the corresponding processes, which indicates that they are equivalent measures.

Consider a scalar multiplicative functional $\upsilon(t\, s; \xi(\cdot))$ of the process $\xi(t)$ defined by a linear stochastic equation

$$\begin{aligned} &d_s \upsilon(t, s; \xi(\cdot)) \\ &\quad = (\alpha_{\xi(s)}(s), dw)\, \upsilon(t, s; \xi(\cdot)), \quad \upsilon(t, t; \xi(\cdot)) = 1\,. \end{aligned}$$

Due to Ito's formula, we may easily verify that

$$\upsilon(t, \tau; \xi(\cdot)) = \exp\left\{ \int_\tau^t (\alpha_{\xi(s)}(s)\, dw) - \frac{1}{2} \int_\tau^t \| \alpha_{\xi(s)}(s) \|^2\, ds \right\}.$$

The theorem concerning multiplicative functionals stated in Chapter 4, implies that the evolution family

$$U(\tau, t) f(x) = E\, \upsilon(t, \tau; \xi(\cdot)) f(\xi_{\tau,x}(t))$$

$$= \int_{M_{\tau,x}} f(y(\cdot))\, \upsilon(t, \tau; \xi(\cdot))\, \mu^{\xi}_{\tau,x}(d\, y(\cdot)) \tag{3.11}$$

has the same generator as the family given by (3.9) since (3.8) holds. Therefore, the function $u(\tau, x) = U(\tau, t) f(x)$ gives a solution of the Cauchy problem (3.10). Comparing (3.9) and (3.11), we obtain the relation

$$E\, F(\eta_{\tau,x}(t)) = E\, F(\xi_{\tau,x}(t))\, \upsilon(t, \tau; \xi(\cdot)) \tag{3.12}$$

for functionals of the form $F(y(\cdot)) = f(y(\tau))$, as the solution of the Cauchy problem (3.10) is unique.

Making use of the Markovian property of the considered processes, we may extend the above relation to a richer class of functionals like $F(y(\cdot)) = \Pi_j f_j(y(t_j))$ and as a consequence, their linear combinations. Indeed, for example, setting

$$F(y) = f_1(y(s)) f_2(y(t)) \text{ for } \tau < s < t,$$

we obtain

$$E\, F(\eta_{\tau,x}(t))$$

$$= E f_1(\eta_{\tau,x}(s))\, E_s f_2(\eta_{s,\eta(s)}(t))$$

$$= E f_1(\xi_{\tau,x}(s))\, \upsilon(\tau, s; \xi(\cdot))\, E_s f_2(\xi_{s,\xi(s)}(t))\, \upsilon(t, s; \xi(\cdot))$$

$$= E\, E_s\, \upsilon(t, s; \xi(\cdot))\, \upsilon(s, \tau; \xi(\cdot)) f_1(\xi_{\tau,x}(s)) \times$$

$$\times f_2(\xi_{s,\xi(s)}(t)) = E\, \upsilon(t, \tau; \xi(\cdot)) f(\xi_{\tau,x}(t)).$$

In a similar way, one may compute the corresponding relation for a larger number of factors.

Now we may calculate that for a large enough collection of functionals F over the trajectory space $M_{\tau,x}$, the equality

$$\int_{M_{\tau,x}} F(y(\cdot))\, \mu^{\xi}_{\tau,x}(d\, y(\cdot))$$

$$= \int_{M_{\tau,x}} F(y(\cdot))\, \upsilon(t, \tau; y(\cdot))\, \mu^{\xi}_{\tau,x}(d\, y(\cdot))$$

is valid. This equality shows that μ^{η} and μ^{ξ} are absolutely continuous measures and the relation

$$\frac{\mu^{\eta}(d\, y)}{\mu^{\xi}(d\, y)} = \upsilon(t, \tau; y(\cdot))$$

$$= \exp\left\{ \int_{\tau}^{t} (\alpha_{\xi(s)}(s)\, d\, w) - \frac{1}{2} \int_{\tau}^{t} \| \alpha_{\xi(s)}(s) \|^2\, d\, s \right\} \tag{3.13}$$

holds.

In this way, we come to the following assertion.

THEOREM 3.2. *Let the coefficients of* (3.6) *and* (3.7) *satisfy* (3.8). *Then the probability distributions* μ^{η} *and* μ^{ξ} *are equivalent and* (3.13) *holds.*

COROLLARY 3.1. *Under the above conditions, the transition probabilities* $p^{\xi}(\tau, x, t, \cdot)$ *and* $p^{\eta}(\tau, x, t, \cdot)$ *are equivalent as well and the relation*

$$\frac{p^{\eta}(\tau, x, t, \mathrm{d}\, y)}{p^{\xi}(\tau, x, t, \mathrm{d}\, y)} = E\left\{\upsilon(t, \tau; \xi_{\tau,x}(\cdot))/\xi_{\tau,x}(t) = y\right\} \tag{3.14}$$

holds.

Let, in particular, $X = G$ be a Lie group. Assume that the coefficients of (3.6) are left invariant and the process $\eta(t)$ is given by

$$\eta(t) = \overset{\curvearrowright}{\exp} \int_0^t \varphi(s)\, \mathrm{d}\, s. \tag{3.15}$$

Here $\varphi(s)$ is a curve in the Lie algebra $\mathcal{G}$. As results from Section 2 $\eta(t)$ solve (3.7) if

$$b_x(t) = a_x(t) + \mathrm{Ad}\, x^{-1}\, \varphi(t)\,. \tag{3.16}$$

Assume $(\mathcal{G}, H, i)$ is Hilbert–Schmidt structure if the Lie group G and H is an ideal in $\mathcal{G}$. Let $w(t)$ be the Wiener process connected with H in the canonic way and $A \equiv I_H$ in (3.6). If $\varphi(t)$ is an H-valued curve, then

$$\alpha(t, x) = \mathrm{d}\, L_x(\mathrm{Ad}\, x^{-1})\, \varphi(t) \in \mathrm{d}\, L_x H \tag{3.17}$$

and (3.8) holds.

As a result, in the considered case the multiplication (3.15) leads to an absolutely continuous transformation of the measure μ on the space of G-valued continuous path.

Notice that the set $C([t, \tau], G)$ is an infinite dimensional group though G may have a finite dimension.

In this case $H = \mathcal{G}$ and (3.17) is no longer a restriction. In this way, we obtain a large enough collection of shifts (3.15) which transforms the measure into an absolutely equivalent one.

In the infinite dimensional case $(\dim G = \infty)$, the most interesting objects are transition probabilities which give measures on G rather than on trajectory space. It follows from the above results that the measure $p(\tau, e\; t, \cdot)$ is transformed into an equivalent one under transformations like $\ell_{\exp \varphi(t)}$ for $\varphi(t) \in H$, i.e. under shifts giving elements of invariant subgroups generated by the ideal H.

3.5. ADMISSIBLE TRANSFORMATIONS OF SMOOTH MEASURES

Let B be a Banach space with a Hilbert–Schmidt structure $B \supset H \supset B^*$. Consider a Borel measure μ with a continuous vector logarithmic derivative $\lambda : B \to B$.

A measurable mapping $f: B \to B$ is called an admissible mapping for the measure μ if $\mu^f = \mu \circ f^{-1}$ is absolutely continuous with respect to μ. Let f be an invertible mapping with the inverse $f^{-1}: y \mapsto x = y + F(y)$. Suppose that the following assumptions hold:

(a) $F: B \to B^*$ is a smooth mapping and both $F(x)$ and $F'(x)$ are bounded over B.

(b) Linear operator $I + tF'(x)$ possesses an invertible restriction to H for $(t, x) \in [0, 1] \times B$ and $\sup_{t,x} \| (I + tF(x))^{-1} \| < \infty$.

Due to Theorem 3.1, we succeed in investigating whether f is an admissible mapping or not under the above assumptions and to derive the explicit expression for the density $\mu^f(dx)/\mu(dx)$. Let us consider a deformation of the mapping f putting $S^{-1}(t, y) = y + tF(y)$ and taking $S(0, y) = y$ and $S(1, y) = f(y)$. Next, construct a vector field which will state for a generator of the above evolution family of mappings.

By differentiation, the relation $x = y(\tau) + \tau F(y(\tau))$ with respect to τ, where $y(t) = S(t, x)$, we obtain

$$0 = y'(\tau) + F(y(\tau)) + \tau F'(y(\tau))\, y'(\tau)$$

or

$$y'(\tau) = -\left(I + \tau F'(y(\tau))\right)^{-1} F(y(\tau)).$$

Hence, the vector field which we had been looking for may be presented in the form

$$z(t, x) = -(I + tF'(x))^{-1} F(x).$$

Let us now derive the expression for the vector field υ, $S(\tau, \cdot)$-connected with $z(\tau, \cdot)$

$$z(\tau, y) = S'(\tau, x)\, \upsilon(x), \qquad y = S(\tau, x).$$

Since $(S^{-1})'(t, y) = I + tF'(y)$, then

$$\upsilon(x) = -(I + tF'(y))\,(I + tF'(y))\, F(y) = -F(y).$$

In addition, $\upsilon'(x)\,(I + tF'(y)) = -F'(y)$ and, thus,

$$\upsilon'(x) = -F'(y)\,(I + tF'(y))^{-1}.$$

Taking the above relations into account, one may derive the expression for the logarithmic derivative of the measure $\mu_t = S(t, \cdot)\,\mu$ along the vector field $z(t, \cdot)$

$$\rho(\mu_t, z(t, \cdot), y) = \rho(\mu, \upsilon, x) = \langle \lambda(x), \upsilon(x) \rangle + \operatorname{tr} \upsilon'(x)$$

$$= -\langle \lambda(y + tF(y)), F(y) \rangle - \operatorname{tr} F'(y)\,(I + tF'(y))^{-1}. \tag{3.18}$$

Introduce the notation

$$Q\,(y, h) = \int_0^1 \lambda\,(y + t\,h)\ \mathrm{d}\,t$$

and use the well-known formula

$$\int_0^1 \operatorname{tr}\,(I + t\,K)^{-1}\,K\ \mathrm{d}\,t = \ln\ \det\,(I + K)$$

for a nuclear operator $K = F'(y)$. It results in changing (3.34) to the following

$$\rho\,(\mu_t\,, z\,(t, \cdot), y) = \langle Q\,(y, F\,(y)),\ F\,(y)\rangle - \ln\ \det\,(I + F'\,(y)). \tag{3.19}$$

Due to Theorem 3.1, we may now state the following assertion.

THEOREM 3.3. *Let* B *be a Banach space equipped with a Hilbert–Schmidt structure and* μ *be a smooth Borel measure on* B *possessing logarithmic derivative* $\rho\,(\mu, h, x) = \langle \lambda\,(x), h\rangle$ *along* $h \in B^*$ *with* $\lambda : B \to B$ *being a continuous mapping. The mapping* $f : B \to B$ *is admissible if it satisfies the above conditions* (a) *and* (b). *In addition, measures* μ *and* μ^f *come to be equivalent and the density* $\mu^f\,(\mathrm{d}\,y)/\mu\,(\mathrm{d}\,y)$ *may be presented in the form*

$$\frac{\mu^f\,(\mathrm{d}\,y)}{\mu\,(\mathrm{d}\,y)}$$

$$= \det\,(I + F'\,(y))\ \exp\left\{\left\langle \int_0^1 \lambda\,(y + t\,F\,(y))\ \mathrm{d}\,t, F\,(y)\right\rangle\right\}. \tag{3.20}$$

Remark 1. If μ is the canonical Gaussian measure with the identity correlation operator in H, then (3.36) is transformed into a well-known relation

$$\frac{\mu^f\,(\mathrm{d}\,y)}{\mu\,(\mathrm{d}\,y)} = \det\,(I + F'\,(y))\ \exp\left\{ -\langle y, F\,(y)\rangle - \frac{1}{2}\,\| F\,(y)\|_H^2 \right\}.$$

Notice that, when adapted to this case, the above proof comes to be more simple from a technical point of view than the corresponding proof of the general case.

Remark 2. Each of the two terms in the right-hand side of (3.18) only make sense for $F : B \to B^*$. Nevertheless, if a derivative $\lambda'\ (x) : H \to H$ does exist and $\sup_x \|\lambda'\ (x)\| < \infty$, then the logarithmic derivative $\rho\,(\mu, z, x)$ may be extended to vector fields $z : B \to H$ as has been shown in Chapter 1. Thus, it results that the logarithmic derivative (3.18) makes sense, although each summand has lost its respective derivative. That is why Theorem 3.8 may be extended to the case when the condition (a) is weakened in a corresponding way. Formula (3.20) may be written in a form admitting an extension to the considered case. For this purpose, we introduce a regularized determinant which is defined for a nuclear operator K by the relation

$$\widetilde{\det}\,(I + K) = \det\,(I + K)\ e^{-\operatorname{tr} K}$$

and still makes sense in the case when K is a Hilbert–Schmidt operator. Consider a new differential expression

$$\beta(t, F, y) = \langle \lambda(y + tF(y)),\ F(y)\rangle + \operatorname{tr} F'(y)$$

which differs from (3.18) by the term $\operatorname{tr}[K(I + tK)^{-1} - K]$ and notice that this term makes sense, even in the case when $F'(y)$ is the Hilbert–Schmidt operator. As a result, one may rewrite (3.20) in the form

$$\frac{\mu^f(d\,y)}{\mu\,(d\,y)} = \widetilde{\det}\,(I + F'(y))\ \exp \int_0^1 \beta(t, F, y)\ d\,t$$

and this expression is valid for the considered more general case.

Historical Comments

Chapters 1 and 2. The reader may find an exposition of differential calculus in Banach spaces in a number of books. The closest in style and choice of material to our book is [28], where the reader may also find a more detailed list of references.

The main notions of differential geometry of Banach manifolds may be found in [58]. The way we prefer for developing the theory of connections on Banach manifolds is based on [36].

Differential calculus for measures in Banach spaces was developed by S.V. Fomin and his collaborators [28]. Another approach to this theory may be found in A.V. Skorohod's works [71].

Measure differentiation along vector fields in linear spaces and smooth manifolds have been introduced in [29].

Integration by part relations and logarithmic derivative second moment estimates for a Gaussian measure have been derived in the course of investigations concerning extended stochastic integrals. The very existence of a logarithmic derivative of a Gaussian measure smooth image has been proved in [24].

Chapter 3. A number of books such as I.I. Gichman and A.V. Skorohod [45]–[46], N. Ikeda and S. Watanabe [48], K. Ito [50], H. McKean [64] and others, are devoted to stochastic differential equation theory in the finite-dimensional case.

The stochastic differential equation theory for an infinite-dimensional Hilbert space has been developed as well (see detailed references in [28]). This theory has been extended to the case of Banach spaces by the authors [9]. Operator multiplicative functionals have been introduced by M. Pinsky [69] and Yu. Dalecky [28].

A number of authors have studied the smoothness properties of a random flow generated by the solution of a stochastic differential equation (see [48]).

The extension of those authors' approaches to the infinite-dimensional case meets some obstacles. Among them is the absence of a direct analogue of the Kolmogorov theorem about path continuity with respect to the initial data. To avoid dealing with those difficulties, we have chosen a weaker notion of smoothness.

Chapter 4. We have pointed out in the introduction, the important part played by the original works by K. Ito [49], H. McKean [62], R. Gangolli [41], see as well our own work [30].

Detailed expositions of diffusion theory on manifolds exist in [48], [37]. The infinite-dimensional situation was discussed in the book by K. Elworthy and in some of the works of H. Kuo [56]. Our way of exposition follows [12]. In the introduction, we have discussed the crucial principles which distinguish our approach. Tensor diffusion was originally investigated by K. Ito [51] and then by E.B. Dynkin [32]. Their results have been extended to infinite-dimensional general bundles by Ya. I. Belopolskaya [8].

Chapter 5. Parabolic diffusion equations in infinite-dimensional spaces have been treated independently by L. Gross and Yu. L. Dalecki and their collaborators (detailed results and comments may be seen in [28]). Parabolic equations on infinite-dimensional smooth manifolds and systems of those equations in sections of vector bundles have been investigated by the authors [12]. We follow this paper in Section 1.

H. McKean [63] and M.I. Freidlin [39] were the first to treat quasilinear parabolic equations from the point of view of stochastic equations. We have extended their approach to the infinite-dimensional case [10] and to nonlinear parabolic equations [13].

The forward Kolmogorov equation for a diffusion process in infinite-dimensional space treated in a weak sense, has been investigated by Yu. L. Dalecky and S.V. Fomin [28]. Many works connected with Malliavin calculus are devoted to studying stochastic diffusion equation transition probability properties. The investigations are based on treating the transition probability as a smooth image of a white noise distribution (see [59]–[61], [18], [73]).

This approach has been extended to the infinite-dimensional case by Yu. L. Dalecky [24].

Chapter 6. A construction of Brownian motion on uniform spaces and Lie groups has been given by K. Iosida and H. McKean for the finite-dimensional case. Various examples of quasi-invariant measures on infinite-dimensional groups have been constructed in [1].

In [30], we have constructed a transition probability quasi-invariant with respect to invariant subgroup actions. Under some additional conditions, the connection between smoothness and equivalence of measures has been studies by A.V. Skorohod [71].

For what concern measure behavior under shifts along vector fields, the reader is referred to [25]. The conditions given in this article may be weakened, thanks to the work of D. Bell [6] in which shifts in linear spaces are discussed. Our exposition follows [21]. The result stated in Theorem 3.3 was obtained by Yu. L. Dalecky and G.A. Sochadze (see *Functional analysis and its applications* **22** (1988), 77–78).

References

1. S. Albeverio, R. Hoegh–Krohn, D. Testard, and A.Vershik, Factorial representation of paths groups, *Funct. Anal.* **51** (1983), 115–131.
2. V.I. Arnold, Sur la géometric différentielle des groupes de Lie de dimension infinie et ses applications à la hydrodynamique des fluides parfaits, *Ann. Inst. Fourier, (Grenoble)* **16**, 319-361.
3. M. Asorey, and P. Mitter, Regularized continuum Yang processes and Feynmann–Kac functional integrals, *Comm. Math. Phys.* **80** (1981), 43–58.
4. V.I. Averbuch, O.G. Smoljanov, and S.V. Fomin, Generalized functions and differential equations in linear spaces, *Trans Moscow Math. Soc.* **24** (1971), 140–184.
5. P. Baxendale, Wiener processes on manifolds of maps, *Proc. Royal Soc. Edinburgh, Sect. A.* **87** (1980), 127–152.
6. D. Bell, A quasi–invariance theorem for measures on Banach spaces, *Trans. Amer. Math. Soc.* **290** (1985), 851–855.
7. Ya. I. Belopolskaya, Diffusion processes in Banach spaces and manifolds, *Teor. Veroyatnost Mat. Statis*, No. 9 (1973), pp. 27–36.
8. Ya. I. Belopolskaya, Diffusion processes in vector bundels, *Teor. Veroyatnost Mat. Statis*, No. 12 (1975), pp. 3–13, 170 (Eng. Summary).
9. Ya. I. Belopolskaya and Yu. L. Dalecky, Diffusion processes in smooth Banach spaces and manifolds, *Trans. Moscow Math. Soc.* **37** (1978), 113–150.
10. Ya. I. Belopolskaya and Yu. L. Dalecky, Investigation of a Cauchy problem for quasilinear parabolic systems with finite and infinte dimensional arguments with the help of Markov random processes, *Izv. Vyssh. Uchebn. Zaved. Math.* **2**:*6* (1978), 5–17 (in Russian).
11. Ya. I. Belopolskaya and Yu. L. Dalecky, Markov processes connected with nonlinear parabolic systems, *Soviet Math. Dokl.* **21** (1980), 99–103.
12. Ya. I. Belopolskaya and Yu. L. Dalecky, Ito's equations and differential geometry, *Russian Math Surveys* **37**:*3* (1982), 95–142.
13. Ya. I. Belopolskaya and Yu. L. Dalecky, Cauchy problem for a nonlinear parabolic system via operator–valued multiplicative functionals of Markov processes, *J. Soviet Math.* **21** (1983), 653–679.
14. Ya. I. Belopolskaya and Yu. L. Dalecky, Probability methods in some global analysis problems, *2nd Intern. Vilius Conf. on Prob. Theory and Math. Stats.* Vol. 1 (1977), pp. 27–30.
15. Ya. I. Belopolskaya and Yu. L. Dalecky, Stochastic equations and differential geometry, *Springer Lecture Notes in Math.* **1214** (1986), 131–158.
16. J.–M. Bismut, Méchanique aléatoire, *Springer Lecture Notes in Math.* **866** (1986).
17. J.–M. Bismut, The Atiyah–Singer theorems: a probabilistic approach 1, 2, *J. Funct. Anal.* **57** (1984), 56–99; 329–348.
18. J.–M. Bismut, Martingales, the Malliavin calculus and hypoellipticity under general Hörmander conditions, *Z. Wahrsch. Verw. Geb.* **56** (1981), 469–505.
19. R. Bishop and R. Crittenden (1964), *Geometry of Manifolds*, Academic Press, N.Y., London.
20. Yu. N. Blagoveschensky and M.I. Freidlin, Certain properties of diffusion processes depending on a certain parameter, *Soviet Math. Dokl.* **2** (1961), 633–636.
21. D.D. Bleecker, Physics from the point of *G*–bundles, *J.Theor. Phys.* **23**:*8* (1984), 735–750.
22. Yu. L. Dalecky, Multiplicative operator functionals of Markov processes and their applications, *Springer Lecture Notes in Control and Inform. Sci.* **25** (1980), 38–49.
23. Yu. L. Dalecky, Infinite dimensional elliptic operators and parabolic equations associated with them, *Russian Math. Surveys* **22**:*4* (1967), 1–53.
24. Yu. L. Dalecky, Stochastic differential geometry, *Russian Math. Surveys* **38**:*3* (1983), 97–125.

25. Yu. L. Dalecky, Absolute continuity of measures generated by diffusion processes on smooth manifolds, *3rd Intern. Vilnius Conf. Prob. Theory Math. Stat.*, Vol. 1 (1981), pp. 167–170.
26. Yu. L. Dalecky, Multiplicative operators of diffusion processes and differential equations in sections of vector bundles, *Uspekli Mat. Nauk.* **30:2** (1975), 209–210 (in Russian).
27. Yu. L. Dalecky and S.V. Fomin, Generalized measures in a Hilbert space and a forward Kolmogorov equation, *Soviet Math. Dokl.* **13** (1972), 993–997.
28. Yu. L. Dalecky and S.V. Fomin, *Measures and Differential Equations in Infinite Dimensional Spaces*, Nauka Moscow (1983).
29. Yu. L. Dalecky and B.D. Marjanin, Smooth measures on infinite dimensional manifolds, *Soviet Math. Dokl.* **32:3** (1985), 863–866.
30. Yu. L. Dalecky and Ya. I. Shneiderman, Diffusion and quasi–invariant measures on infinite dimensional Lie groups, *Funct. Anal. Appl.* **3** (1975), 156–168.
31. B.A. Dubrovin, S.P. Novikov, and A.T. Fomenko, *Modern Geometry*, Springer, New York (1984) (Transl. from Russian).
32. E.B. Dynkin, Diffusion of tensors, *Soviet Math. Dokl.* **9** (1968), 532–535.
33. D. Ebin and J. Marsden, Groups of diffeomorphisms and the motion of an incompressible fluid, *Ann. of Math.* **92:1** (1979), 102–163.
34. J. Eells and K.D. Elworthy, Wiener integration on certain manifolds in *Problems of Nonlinear Analysis*, Cremonese, Rome (1971), pp. 67–94.
35. J. Eells, A setting for global analysis, *Bull. Amer. Math. Soc.* **72** (1966), 751–807.
36. H. Eliasson, Geometry of manifolds of maps, *J. Diff. Geom.* **1** (1967), 164–194.
37. K.D. Elworthy, *Stochastic Differential Equation on Manifolds*, London Math. Soc. Lecture Notes (1982).
38. K.D. Elworthy, Stochastic dynamical systems and their flows, *Stoch.Anal.* (1978), 79–95.
39. M.I. Freidlin, Quasilinear parabolic equations and measures in function space, *Funct. Anal. Appl.* **1:3** (1967), 237–240.
40. M.I. Freidlin, On the existence in the whole of degenerated quasilinear equation smooth solutions, *Math. Sb.* **70:3** (1969), 332–348.
41. R. Gangolli, On the construction of certain diffusions on a differentiable manifold, Z. *Wahrsch. Verw. Geb.* **2** (1964), 406–409.
42. B. Gaveau and P. Trauber, Une construction de la qualification euclidiénne du champ de Yang–Mills regularisé, *J. Funct. Anal.* **42** (1981), 356–357.
43. I.I. Gihman, On certain differential equations with random coefficients, *Ukr. Math. J.* **2:3** (1950), 45–69.
44. I.I. Gihman, On the theory of differential equations with random coefficients, *Ukr. Math. J.* **2:4** (1950), 37–63.
45. I.I. Gihman and A.V. Skorohod, *Introduction to the Theory of Random Processes*, Springer, New York (1979).
46. I.I. Gihman and A.V. Skorohod, *Theory of Stochastic Processes*, Nauka, Moscow, Vol. 1 (1971), Vol. 2 (1973), Vol. (1975) (in Russian); Eng. trans.: Springer, New York (1974).
47. Yu. E. Gliklich, Riemannian parallel displacement, Ito's integral and stochastic equations on manifolds in *Analysis on Manifolds and Differential Equations*, Voronež (1986), pp. 25–45.
48. N. Ikeda and S. Watanabe, *Stochastic Differential Equations and Diffusion Processes*, North–Holland, Amsterdam (1981).
49. K. Ito, On stochastic differential equations on a differential manifold 1, *Nagoya Math. J.* **1** (1950), 35–47.
50. K. Ito, Stochastic differential equations on a differentiable manifold, *Mem. Coll Sci. Uni. Kyoto Math.* **28** (1953), 81–85.
51. K. Ito, The Brownian motion and tensor fields on Riemannian manifold in *Proc. Intern. Congr. Math. Stockholm* (1963), pp. 536–539.
52. K. Ito, Stochastic parallel displacement probabilistic methods in differential equations, *Springer Lecture Notes in Math.* **451** (1975), 213–229.
53. W.S. Kendall, Brownian motion, negative curvature and harmonic maps, *Springer Lecture Notes in Math.* **851** (1981), 479–491.
54. S. Kobayashi and K. Nomizu, *Foundations of Differential Geometry*, Interscience, N.Y., London (1963/69) (2 Vols.).

55. H. Kuo, Gaussian measure in Banach spaces, *Springer Lecture Notes in Math* **463** (1975).
56. H. Kuo, Diffusion and Brownian motion on infinite dimensional manifolds, *Trans. Amer. Math. Soc.* **169** (1972), 439–459.
57. S. Kwapien, Some applications of the theory of absolutely summing operators, Aarhus Univ. Math. Inst. Prepr. 1968/1969, No. 37.
58. S. Lang, *Introduction to Differentiable Manifolds*, Interscience, N.Y. (1962).
59. P. Malliavin, *Géometrie différentielle stochastique*, Les Presses de l'Univ. de Montréal, Montréal (1978).
60. P. Malliavin, *C^k–Hypoellipticity with Degeneracy*, in *Stoch. Anal.* Academic Press, N.Y. (1978), pp. 199–214; 327–370.
61. P. Malliavin, Stochastic calculus of variation and hypoelliptic operators, *Proc. Intern. Symp. SDE Kyoto 1976*, Wiley, N.Y. (1978), pp. 195–263.
62. H.P. McKean, Brownian motion on the 3–dimensional rotation group, *Mem. Coll Sci. Kyoto Univ.* **33** (1960), 25–38.
63. H.P. McKean, A class of Markov processes associated with nonlinear parabolic equations, *Proc. Nat. Acad. Sci. USA* **59**:*6* (1966), 1907–1911.
64. H.P. McKean, *Stochastic Integrals*, Academic Press, N.Y. (1969).
65. P. Meyer, A differential geometric formalism for the Ito calculus, *Springer Lecture Notes in Math* **851** (1981), 256–270.
66. S.A. Molchoanov, Diffusion processes and Riemannian geometry, *Russian Math. Surveys* **30**:*1* (1975), 1–63.
67. E. Mourier, Elements aléatoires dans un espace de Banach, *Ann. Inst. Henri Poincaré* **13** (1953), 161–244.
68. H. Omori, Infinite dimensional Lie transformation groups, *Springer Lecture Notes in Math.* **427** (1974), 249.
69. M. Pinsky, Multiplicative operator functional of Markov process, *Bull. Amer. Math. Soc.* **77** (1971), 377.
70. M. Pinsky and A. Friedman, *Stochastic Analysis*, Academic Press, N.Y. (1978).
71. A.V. Skorohod, *Integration in Hilbert Space*, Springer, New York (1974).
72. A.V. Skorohod, On admissible translations of measures in Hilbert space, *Theor. Probab. Appl.* **15** (1970), 557–580.
73. D. W. Stroock, The Malliavin calculus: a functional analytic approach, *J. Funct. Anal.* **44** (1981), 212–217.
74. M.I. Vishik and A.B. Marchenko, Boundary value problems for second order elliptic and parabolic operators on infinite dimensional manifolds with boundary, *Math. USSR. Sb.* **19** (1973), 325-364.
75. A. D. Wentzel and M.I. Freidlin, *Fluctuations in Dynamical Systems*, Nauka, Moscow (1979).
76. Y. Yamato, Stochastic differential equations and nilpotent Lie algebras, *Z. Wahrsch. Verw. Geb.* **47** (1979), 213–229.

Index